T0245433

CAMBRIDGE LIBRARY COLLECTION

Books of enduring scholarly value

Botany and Horticulture

Until the nineteenth century, the investigation of natural phenomena, plants and animals was considered either the preserve of elite scholars or a pastime for the leisured upper classes. As increasing academic rigour and systematisation was brought to the study of 'natural history', its subdisciplines were adopted into university curricula, and learned societies (such as the Royal Horticultural Society, founded in 1804) were established to support research in these areas. A related development was strong enthusiasm for exotic garden plants, which resulted in plant collecting expeditions to every corner of the globe, sometimes with tragic consequences. This series includes accounts of some of those expeditions, detailed reference works on the flora of different regions, and practical advice for amateur and professional gardeners.

Hortus Kewensis

Trained as a gardener in his native Scotland, William Aiton (1731–93) had worked in the Chelsea Physic Garden prior to coming to Kew in 1759. He met Joseph Banks in 1764, and the pair worked together to develop the scientific and horticultural status of the gardens. Aiton had become super-intendent of the entire Kew estate by 1783. This important three-volume work, first published in 1789, took as its starting point the plant catalogue begun in 1773. In its compilation, Aiton was greatly assisted with the identification and scientific description of species, according to the Linnaean system, by the botanists Daniel Solander and Jonas Dryander (the latter contributed most of the third volume). Aiton added dates of introduction and horticultural information. An important historical resource, it covers some 5,600 species and features a selection of engravings. Volume 2 continues to catalogue the plants, covering Octandria to Monadelphia.

Cambridge University Press has long been a pioneer in the reissuing of out-of-print titles from its own backlist, producing digital reprints of books that are still sought after by scholars and students but could not be reprinted economically using traditional technology. The Cambridge Library Collection extends this activity to a wider range of books which are still of importance to researchers and professionals, either for the source material they contain, or as landmarks in the history of their academic discipline.

Drawing from the world-renowned collections in the Cambridge University Library and other partner libraries, and guided by the advice of experts in each subject area, Cambridge University Press is using state-of-the-art scanning machines in its own Printing House to capture the content of each book selected for inclusion. The files are processed to give a consistently clear, crisp image, and the books finished to the high quality standard for which the Press is recognised around the world. The latest print-on-demand technology ensures that the books will remain available indefinitely, and that orders for single or multiple copies can quickly be supplied.

The Cambridge Library Collection brings back to life books of enduring scholarly value (including out-of-copyright works originally issued by other publishers) across a wide range of disciplines in the humanities and social sciences and in science and technology.

Hortus Kewensis

*Or, a Catalogue of the Plants Cultivated
in the Royal Botanic Garden at Kew*

VOLUME 2:
OCTANDRIA TO MONADELPHIA

WILLIAM AITON

CAMBRIDGE
UNIVERSITY PRESS

University Printing House, Cambridge, CB2 8BS, United Kingdom

Published in the United States of America by Cambridge University Press, New York

Cambridge University Press is part of the University of Cambridge.
It furthers the University's mission by disseminating knowledge in the pursuit of
education, learning and research at the highest international levels of excellence.

www.cambridge.org
Information on this title: www.cambridge.org/9781108069687

© in this compilation Cambridge University Press 2014

This edition first published 1789
This digitally printed version 2014

ISBN 978-1-108-06968-7 Paperback

This book reproduces the text of the original edition. The content and language reflect
the beliefs, practices and terminology of their time, and have not been updated.

Cambridge University Press wishes to make clear that the book, unless originally published
by Cambridge, is not being republished by, in association or collaboration with, or
with the endorsement or approval of, the original publisher or its successors in title.

HORTUS KEWENSIS;

OR, A

CATALOGUE

OF THE

PLANTS

CULTIVATED IN THE

ROYAL BOTANIC GARDEN AT KEW.

BY WILLIAM AITON,

GARDENER TO

HIS MAJESTY.

IN THREE VOLUMES.

VOL. II.

OCTANDRIA — MONADELPHIA.

LONDON:

PRINTED FOR GEORGE NICOL, BOOKSELLER
TO HIS MAJESTY, PALL MALL.

M.DCC.LXXXIX.

Claſſis VIII.

OCTANDRIA

MONOGYNIA.

TROPÆOLUM. *Gen. pl.* 466.

Cal. 1-phyllus, calcaratus. *Petala* 5, inæqualia.
 Baccæ 3, ſiccæ.

1. T. foliis peltatis repandis, petalis acuminato-ſetaceis. *minus.*
 Syſt. veget. 357.
 Small Indian Creſs, or Tropæolum.
 Nat. of Peru.
 Cult. 1596, by Mr. John Gerard. *Hort. Ger.*
 Fl. June——October.   H. ☉.

2. T. foliis peltatis ſubquinquelobis, petalis obtuſis. *majus.*
 Syſt. veget. 357. *Curtis magaz.* 23.
α flore ſimplici.
 Great Indian Creſs, or Tropæolum.
β flore pleno.
 Double-flower'd Indian Creſs, or Tropæolum.
 Nat. of Peru.
 Introd. 1686, by Dr. Lumley Lloyd. *Collinſ. mſcr.*
 Fl. June——October. α. H. ☉. β. Gr. H. ♃.

3. T. foliis palmatis, petalis multifidis. *Sp. pl.* 490. *peregri-*
 Fringed-flower'd Tropæolum.   *num.*
 Nat. of Peru.
 Introd. 1775, by Benjamin Bewick, Eſq.
 Fl. September and October.   S. ☉.

Vol. II.   B   RHEXIA.

RHEXIA. *Gen. pl.* 468.

Cal. 4-fidus. *Petala* 4, calyci inserta. *Antheræ* declinatæ. *Capsula* 4-locularis, intra ventrem calycis.

virgini-   1. R. foliis sessilibus serratis, calycibus glabris. *Sp. pl.*
ca.   491.
        Hairy-leav'd Rhexia.
        Nat. of North America.
        Cult. 1759, by Mr. Ph. Miller. *Mill. dict. edit.* 7. *n.* 1.
        Fl. July and August.   H. ♃.

ŒNOTHERA. *Gen. pl.* 469.

Cal. 4-fidus. *Petala* 4. *Capf.* cylindrica, infera, *Sem.* nuda.

biennis.   1. Œ. foliis ovato-lanceolatis planis, caule muricato
        subvilloso. *Sp. pl.* 492.
        Broad-leav'd Tree Primrose, or Œnothera.
        Nat. of North America.
        Cult. 1629. *Park. parad.* 263. *f.* 6.
        Fl. June——September.   H. ♂.

grandi-   2. Œ. foliis ovato-lanceolatis, staminibus declinatis,
flora.   caule fruticoso. *L'Herit. stirp. nov. tom.* 2. *tab.* 4.
        Great-flower'd Œnothera.
        Nat. of North America.
        Introd. 1778, by John Fothergill, M. D.
        Fl. July and August.   G. H. ♂.

parviflo-   3. Œ. foliis ovato-lanceolatis planis, caule lævi subvillo-
ra.   so. *Sp. pl.* 492.
        Small-flower'd Œnothera.
        Nat. of North America.

                                   Introd.

Introd. 1775, by John Earl of Bute.
Fl. June——Auguſt. H. ♂.

4. Œ. foliis denticulatis, caulibus ſimplicibus piloſis, *longiflo-*
petalis diſtantibus bilobis. *Syſt. veget.* 358. *Jacqu.* *ra.*
hort. 2. *p.* 81. *t.* 172.
Long-flower'd Œnothera.
Nat. of Buenos Ayres.
Introd. 1776, by Chevalier Murray.
Fl. July——September. H. ♂.

5. Œ. foliis lanceolatis undulatis. *Sp. pl.* 492. *molliſſi-*
Soft Œnothera. *ma.*
Nat. of Buenos Ayres.
Cult. 1732, by James Sherard, M. D. *Dill. elth.* 297.
t. 219. *f.* 286.
Fl. June——October. H. ♂.

6. Œ. foliis ovatis dentatis: inferioribus lyratis, capſulis *roſea.*
clavatis.
Œnothera roſea. *L'Herit. ſtirp. nov. tom.* 2. *tab.* 6.
Roſe-flower'd Œnothera.
Nat. of Peru.
Introd. 1783, by Monſ. Thouin.
Fl. moſt part of the Summer. G. H. ♃.

7. Œ. foliis dentato-ſinuatis, capſulis priſmaticis. *ſinuata.*
Œnothera ſinuata. *Linn. mant.* 228. *Murray in nov.*
comment. getting. 5. *pag.* 44. *tab.* 9.
Œnothera repanda. *Medicus in act. palat. vol.* 3. *phyſ.*
pag. 198. *tab.* 8.
Scollop-leav'd Œnothera.
Nat. of North America.
Introd. 1770, by Monſ. Richard.
Fl. July. H. ☉.
8. Œ.

fruticofa.   8. Œ. foliis lanceolatis fubdentatis, capfulis pedicellatis acutangulis, racemo pedunculato. *Sp. pl.* 492.
L' Herit. ftirp. nov. tom. 2. *tab.* 5.
Shrubby Œnothera.
Nat. of Virginia.
Cult. 1739, by Mr. Philip Miller. *Rand. chel.* Onagra 5.
Fl. June——Auguft.   H. ♃.

pumila.   9. Œ. foliis lanceolatis obtufis glabris fubpetiolatis, caulibus proftratis, capfulis acutangulis. *Sp. pl.* 493.
Dwarf Œnothera.
Nat. of North America.
Cult. 1757, by Mr. Ph. Miller. *Mill. ic.* 125. *t.* 188.
Fl. May——September.   H. ♃.

G A U R A.   *Gen. pl.* 470.

Cal. 4-fidus, tubulofus.   *Cor.* 4-petala, adfcendens verfus latus fuperius.   *Nux* infera, 1-fperma, 4-angula.

biennis.   1. GAURA. *Sp. pl.* 493.
Biennial Gaura.
Nat. of Penfylvania and Virginia.
Introd. 1762, by Mr. James Gordon.
Fl. Auguft——October.   H. ♂.

E P I L O B I U M.   *Gen. pl.* 471.

Cal. 4-fidus. *Petala* 4. *Capf.* oblonga, infera. *Sem.* pappofa.

* *Staminibus declinatis.*

angufti-
folium.   1. E. foliis fparfis lineari-lanceolatis integerrimis venofis, floribus inæqualibus.

Epilobium

Epilobium anguftifolium. *Sp. pl.* 493. *Curtis lond.*
Narrow-leav'd Epilobium, or Rofe-bay Willow-herb.
Nat. of Britain.
Fl. July and Auguft.   H. ♃.

2. E. foliis fparfis linearibus obfolete denticulatis ave-   *anguftif-*
   niis, petalis æqualibus integerrimis.   *fimum.*
   Epilobium anguftiffimum. *Weber dec. pl. min. cogn.*
   p. 3.
   Epilobium anguftifolium γ. *Sp. pl.* 494.
   Epilobium flore difformi, foliis linearibus. *Hall.*
   hift. 1001.
   Linear-leav'd Epilobium.
   Nat. of Switzerland.
   Introd. 1775, by the Doctors Pitcairn and Fothergill.
   Fl. July and Auguft.   H. ♃.

   * * *Staminibus erectis regularibus, petalis bifidis.*

3. E. foliis ovato-lanceolatis femiamplexicaulibus hir-   *hirfutum.*
   futis, caule ramofiffimo, radice repente. *Curtis lond.*
   Epilobium hirfutum α. *Sp. pl.* 494.
   Epilobium ramofum. *Hudf. angl.* 162.
   Epilobium grandiflorum. *Allion. pedem.* 1. *p.* 279.
   Large-flower'd Epilobium.
   Nat. of Britain.
   Fl. July and Auguft.   H. ♃.

4. E. foliis oblongo-lanceolatis dentatis pubefcentibus,   *villofum.*
   caule tereti villofo. *Curtis lond.*
   Epilobium hirfutum β. *Sp. pl.* 494.
   Epilobium hirfutum. *Hudf. angl.* 161. *Allion. pedem.* 1.
   p. 279.
   Hoary Epilobium.
   Nat. of Britain.
   Fl. July and Auguft.   H. ♃.

monta-num.

5. E. foliis oppofitis ovatis dentatis. *Sp. pl.* 494. *Curtis lond.*
Mountain Epilobium.
Nat. of Britain.
Fl. June.   H. ♂.

tetrago-num.

6. E. foliis lanceolatis denticulatis : imis oppofitis, caule tetragono. *Sp. pl.* 494. *Curtis lond.*
Square-ftalk'd Epilobium.
Nat. of Britain.
Fl. July.   H. ♃.

paluftre.

7. E. foliis oppofitis lanceolatis integerrimis, petalis emarginatis, caule erecto. *Sp. pl.* 495.
Marfh Epilobium.
Nat. of Britain.
Fl. July.   H. ♃.

alpinum.

8. E. foliis oppofitis ovato-lanceolatis integerrimis, filiquis feffilibus, caule repente. *Sp. pl.* 495.
Alpine Epilobium.
Nat. of Britain.
Fl. July.   H. ♃.

MELICOCCA. *Gen. pl.* 472.

Cal. 4-partitus. *Petala* 4, reflexa infra calycem. *Stigm.* fubpeltatum. *Drupa* coriacea.

bijuga.

1. MELICOCCA. *Sp. pl.* 495.
Winged-leav'd Melicocca.
Nat. of Jamaica.
Introd. 1778, by Thomas Clark, M.D.
Fl.
       S. ♄.

KŒLREU-

KŒLREUTERIA. *Laxmann.*

Cal. 5-phyllus. *Cor.* 4-petala, irregularis. *Capf.*
   3-locularis. *Sem.* bina.

1. KOELREUTERIA. *Laxmann in nov. comm. petrop.* panicula-
   16. *p.* 561. *tab.* 18.   ta.
   Kœlreuteria paullinioides. *L'Herit. fert. angl. tab.* 19.
   Sapindus chinenfis. *Linn. fuppl.* 228.
   Panicled Kœlreuteria.
   Nat. of China.
   Introd. about 1763, by George W. Earl of Coventry.
   Fl. Auguft.   H. ♄.

G U A R E A. *Linn. mant.* 150.

Cal. 4-dentatus. *Petala* 4. *Nectarium* cylindricum,
   ore Antheras gerens. *Capf.* 4-locularis, 4-valvis.
   Sem. folitaria.

1. GUAREA. *Syft. veget.* 360.   trichilioi-
   Trichilia Guara. *Sp. pl.* 551.   des.
   Afh-leav'd Guarea.
   Nat. of South America.
   Cult. 1770, by Mr. James Gordon.
   Fl.   G. H. ♄.

X I M E N I A. *Gen. pl.* 477.

Cal. 4-fidus. *Petala* 4, pilofa, revoluta. *Drupa* monofperma.

1. X. foliis oblongis, pedunculis multifloris. *Sp. pl.* 497.   america-
   American Ximenia.   na.
   Nat. of the Weft Indies.
   Cult. 1759, by Mr. Ph. Miller. *Mill. dict. edit.* 7. *n.* 1.
   Fl.   S. ♄.

F U C H S I A. *Gen. pl.* 128.

Cal. 1-phyllus, coloratus, corollifer, maximus. *Petala* 4, parva. *Bacca* infera, 4-locularis, polyſperma.

coccinea.   1. F. foliis oppoſitis ovatis denticulatis, petalis obovatis obtuſis.
Thilco. *Feuillée it.* 3. *p.* 64. *t.* 47.
Scarlet-flower'd Fuchſia.
Nat. of Chili.
Introd. 1788, by Captain Firth.
Fl. May——July.   S. ♄.

C H L O R A. *Linn. mant.* 10.

Cal. 8-phyllus. *Cor.* 1-petala, 8-fida. *Capſ.* 1-locularis, 2-valvis, polyſperma.

perfolia-   1. C. foliis perfoliatis. *Syſt. veget.* 361.
ta.   Gentiana perfoliata. *Sp. pl.* 335.
Yellow Wort, or perfoliate Centory.
Nat. of Britain.
Fl. June and July.   H. ☉.

M I C H A U X I A. *L'Heritier monogr.*

Cal. 16-partitus. *Cor.* rotata, 8-partita. *Nect.* 8-valve, ſtaminiferum. *Capſ.* 8-locularis, polyſperma.

campanu-   1. Michauxia. *L'Heritier monogr.*
loides.   Rough leav'd Michauxia.
Nat. of the Levant.
Introd. 1787, by Monſ. L'Heritier.
Fl. June——Auguſt.   G. H. ♂.

DODONÆA.

DODONÆA. *Linn. mant.* 149.

Cal. 4-phyllus. *Cor.* 0. *Capf.* 3-locularis, inflata.
Sem. bina.

1. D. foliis oblongis. *Linn. fuppl.* 218.   *vifcofa.*
Broad-leav'd Dodonæa.
Nat. of the Countries between the Tropics.
Introd. 1690, by Mr. Bentick. *Br. Muf. Sloan. mff.*
   3370.
Fl. June and July.   S. ♄.

2. D. foliis linearibus. *Linn. fuppl.* 218.   *angufti-*
Narrow-leav'd Dodonæa.   *folia.*
Nat. of the Cape of Good Hope.
Cult. 1758, by Mr. Philip Miller.
Fl. May——Auguft.   G. H. ♄.

LAWSONIA. *Gen. pl.* 482.

Cal. 4-fidus. *Petala* 4. *Stamina* 4 parium. *Capf.*
   4-locularis, polyfperma.

1. L. ramis inermibus. *Sp. pl.* 498.   *inermis.*
Smooth Lawfonia.
Nat. of Egypt.
Cult. 1739, by Mr. Philip Miller. *Mill. dict. vol.* 2.
   Liguftrum 2.
Fl.   S. ♄.

2. L. ramis fpinofis. *Sp. pl.* 498.   *fpinofa.*
Prickly Lawfonia.
Nat. of the Eaft Indies.
Cult. 1759, by Mr. Ph. Miller. *Mill. dict. edit.* 7. *n.* 2.
Fl.   S. ♄.

VACCI-

VACCINIUM. *Gen. pl.* 483.

Cal. fuperus. *Cor.* 1-petala. *Filamenta* receptaculo
inferta. *Bacca* 4-locularis, polyfperma.

* *Foliis annotinis f. deciduis.*

Myrtil-
lus.
1. V. pedunculis unifloris, foliis ferratis ovatis deciduis,
caule angulato. *Sp. pl.* 498.
Common Blea-berry, or Whortle-berry.
Nat. of Britain.
Fl. April.   H. ♃.

pallidum.
2. V. racemis braĉteatis, corollis cylindraceo-campanu-
latis, foliis ovatis acutis ferrulatis glabris deciduis.
Pale Whortle-berry.
Nat. of North America.   *Samuel Martin*, M. D.
Introd. 1772.
Fl. May and June.   H. ♃.

ftamine-
um.
3. V. pedunculis folitariis nudis unifloris, antheris corol-
la longioribus, foliis oblongo-ovatis acutis integer-
rimis fubtus fubglaucis.
Vaccinium ftamineum. *Sp. pl.* 498.
Green-wooded Whortle-berry.
Nat. of North America.
Introd. 1772, by Mr. William Young.
Fl. May and June.   H. ♃.

uligino-
fum.
4. V. pedunculis unifloris, foliis integerrimis deciduis
ovalibus bafi attenuatis lævibus.
Vaccinium uliginofum. *Sp. pl.* 499.
Marfh Bilberry, or Whortle-berry.
Nat. of Britain.
Fl. April and May.   H. ♃.

5. V.

5. V. pedunculis folitariis nudis unifloris, foliis ovatis *diffufum.*
 acutis obfolete ferratis fubtus villofiufculis.
Shining-leav'd Whortle-berry.
Nat. of South Carolina. Mr. *John Cree.*
Introd. 1765.
Fl. May——July.   H. ♄.

6. V. pedunculis folitariis unifloris, foliis elliptico-lan- *angufti-*
 ceolatis glabris obfolete ferrulatis.   *folium.*
Narrow-leav'd Whortle-berry.
Nat. of Newfoundland and Labrador.
Introd. about 1776, by Benjamin Bewick, Efq.
Fl. April and May.   H. ♄.

7. V. racemis nudiufculis, corollis cylindrico-ovatis, ca- *fufcatum.*
 lycibus acutis, foliis ellipticis acutis integerrimis :
 venis fubtus villofiufculis.
Clufter-flower'd Whortle-berry.
Nat. of North America. Mr. *William Young.*
Introd. 1770.
Fl. May and June.   H. ♄.

8. V. racemis bracteatis, pedicellis bracteolatis, corollis *frondo-*
 fubcampanulatis, foliis obovato-oblongis integerri- *fum.*
 mis deciduis.
Vaccinium frondofum. *Sp. pl.* 499.
Obtufe-leav'd Whortle-berry.
Nat. of North America.
Introd. 1770, by Mr. William Young.
Fl. June.   H. ♄.

9. V. racemis bracteatis, pedicellis bracteolatis, corollis *venuftum.*
 fubcampanulatis, foliis ellipticis acutis integerrimis
 deciduis glabris.
Red-twigg'd Whortle-berry.
          Nat.

Nat. of North America. Mr. *William Young.*
Introd. 1770.
Fl. May and June.   H. ♄ .

refinofum.   10. V. racemis bracteatis, corollis ovatis, foliis ellipticis
acutiufculis integerrimis deciduis atomis refinofis
irroratis.
Clammy Whortle-berry.
Nat. of North America.   *Samuel Martin,* M. D.
Introd. 1772.
Fl. May and June.   H. ♄ .

amœnum.   11. V. racemis bracteatis, corollis fubcylindraceis, foliis
ellipticis fubferrulatis deciduis : venis fubtus vil-
lofiufculis.
Broad-leav'd Whortle-berry.
Nat. of North America.   Mr. *John Cree.*
Introd. 1765.
Fl. May and June.   H. ♄ .

virgatum.   12. V. racemis feffilibus, corollis fubcylindraceis, foliis
oblongo-ellipticis ferrulatis deciduis utrinque gla-
bris, ramis floriferis elongatis.
Privet-leav'd Whortle-berry.
Nat. of North America.   Mr. *William Young.*
Introd. 1770.
Fl. April and May.   H. ♄ .

tenellum.   13. V. racemis bracteatis feffilibus, corollis ovato-cylin-
draceis, foliis oblongo-ellipticis fubcuneiformibus
ferrulatis deciduis fubglabris.
Gale-leav'd Dwarf Whortle-berry.
Nat. of North America.   Mr. *William Young.*
Introd. 1772.
Fl. May and June.   H. ♄ .
                                                        14. V.

Ehret. del.

Vaccinium macrocarpon.

M. Kenzie sc.

Tab.7. Vol.2. Page 13.

14. V. floribus racemosis, foliis crenulatis ovatis acutis, *Arctosta-*
    caule arboreo. *Sp. pl.* 500. *phylos.*
    Madeira Whortle-berry.
    Nat. of Madeira and the Levant.
    Introd. 1777, by Mr. Francis Masson.
    Fl. June and July. G. H. ♄.

          ** *Foliis sempervirentibus.*
15. V. racemis terminalibus nutantibus, foliis obovatis *Vitis*
    revolutis integerrimis subtus punctatis. *Sp. pl.* *idæa.*
    500.
    Red Bilberry, or Whortle-berry.
    Nat. of Britain.
    Fl. April and May. H. ♄.

16. V. foliis integerrimis ovatis acutis margine revo- *Oxycoc-*
    lutis, caulibus repentibus filiformibus. *cos.*
    Vaccinium Oxycoccos. *Sp. pl.* 500.
    Common Cranberry.
    Nat. of Britain.
    Fl. May and June. H. ♄.

17. V. foliis integerrimis ovali-oblongis obtusis planis, *macro-*
    caulibus repentibus filiformibus. TAB. 7. *carpon.*
    Vitis idæa palustris americana, oblongis splendenti-
    bus foliis, fructu grandiore rubro plurimis intus
    acinis referto. *Pluk. alm.* 392. *t.* 320. *f.* 6. mala.
    American Cranberry.
    Nat. of North America.
    Cult. 1760, by Mr. James Gordon.
    Fl. May. H. ♄.

18. V. foliis ovato-oblongis acutis serratis perennanti- *meridio-*
    bus planis lucidis, racemis terminalibus erectis, *nale.*
    corollis prismaticis.
                                    Vaccinium

Vaccinium meridionale. *Swartz prodr.* 62.
Jamaica Whortle-berry.
Nat. of Jamaica. *Thomas Clark*, M. D.
Introd. 1778.
Fl.   ·   G. H. ♄.

E R I C A. *Gen. pl.* 484.

Cal. 4-phyllus. *Cor.* 4-fida. *Filamenta* receptaculo
inferta. *Antheræ* bifidæ. *Capf.* 4-locularis.

* *Antheris ariftatis, foliis oppofitis.*

vulgaris.	1. E. antheris ariftatis, corollis campanulatis fubæqua-
	libus, calycibus duplicatis, foliis oppofitis fagitta-
	tis. *Syft. veget.* 363. *Curtis lond.*
purpuraf- cens.	α floribus purpurafcentibus.
	Common Heath.
alba.	β floribus albis.
	White-flower'd Common Heath.
	Nat. of Britain.
	Fl. June——Auguft. H. ♄

lutea.	2. E. antheris ariftatis, corollis ovatis acuminatis,
	floribus congeftis, foliis oppofitis linearibus. *Syft.*
	veget. 364.
	Yellow Heath.
	Nat. of the Cape of Good Hope.
	Introd. 1774, by Mr. Francis Maffon.
	Fl. G. H. ♄.

——— *foliis ternis.*

halicaca- ba.	3. E. antheris ariftatis, corollis ovatis inflatis, ftylo in-
	clufo, floribus folitariis, foliis ternis. *Syft. veget.*
	364.
	Purple-ftalk'd Heath.
	Nat. of the Cape of Good Hope.

Introd.

Introd. about 1780.
Fl. May and June. G. H. ♄.

4. E. foliis ternis, ftylo inclufo, corolla oblonga inflata, *Monfo-*
calyce calyculato, floribus terminalibus ramulorum *niana.*
obtuforum. *Linn. fuppl.* 223.
Bladder-flower'd Heath.
Nat. of the Cape of Good Hope.
Introd. 1787, by Mr. Francis Maffon.
Fl. G. H. ♄.

5. E. antheris ariftatis, corollis fubglobofis mucofis, ftylo *mucofa.*
inclufo, foliis ternis. *Syft. veget.* 364.
Mucous Heath.
Nat. of the Cape of Good Hope.
Introd. 1787, by Mr. Francis Maffon.
Fl. G. H. ♄.

6. E. antheris ariftatis, corollis ovato-conicis villofis, *urceola-*
ftylo inclufo, calycibus lanceolatis, floribus umbel- *ris.*
latis, foliis ternis.
Erica urceolaris. *Berg. cap.* 107. *Thunb. Erica,*
n. 55.
Hairy-flower'd Heath.
Nat. of the Cape of Good Hope.
Introd. about 1778, by Mr. James Gordon.
Fl. May and June. G. H. ♄.

7. E. antheris ariftatis, corollis ovato-conicis, ftylo me- *marifolla.*
diocri, foliis ternis ovatis pubefcentibus fubtus al-
bidis.
Marum-leav'd Heath.
Nat. of the Cape of Good Hope. Mr. *Francis Maffon.*
Introd. 1773.
Fl. May and June. G. H ♄.

8. E.

fcoparia.   8. E. antheris ariftatis, corollis campanulatis, ftigmate exferto peltato, foliis ternis. *Syft. veget.* 364.

Small green-flower'd Heath.

Nat. of the South of Europe.

Introd. about 1770.

Fl. April and May.   G. H. ♄.

arborea.   9. E. antheris ariftatis, corollis campanulatis, ftylo exferto, foliis·ternis, ramulis incanis. *Syft. veget.* 365.

Tree Heath.

Nat. of the South of Europe and Madeira.

Introd. about 1748.

Fl. February——May.   H. ♄.

cruenta.   10. E. antheris ariftatis, corollis cylindraceis incurvis, ftylo exferto, calycibus fubulatis bafi dilatatis, bracteis remotis, foliis glabris.

Bloody-flower'd Heath.

Nat. of the Cape of Good Hope. Mr. *Francis Maffon.*

Introd. 1774.

Fl. at various Seafons.   G. H. ♄.

DESCR. *Rami* teretes, glabri. *Ramuli* pubefcentes. *Folia* terna, petiolata, lineari-fubulata, fulco exarata, patentia, glabra, femuncialia. *Petioli* adpreffi, vix femilineares. *Flores* axillares. *Pedunculi* vix femunciales, interdum bi- vel trifidi. *Braɛteæ* tres, fubulatæ, triquetræ. *Calycis* foliola e lata bafi fubulata, carinata, glabra, bilinearia. *CoroƖla* intenfe punicea, cylindracea, uncialis, parum incurva, glabra, fubpellucida, fuperne parum ventricofa, ore quadrifido : *laciniæ* latæ, acutiufculæ, fubereɛtæ. *Filamenta* albida. *Antheræ* inclufæ, oblongæ, brunneæ, bifidæ, ariftatæ . *ariftæ* fubulato-capillares, longitudine antherarum. *Stylus* corolla paulo longior, rubicundus. *Stigma* exfertum, incraffatum, atropurpureum.

         ———— *foliis*

——— *foliis quaternis.*

11. E. antheris ariſtatis, corollis globoſis, ſtylo incluſo,   *ramenta-*
ſtigmate duplicato, foliis quaternis ſetaceis. *Syſt.*   *cea.*
veget. 365.
Slender-branched Heath.
Nat. of the Cape of Good Hope.
Introd. 1786, by Mr. Francis Maſſon.
Fl. July.   G. H. ♄.

12. E. antheris ariſtatis, corollis campanulatis, ſtylo inclu-   *perſoluta.*
ſo, calycibus ciliatis, foliis quaternis. *Syſt. veget.* 365.
Bluſh-flower'd Heath.
Nat. of the Cape of Good Hope.
Introd. 1774, by Mr. Francis Maſſon.
Fl. February——May.   G. H. ♄.

13. E. antheris ariſtatis, corollis campanulatis glabris,   *ſtrigoſa.*
ſtylo exſerto, foliis quaternis pubeſcentibus ciliatis.
Dwarf Downy Heath.
Nat. of the Cape of Good Hope. Mr. *Francis Maſſon.*
Introd. 1775.
Fl. March and April.   G. H. ♄.
DESCR. *Rami* villoſiuſculi. *Folia* quaterna, petiolata, linearia, acuta, patentia, villoſiuſcula, pilis longis raris apice glanduloſis ciliata, trilinearia. *Petioli*
ſemilineares. *Flores* in ultimis ramulis axillares.
Pedunculi ſeſquilineares. *Bracteæ* tres, lineares,
minutæ, caducæ. *Calycis* foliola lanceolata, villoſa, adpreſſa, lineam longa. *Corolla* campanulata,
pallide rubens, glabra, calyce fere duplo longior:
tubus parum ventricoſus ; *limbi* laciniæ obtuſæ, erec-
tæ. *Filamenta* alba, calyce paulo breviora. *Antheræ* incluſæ, oblongæ, nigricantes, bipartitæ, bia-
riſtatæ : *ariſtæ* ſubulatæ, breves, parum divaricatæ.
Stylus corolla paulo longior, vireſcens. *Stigma* in-
craſſatum, convexum, nigro-purpureum.

Tetralix. 14. E. antheris ariſtatis, corollis ovatis, ſtylo incluſo, foliis quaternis ciliatis, floribus capitatis. *Syſt. veget.*
365. *Curtis lond.*

rubeſ-   α floribus rubeſcentibus.
cens.   Croſs-leav'd Heath.
alba.   β floribus albis.
           White-flower'd croſs-leav'd Heath.
           Nat. of Britain.
           Fl. June and Auguſt.   H. ♄ .

abietina. 15. E. antheris ariſtatis, corollis groſſis, ſtylo incluſo,
foliis quaternis, floribus feſſilibus. *Syſt. veget.* 365.
Fir Heath.
Nat. of the Cape of Good Hope.
Introd. 1774, by Mr. Francis Maſſon.
Fl. June and July.   G. H. ♄ .

         * * *Antheris criſtatis, foliis ternis.*

triflora. 16. E. antheris criſtatis, corollis globoſo-campanulatis,
ſtylo incluſo, foliis ternis, floribus terminalibus.
Syſt. veget. 366.
Three-flower'd Heath.
Nat. of the Cape of Good Hope.
Introd. 1787, by Mr. Francis Maſſon.
Fl.   G. H. ♄ .

baccans. 17. E. antheris criſtatis, corollis globoſo-campanulatis
tectis, ſtylo incluſo, foliis ternis imbricatis. *Syſt.*
veget. 366.
Arbutus-flower'd Heath.
Nat. of the Cape of Good Hope.
Introd. 1774, by Mr. Francis Maſſon.
Fl. April and May.   G. H. ♄ .

corifolia. 18. E. antheris criſtatis, corollis ovatis, ſtylo incluſo, ca-
                                       lycibus

lycibus turbinatis, foliis ternis, floribus umbellatis.
Syft. veget. 366.
Slender-twig'd Heath.
Nat. of the Cape of Good Hope.
Introd. 1774, by Mr. Francis Maffon.
Fl. Auguft. G. H. ♄.

19. E. antheris criftatis, corollis ovatis, ftylo fubexferto,   *cinerea.*
    foliis ternis, ftigmate capitato. *Syft. veget.* 366.
    Curtis lond.
α floribus rubentibus.   rubens.
  Fine-leav'd Heath.
β floribus albis.   alba.
  White-flower'd fine-leav'd Heath.
  Nat. of Britain.
  Fl. June and July.   H. ♄.

20. E. antheris criftatis, corollis campanulatis, ftylo ex- *panicula-*
    ferto, foliis ternis, floribus minutis. *Syft. veget.* *ta,*
    366.
  Panicled Heath.
  Nat. of the Cape of Good Hope.
  Introd. 1774, by Mr. Francis Maffon.
  Fl. February——April.   G. H. ♄.

21. E. antheris criftatis, corollis cylindricis, ftylo ex- *auftralis.*
    ferto, foliis ternis patentibus. *Syft. veget.* 366.
  Spanifh Heath.
  Nat. of Spain and Portugal.
  Introd. 1769, by George W. Earl of Coventry.
  Fl. April and May.   H. ♄.

——— *foliis quaternis.*

22. E. antheris criftatis, corollis ovatis, foliis quaternis, *empetri-*
    floribus feffilibus lateralibus. *Syft. veget.* 366. *folia.*
                    C 2   Crow-

Crow-berry-leav'd Heath.
Nat. of the Cape of Good Hope.
Introd. 1774, by Mr. Francis Maſſon.
Fl. April and May.   G. H. ♃.

margari-   23. E. antheris criſtatis, corollis urceolato-campanulatis,
tacea.   ſtylo mediocri.
         Pearl-flower'd Heath.
         Nat. of the Cape of Good Hope.   Mr. *Fr. Maſſon.*
         Introd. 1775.
         Fl. May and June.   G. H. ♃.
         DESCR. *Flores* terminales, quatuor vel octo, e ſupre-
         mis axillis. *Pedunculi* filiformes, foliis breviores.
         Bracteæ tres, lineares. *Calycis* foliola e lata baſi
         ſubulata, carinata: carina ſulco exarata, lævia,
         ſeſquilinearia. *Corolla* alba: *tubus* urceolatus, ca-
         lyce paulo longior; *limbi* laciniæ obtuſæ, patulæ.
         Filamenta alba, tubo breviora, apice inflexa. *An-*
         theræ ovatæ, compreſſæ, brunneæ, bipartitæ, criſ-
         tatæ: *laminæ* oblongæ, acutæ, extus inciſo-ſerratæ,
         albæ, longitudine antherarum. *Germen* ſuperne
         octonoduloſum, rubicundum. *Stylus* longitudine
         corollæ, albicans. *Stigma* capitatum, ſubtus pla-
         num, ſupra convexum, ſubquadrilobum.

         * * * *Antheris muticis incluſis, foliis ternis.*
ſpumoſa.   24. E. antheris muticis incluſis, corollis ternis calyce
         communi obtectis, ſtylo exſerto, foliis ternis. *Syſt.*
         veget. 367.
         Six-angled Heath.
         Nat. of the Cape of Good Hope.
         Introd. 1774, by Mr. Francis Maſſon.
         Fl.   G. H. ♃.

capitata.   25. E. antheris muticis mediocribus, corollis tectis calyce
                               lanato,

lanato, foliis ternis, floribus feffilibus. *Syft. veget.*
367.
Woolly Heath.
Nat. of the Cape of Good Hope.
Introd. 1774, by Mr. Francis Maffon.
Fl. April——July. G. H. ♄.

26. E. antheris muticis inclufis, corollis ovatis groffis, *ciliaris.*
ftylo exferto, foliis ternis, racemis fecundis. *Syft.*
veget. 368.
Ciliated Heath.
Nat. of Spain and Portugal.
Introd. about 1773.
Fl. July——September. H. ♄.

27. E. antheris muticis fubexfertis, corollis ternis ovatis *petiolata.*
longitudine calycis, ftylo exferto, foliis ternis ob-
longis petiolatis.
Erica petiolata. *Thunb. Erica, n.* 7. *tab.* 6.
Rofemary-leav'd Heath.
Nat. of the Cape of Good Hope.
Introd. 1774, by Mr. Francis Maffon.
Fl. March——June. G. H. ♄.

——— *foliis quaternis, pluribufve.*

28. E. antheris muticis inclufis, corollis clavatis groffis, *tubiflora.*
ftylo inclufo, foliis quaternis fubciliatis. *Syft.*
veget. 368.
Tube-flower'd Heath.
Nat. of the Cape of Good Hope.
Introd. 1775, by Meffrs. Kennedy and Lee.
Fl. June——Auguft. G. H. ♄.

29. E. antheris muticis inclufis, corollis clavatis groffis, *curviflo-*
ftylo inclufo, foliis quaternis glabris. *Syft. veget.* *ra.*
368.

Curve-

Curve-flower'd Heath.
Nat. of the Cape of Good Hope.
Introd. 1774, by Mr. Francis Maſſon.
Fl. Auguſt——October.   G. H. ♄.

conſpicua. 30. E. antheris muticis ſubincluſis, corollis cylindricis
curvis longiſſimis piloſis : limbo revoluto, ſtylo
exſerto, foliis quaternis glabris.
Long-tubed yellow Heath.
Nat. of the Cape of Good Hope. Mr. *Fr. Maſſon.*
Introd. 1774.
Fl. May——Auguſt.   G. H. ♄.

cerinthoi- 31. E. antheris muticis incluſis, corollis clavatis groſſis,
des,   ſtigmate incluſo cruciato, foliis quaternis. *Syſt.*
veget. 368.
Honey-wort-flower'd Heath.
Nat. of the Cape of Good Hope.
Introd. 1774, by Mr. Francis Maſſon.
Fl. moſt part of the Year.   G. H. ♄.

viſcaria. 32. E. antheris muticis incluſis, corollis campanulatis
glutinoſis, ſtylo incluſo, foliis quaternis, floribus
racemoſis. *Syſt. veget.* 369.
Clammy-flower'd Heath.
Nat. of the Cape of Good Hope.
Introd. 1774, by Mr. Francis Maſſon.
Fl. March.   G. H. ♄.

comoſa, 33. E. antheris muticis incluſis, corollis ovato-oblongis,
ſtylo incluſo, foliis quaternis, floribus congeſtis.
Syſt. veget. 369.
Tufted-flower'd Heath.
Nat. of the Cape of Good Hope.
Introd. 1787, by Mr. Francis Maſſon.
Fl.   G. H. ♄.
34. E.

34. E. antheris muticis inclusis, corollis cylindricis basi *concinna.* attenuatis; floribus terminalibus umbellatis, foliis subsenis glabris.

Flesh-colour'd Heath.

Nat. of the Cape of Good Hope. Mr. *Fr. Masson.*
Introd. 1773.

Fl. September and October.   G. H. ♄.

DESCR. *Rami* glabri. *Folia* ramorum sena; ramulorum quaterna, petiolata, erecta, acerosa, glabra, quadrilinearia. *Petioli* glabri, vix semilineares. *Flores* terminales, umbellati: *umbellæ* 3—6 floræ. *Pedunculi* filiformes, pilosiusculi, bilineares. *Bracteæ* tres, subulatæ, ciliatæ, calyci adpressæ. *Calycis* foliola e lata basi subulata, ciliata, dorso sulco exarata, bilinearia. *Corolla* carnea, extus villosiuscula, cylindracea, vix uncialis, crassitie pennæ gallinæ. *Filamenta* glabra, tubo corollæ paulo breviora. *Antheræ* oblongæ, basi acuminatæ, apice ad medium bipartitæ, ubi filamentis affixæ, muticæ, brunneæ, inclusæ. *Germen* turbinatum, supra concavum, margine crenulatum. *Stylus* rubicundus, longitudine staminum. *Stigma* subcapitatum, atrorubens.

35. E. antheris muticis inclusis, corollis cylindricis *Massoni.* grossis, floribus capitatis, foliis octofariis imbricatis pubescentibus. *Linn. suppl.* 221.

Tall downy Heath.

Nat. of the Cape of Good Hope.
Introd. 1787, by Mr. Francis Masson.
Fl.   G. H. ♄.

* * * * *Antheris muticis exsertis, foliis ternis.*

36. E. antheris muticis longissimis exsertis, corollis cy- *Plukene-* lindricis, stylo exserto, calycibus simplicibus, foliis *tii.* ternis. *Syst. veget.* 369.

C 4   Smooth-

Smooth-twig'd pencil-flower'd Heath.
Nat. of the Cape of Good Hope.
Introd. 1774, by Mr. Francis Maffon.
Fl.   G. H. ♄.

Petiveri. 37. E. antheris muticis longiffimis exfertis, corollis acu-
tis, ftylo exferto, calycibus imbricatis, foliis ternis.
Syft. veget. 369.
Downy-twig'd pencil-flower'd Heath.
Nat. of the Cape of Good Hope.
Introd. 1774, by Mr. Francis Maffon.
Fl. January——March.   G. H. ♄.

———— *foliis quaternis, pluribufve.*

herbacea. 38. E. antheris muticis exfertis, corollis oblongis, ftylo
exferto, foliis quaternis, floribus fecundis. *Syft.*
veget. 370. *Curtis magaz.* 11.
Erica carnea. *Sp. pl.* 504. *Jacqu. auftr.* 1. *p.* 21.
t. 32.
Early-flowering Dwarf Heath.
Nat. of Auftria and Switzerland.
Introd. 1763, by the Earl of Coventry.
Fl. January——April.   H. ♄.

multiflo- 39. E. antheris muticis exfertis, corollis cylindricis, ftylo
ra.   exferto, foliis quinis, floribus fparfis. *Syft. veget.*
370.
Many-flower'd Heath.
Nat. of the South of Europe.
Fl. June——November.   H. ♄.

mediter- 40. E. antheris muticis exfertis, corollis ovatis, ftylo
ranea.   exferto, foliis quaternis patentibus, floribus fparfis. *Syft. veget.* 370.
Mediterranean Heath.

Nat.

Nat. of the South of Europe.
Introd. about 1765, by Mr. Joſhua Brooks.
Fl. March——May.   H. ♄.

41. E. antheris muticis exſertis, corollis cylindraceis ſub-   *grandi-*
incurvis glabris, ſtylo elongato, floribus axillaribus   *flora.*
pedunculatis, foliis ſubſenis aceroſis glabris.
Erica grandiflora. *Linn. ſuppl.* 223.
Great-flower'd Heath.
Nat. of the Cape of Good Hope. Mr. *Fr. Maſſon.*
Introd. 1775.
Fl. May —— July.   G. H. ♄.

D A P H N E. *Gen. pl.* 485.

Cal. 0. *Cor.* 4-fida, corollacea, marceſcens, ſtamina
includens. *Bacca* 1-ſperma.

* *Floribus lateralibus.*

1. D. floribus ſeſſilibus ternis caulinis, foliis lanceolatis   *Mezere-*
deciduis. *Sp. pl.* 509.   *um.*
α floribus rubris.   rubrum.
Common Spurge Olive, or Mezerion.
β Thymelæa Lauri folio deciduo, flore albido, fructu   album.
flaveſcente. *Du Hamel arb.* 2. *p.* 325. *n.* 4. *Du*
Roi hort. harbecc. 1. *p.* 213.
White-flower'd Mezerion.
Nat. of England.
Fl. February and March.   H. ♄.

2. D. floribus ſeſſilibus aggregatis axillaribus, foliis ova-   *Tarton-*
tis utrinque pubeſcentibus nervoſis. *Sp. pl.* 510.   *raira.*
Silvery-leav'd Daphne.
Nat. of the South of France.
Cult. 1739. *Mill. dict. vol.* 2.   Thymelæa 16.
Fl.   H. ♄.
                                                 3. D.

alpina. 3. D. floribus feffilibus aggregatis lateralibus, foliis lan-
ceolatis obtufiufculis fubtus tomentofis. *Sp. pl.* 510.
Alpine Daphne.
Nat. of the Alps of Italy, Switzerland, and Auftria.
Cult. 1759, by Mr. Ph. Miller. *Mill. dict. edit.* 7. *n.* 5.
Fl. May and June.   H. ♄.

Laureola. 4. D. racemis axillaribus quinquefloris, foliis glabris
lanceolatis. *Syft. veget.* 371. *Jacqu. auftr.* 2.
p. 49. *t.* 183.
Common Daphne, or Spurge-laurel.
Nat. of Britain.
Fl. January——March.   H. ♄.

** *Floribus terminalibus.*

odora. 5. D. capitulo terminali fubfeffili multifloro, foliis fparfis
oblongo-lanceolatis glabris.
Daphne odora. *Thunb. japon.* 159. *Ic. Kæmpfer.*
tab. 16. *L'Herit. ftirp. nov. tom.* 2. *tab.* 7.
Sweet-fcented Daphne.
Nat. of China and Japan.
Introd. 1771, by Benjamin Torin, Efq.
Fl. December——March.   G. H. ♄.

Cneorum. 6. D. floribus congeftis terminalibus feffilibus, foliis
lanceolatis nudis mucronatis. *Syft. veget.* 371.
Jacqu. auftr. 5. *p.* 12. *t.* 426.
Trailing Daphne.
Nat. of Switzerland and Auftria.
Cult. 1739. *Mill. dict. vol.* 2. Thymelæa 2.
Fl. April——September.   H. ♄.

Gnidium. 7. D. panicula terminali, foliis lineari-lanceolatis acuminatis. *Sp. pl.* 511.
Flax-leav'd Daphne.

Nat. of Spain and Italy.
Cult. 1597, by Mr. John Gerard. *Ger. herb.* 1217.
  f. 2.
Fl. June and July.   H. ♄.

D I R C A. *Gen. pl.* 486.

Cal. o. *Cor.* tubulofa limbo obfoleto. *Stam.* tubo
        longiora. *Bacca* 1-fperma.

1. DIRCA. *Sp. pl.* 512.   *paluſtris.*
   Marſh Leather-wood.
   Nat. of Virginia.
   Cult. 1750, by Archibald Duke of Argyle.
   Fl. March and April.   H. ♄.

G N I D I A. *Gen. pl.* 487.

Cal. infundibuliformis, 4-fidus. *Petala* 4, calyci in-
       ferta. *Sem.* 1, fubbaccatum.

1. G. foliis omnibus linearibus acutis, floribus termina- *ſimplex.*
   libus feſſilibus. *Linn. mant.* 67.
   Flax-leav'd Gnidia.
   Nat. of the Cape of Good Hope.
   Introd. 1786, by Mr. Francis Maſſon.
   Fl.   G. H. ♄.

2. G. foliis ovatis tomentofis : floralibus quaternis, caule *ſericea.*
   hirfuto, florum corona fetis octo. *Syſt. veget.* 373.
   Paſſerina fericea. *Sp. pl.* 513.
   Silky Gnidia.
   Nat. of the Cape of Good Hope.
   Introd. 1786, by Mr. Francis Maſſon.
   Fl.   G. H. ♄.

S T E L-

STELLERA. *Gen. pl.* 488.

Cal. o. *Cor.* 4-fida. *Stam.* breviffima. *Sem.* 1, roftratum.

Pafferi-na.
1. S. foliis linearibus, floribus quadrifidis. *Sp. pl.* 512.
Jacqu. ic. collect. 1. *p.* 65.
Flax-leav'd Stellera.
Nat. of the South of Europe.
Cult. 1759, by Mr. Philip Miller.
Fl. July and Auguft. H. ☉.

PASSERINA. *Gen. pl.* 489.

Cal. o. *Cor.* 4-fida. *Stam.* tubo impofita. *Sem.* 1, çorticatum.

filiformis.
1. P. foliis linearibus convexis quadrifariam imbricatis, ramis tomentofis. *Sp. pl.* 513.
African Sparrow-wort.
Nat. of the Cape of Good Hope.
Cult. 1752, by Mr. Ph. Miller. *Mill. dict. edit.* 6. *n.* 1.
Fl. June——Auguft. G. H. ♃.

LACHNÆA. *Gen. pl.* 490.

Cal. o. *Cor.* 4-fida: limbo inæquali. *Sem.* 1, fub-baccatum.

conglome-rata.
1. L. capitulis confertis, foliis laxis. *Sp. pl.* 514.
Clufter-headed Lachnæa.
Nat. of the Cape of Good Hope.
Introd. 1773, by Meffrs. Kennedy and Lee.
Fl. G. H. ♃.

DIGYNIA.

DIGYNIA.

GALENIA. *Gen. pl.* 492.

Cal. 4-fidus. *Cor.* o. *Capf.* fubrotunda, 2-fperma.

1. GALENIA. *Sp. pl.* 515.　　　　　　*africana.*
African Galenia.
Nat. of the Cape of Good Hope.
Cult. 1752, by Mr. Philip Miller. *Mill. dict. edit.* 6.
Fl. June——Auguft.　　　　　　G. H. ♄.

MŒHRINGIA. *Gen. pl.* 494.

Cal. 4-phyllus. *Petala* 4. *Capf.* 1-locularis, 4-valvis.

1. MOEHRINGIA. *Sp. pl.* 515. *Jacqu. auftr.* 5. *p.* 24.　*mufcofa.*
　　t. 449.
Moffy Mœhringia.
Nat. of the South of Europe.
Introd. 1775, by the Doctors Pitcairn and Fothergill.
Fl. June and July.　　　　　　H. ♃.

TRIGYNIA.

POLYGONUM. *Gen. pl.* 495.

Cal. o. *Cor.* 5-partita, calycina. *Sem.* 1, angulatum.

　　＊ Atraphaxoides, *caule frutefcente.*

1. P. caule fruticofo, calycinis foliolis duobus reflexis.　*frutef-*
　　Sp. pl. 516.　　　　　　　　　　　　　　*cens.*
Shrubby Polygonum.
Nat. of Siberia.
Introd. 1770, by Monf. Richard.
Fl. July.　　　　　　H. ♃.
　　　　　　　　　　＊＊ Biftortæ,

** Biftortæ, *fpica unica.*

Biftorta. 2. P. caule fimpliciffimo monoftachyo, foliis ovatis in
petiolum decurrentibus. *Sp. pl.* 516. *Curtis lond.*
Great Biftort, or Snakeweed.
Nat. of Britain.
Fl. May——September.   H. ♃.

vivipa- 3. P. caule fimpliciffimo monoftachyo, foliis lanceolatis.
rum.   *Sp. pl.* 516.
Viviparous Polygonum, or Small Biftort.
Nat. of Britain.
Fl. May——September.   H. ♃.

*** Perficariæ, *piftillo bifido, aut ftamina minus* 8.

virginia- 4. P. floribus pentandris femidigynis, corollis quadrifidis
num.   inæqualibus, foliis ovatis. *Sp. pl.* 516.
Virginian Polygonum.
Nat. of North America.
Cult. 1640, by Mr. John Morrice. *Park. theat.* 857.
f. 6. 7.
Fl. Auguft and September.   H. ♃.

lapathifo- 5. P. floribus hexandris digynis, ftipulis muticis, pedun-
lium.   culis fcabris, feminibus utrinque depreffis. *Curtis*
lond. (fub nomine P. penfylvanici.)
Polygonum Lapathifolium. *Sp. pl.* 517.
α Polygonum penfylvanicum. *Curtis lond.*
Pale-flower'd Polygonum.
β Polygonum penfylvanicum caule maculato. *Curtis*
lond.
Spotted-ftalk'd Polygonum.
Nat. of England.
Fl. September.   H. ☉.

6. P.

6. P. floribus pentandris femidigynis, fpica ovata. *Syft.*   *amphibi-*
   veget. 377. *Curtis lond.*   *um.*
   Amphibious Polygonum, or Perficaria.
   Nat. of Britain.
   Fl. June and July.   H. ♃.

7. P. floribus pentandris trigynis, foliis lanceolatis. *Sp.*   *ocreatum.*
   pl. 517.
   Spear-leav'd Polygonum.
   Nat. of Siberia.
   Introd. 1780, by Peter Simon Pallas, M. D.
   Fl. July.   H. ♃.

8. P. floribus hexandris femidigynis, foliis lanceolatis,   *Hydropi-*
   ftipulis fubmuticis. *Sp. pl.* 517. *Curtis lond.*   *per.*
   Water Polygonum, Pepper, or Arfmart.
   Nat. of Britain.
   Fl. July and Auguft.   H. ☉.

9. P. floribus hexandris digynis, fpicis ovato-oblongis,   *Perfica-*
   foliis lanceolatis, ftipulis ciliatis. *Sp. pl.* 518.   *ria.*
   Curtis lond.
   Spotted Polygonum, or Perficaria.
   Nat. of Britain.
   Fl. Auguft.   H. ☉.

10. P. floribus hexandris fubmonogynis, foliis lineari-   *minus.*
   lanceolatis, caule bafi repente. *Curtis lond.*
   Creeping Polygonum.
   Nat. of England.
   Fl. September.   H. ☉.

11. P. floribus hexandris trigynis, fpicis virgatis, ftipulis   *tinctori-*
   glabris arctis truncatis ciliatis, foliis ovatis acutiuf-   *um.*
   culis glabris.

                        Dyer's

Dyer's Polygonum.
Nat. of China.
Introd. 1776, by John Blake, Efq.
Fl. July and Auguft.   G. H. ♂.

orientale.   12. P. floribus heptandris digynis, foliis ovatis, caule erec-
        to, ftipulis hirtis hypocrateriformibus. *Sp. pl.* 519,
    α floribus carneis.
        Oriental red Perficaria.
    β floribus albis.
        Oriental white Perficaria.
        Nat. of the Eaft Indies.
    α *Cult.* 1707, by the Dutchefs of Beaufort. *Br. Muf.*
        H. S. 137. *fol.* 35.
    β *Introd.* 1781, by Monf. de Wevelinchoven.
        Fl. July——October.   H. ☉.

    **** Polygona, *foliis indivifis, floribus octandris.*

mariti-   13. P. floribus octandris trigynis axillaribus, foliis ovali-
mum.   lanceolatis fempervirentibus, caule fuffrutefcente.
        Sp. pl. 519.
        Sea Polygonum, or Knotgrafs.
        Nat. of England.
        Fl. July.   H. ♃.

avicu-   14. P. floribus octandris trigynis axillaribus, foliis lan-
lare.   ceolatis, caule procumbente herbaceo. *Sp. pl.*
        519. *Curtis lond.*
        Common Knotgrafs.
        Nat. of Britain.
        Fl. June——September.   H. ☉.

divarica-   15. P. floribus octandris trigynis racemofis, foliis lan-
tum.   ceolatis, caule divaricato patulo. *Sp. pl.* 520.
        Divaricated Polygonum.

                                                        Nat.

Nat. of Siberia.
Cult. 1759, by Mr. Philip Miller.
Fl. July and Auguſt. H. ♃.

***** Helxine, *foliis ſubcordatis.*

16. P. foliis ſagittatis, caule aculeato. *Sp. pl.* 521.   *ſagitta-*
     Prickly Polygonum.   *tum.*
     Nat. of North America.
     Cult. 1759, by Mr. Philip Miller.
     Fl. July and Auguſt. H. ☉.

17. P. foliis cordato-ſagittatis, caule inermi erecto, ſe-   *tatari-*
     minibus ſubdentatis. *Sp. pl.* 521.   *cum.*
     Tartarian Polygonum.
     Nat. of Siberia.
     Cult. 1759, by Mr. Philip Miller.
     Fl. July and Auguſt. H. ☉.

18. P. foliis cordato-ſagittatis, caule erectiuſculo inermi,   *Fagopy-*
     ſeminum angulis æqualibus. *Sp. pl.* 522.   *rum.*
     Cultivated Polygonum, or Buck-wheat.
     Nat. of England.
     Fl. July and Auguſt. H. ☉.

19. P. foliis cordatis, caule volubili angulato, floribus   *Convol-*
     obtuſatis. *Syſt. veget.* 379. *Curtis lond.*   *vulus.*
     Black Polygonum, or Bindweed.
     Nat. of Britain.
     Fl. May——September. H. ☉.

20. P. foliis cordatis, caule erecto ſcandente. *Syſt.*   *ſcandens.*
     veget. 379.
     Climbing Polygonum.
     Nat. of North America.
     Cult. 1759, by Mr. Philip Miller.
     Fl. Auguſt and September. H. ♃.

Vol. II.   D   COCCO-

COCCOLOBA. *Gen. pl.* 496.

Cal. 5-partitus, coloratus.   *Cor.* o.   *Bacca* calycina,
monosperma.

Uvifera. 1. C. foliis cordato-subrotundis nitidis. *Sp. pl.* 523.
Round-leav'd Sea-side Grape.
Nat. of the West Indies.
Introd. 1690, by Mr. Bentick. *Br. Mus. Sloan. mss.*
3370.
Fl.   S. ♄.

excoriata. 2. C. foliis ovatis, ramis quasi excorticatis. *Sp. pl.* 524.
Oval-leav'd Sea-side Grape.
Nat. of the West Indies.
Introd. before 1733, by William Houstoun, M.D.
Mill. dict. vol. 2. Guaiabara 1.
Fl.   S. ♄.

punctata. 3. C. foliis lanceolato-ovatis. *Sp. pl.* 523.
Spear-leav'd Sea-side Grape.
Nat. of the West Indies.
Introd. before 1733, by William Houstoun, M.D. *Mill.*
dict. edit. 7. Addenda. Coccolobis 5.
Fl.   S. ♄.

PAULLINIA. *Gen. pl.* 497.

Cal. 5-phyllus. *Petala* 4. *Nectar.* 4-phyllum, in-
æquale. *Caps.* 3, compressæ, membranaceæ, connatæ.

curassa- 1. P. foliis biternatis, petiolulis omnibus marginatis, ra-
vica.   mis inermibus. *Syst. veget.* 380.
Shining-leav'd Paullinia.
Nat. of South America.
                                              Cult.

Cult. 1739, by Mr. Ph. Miller. *Rand. chel.* Cururu 1.
Fl.   S. ♄.

2. P. foliis triternatis, petiolulis nudis. *Syſt. veget.* 380.   *polyphyl-*
Parſley-leav'd Paullinia, or Supple Jack.   *la.*
Nat. of the Weſt Indies.
Cult. 1739, by Mr. Ph. Miller. *Rand. chel.* Serjánia 1.
Fl.   S. ♄.

CARDIOSPERMUM. *Gen. pl.* 498.

Cal. 4-phyllus. *Petala* 4. *Nectarium* 4-phyllum, in-
æquale. *Capſ.* 3, connatæ, inflatæ.

1. C. foliis lævibus. *Sp. pl.* 525.   *Halica-*
Smooth-leav'd Heart-feed.   *cabum.*
Nat. of both the Indies.
Cult. 1594, by Mr. John Gerard. *Ger. herb.* 271. *f.* 2.
Fl. July.   S. ☉.

2. C. foliis ſubtus tomentoſis. *Sp. pl.* 526.   *Corin-*
Parſley-leav'd Heart-feed.   *dum.*
Nat. of Brazil.
Cult. 1759, by Mr. Philip Miller. *Mill. dict. edit.* 7.
n. 3.
Fl. July and Auguſt.   S. ☉.

SAPINDUS. *Gen. pl.* 499.

Cal. 4-phyllus. *Petala* 4. *Capſ.* carnoſæ, connatæ,
ventricoſæ.

1. S. inermis, foliis pinnatis : foliolis lanceolatis; rachi   *Sapona-*
alata.   *ria.*
Sapindus Saponaria. *Sp. pl.* 526. (excluſo ſynonymo
Plukenetii.)
Common Soap-berry.

                        Nat.

Nat. of the Weſt Indies.

Cult. 1697, by the Dutcheſs of Beaufort.   *Br. Muſ.*
> *Sloan. mſſ.* 3357. *fol.* 62.

Fl.   S. ♄ .

rigida.   2. S. inermis, foliis pinnatis : foliolis ovato-oblongis ;
rachi ſimplici, fructibus glabris.

Sapindus rigida.   *Mill. dict.*

Nuciprunifera arbor Americana, fructu ſaponario orbiculato monococco nigro.   *Pluk. alm.* 255. *tab.*
217. *fig.* 7.

Aſh-leav'd Soap-berry.

Nat. of the Weſt Indies.

Cult. 1759, by Mr. Ph. Miller.   *Mill. dict. edit.* 7. *n.* 2.

Fl. July —— September.   G. H. ♄ .

edulis.   3. S. inermis, foliis pinnatis : foliolis lanceolato-oblongis ;
rachi ſimplici, fructibus muricatis.

Lä-tji.   *Oſb. it.* 192, 204.

Litchi.   *Bergius om läckerheter, pag.* 171.

Lêchea.   *Richard hiſt. du Tonquin* 1. *p.* 60.

Liſchia ſive Liſchion Indiæ orientalis.   *Zanon. hiſt.*
147. *tab.* 108.

Li tchi.   *Du Halde chin.* 2. *p.* 144. *fig. in tab. ad*
pag. 154.

Li-ci.   *Boym fl. ſinens. tab.* D.

Chineſe Lee-chee.

Nat. of China, Tunquin, and Cochinchina.

Introd. 1786, by Warren Haſtings, Eſq.

Fl.   S. ♄ .

TETRAGY-

TETRAGYNIA.

P A R I S. *Gen. pl.* 500.
Cal. 4-phyllus. *Petala* 4, anguftiora. *Bacca* 4-
    locularis.

1. PARIS. *Sp. pl.* 526.   *quadrifo-*
    Herb-Paris, True-love, or One-berry.   *lia.*
    Nat. of Britain.
    Fl. May and June.   H. ♃.

A D O X A. *Gen. pl.* 501.
Cal. 2-fidus, inferus. *Cor.* 4-f. 5-fida, fupera. *Bacca*
    4-f. 5-locularis, calyce coalita.

1. ADOXA. *Sp. pl.* 527. *Curtis lond.*   *Mofcha-*
    Tuberous Mofchatel.   *tellina.*
    Nat. of Britain.
    Fl. March and April.   H. ♃.

H A L O R A G I S. *Linn. fuppl.* 34.
Cal. 4-phyllus, fuperus. *Petala* 4. *Drupa* ficca.
    Nux 4-locularis.

1. H. foliis ferratis, floribus verticillatis.   *Cercodia.*
    Haloragis Cercodia. *Medic. bot. beobacht.* 1783. *p.* 73.
    Haloragis Tetragonia. *L'Herit. ftirp. nov. p.* 82.
    Haloragis alata. *Jacqu. ic. mifcell.* 2. *p.* 332.
    Cercodia erecta. *Murray in commentat. gotting.* 1780.
        p. 3. *tab.* 1.
    Tetragonia Ivæfolia. *Linn. fuppl.* 257.
    Whorl'd-flower'd Haloragis.
    Nat. of New Zealand. Sir *Jofeph Banks*, Bart.
    Introd. 1772.
    Fl. moft part of the Summer.   G. H. ♄.
        D 3   *Claffis*

Claſſis IX.

ENNEANDRIA

MONOGYNIA.

LAURUS. *Gen. pl.* 503.

Cal. o. *Cor.* calycina, 6-partita. *Nectarium* glandulis 3, bifetis, germen cingentibus. *Filamenta* interiora glandulifera. *Drupa* 1-fperma.

Cinnamo-   1. L. foliis trinerviis ovato-oblongis : nervis verfus api-
mum.   cem evanefcentibus. *Sp. pl.* 528.
           Cinnamon-tree.
           Nat. of Ceylon.
           Cult. 1768, by Mr. Philip Miller. *Mill. dict. edit.* 8.
           Fl.   S. ♄.

Campho-   2. L. foliis triplinerviis lanceolato-ovatis. *Sp. pl.* 528.
ra.   Camphire-tree.
           Nat. of Japan.
           Cult. 1731, by Mr. Philip Miller. *Mill. act. eau.* 1.
           Camphora.
           Fl.   G. H. ♄.

Chloroxy-   3. L. foliis trinerviis ovatis coriaceis : nervis apicem
lon.   attingentibus. *Sp. pl.* 508.
           Jamaica Laurel.
           Nat. of Jamaica.
           Introd. 1778, by Thomas Clark, M.D.
           Fl.   S. ♄.

4. L.

4. L. foliis venofis lanceolatis perennantibus, floribus *nobilis.*
quadrifidis. *Syft. veget.* 383.
Common Sweet-bay.
Nat. of Italy.
Cult. 1562. *Turn. herb. part* 2. *fol.* 32.
Fl. April and May. H. ♄.

5. L. foliis venofis lanceolatis perennantibus planis, ra- *indica.*
mulis tuberculatis cicatricibus, floribus racemofis.
Sp. pl. 529.
Royal Bay, or Indian Laurel.
Nat. of Madeira.
Cult. 1665. *Rea's flora* 15.
Fl. October and November. G. H. ♄.

6. L. foliis venofis ellipticis acutis perennantibus : axillis *fœtens.*
venarum fubtus villofis, racemis elongatis compo-
fitis paniculæformibus.
Madeira Laurel, or Til.
Nat. of Madeira and the Canary Iflands.
Introd. 1760, by Lord Adam Gordon.
Fl. G. H. ♄.

7. L. foliis ovatis coriaceis tranfverfe venofis perennan- *Perfea.*
tibus, floribus corymbofis. *Syft. veget.* 383.
Alligator Pear.
Nat. of the Weft Indies.
Cult. 1739, by Mr. Philip Miller. *Rand. chel.* Perfea.
Fl. S. ♄.

8. L. foliis lanceolatis perennantibus, calycibus fructus *Borbo-*
baccatis. *Syft. veget.* 383. *nia.*
Broad-leav'd Carolina Bay.
Nat. of Virginia and Carolina.
Cult. 1739, by Mr. Ph. Miller. *Rand. chel. n.* 6.
Fl. April and May. G. H. ♄.

æstiva-
lis.

9. L. foliis venofis oblongis acuminatis annuis fubtus rugofis, ramis fupra-axillaribus. *Syft. veget.* 384.
Willow-leav'd Bay.
Nat. of North America.
Introd. 1765, by Mr. John Cree.
Fl.   G. H. ♄.

Benzoin.

10. L. foliis enerviis ovatis utrinque acutis integris annuis. *Sp. pl.* 530.
Common Benjamin-tree.
Nat. of Virginia.
Cult. 1688, by Bifhop Compton. *Raj. hift.* 2. *p.* 1845.
Fl. April and May.   H. ♄.

Saffafras.

11. L. foliis integris trilobifque. *Sp. pl.* 530.
Saffafras-tree.
Nat. of North America.
Cult. before 1633, by Mr. Wilmot. *Ger. emac.* 1525.
Fl. May and June.   H. ♄.

ANACARDIUM. *Gen. pl.* 520.

Cal. 5 partitus. *Petala* 5, reflexa. *Antheræ* 9: unico caftrato. *Nux* reniformis fupra receptaculum carnofum.

occiden-
tale.

1. ANACARDIUM. *Sp. pl.* 548.
Cafhew-nut.
Nat. of both Indies.
Cult. 1699, by the Dutchefs of Beaufort. *Br. Muf.*
    Sloan. mff. 525 and 3349.
Fl.   S. ♄.

TRIGYNIA.

TRIGYNIA.

RHEUM. *Gen. pl.* 506.

Cal. 0. *Cor.* 6-fida, perſiſtens. *Sem.* 1, triquetrum.

1, R. foliis obtuſis glabris : venis ſubtus piloſiuſculis ;   *Rhapon-*
ſinu baſeos dilatato, petiolis ſupra ſulcatis margine   *ticum.*
rotundatis.
Rheum Rhaponticum. *Sp. pl.* 531.
Rhapontic Rhubarb.
Nat. of Aſia.
Cult. 1629, by Mr. John Parkinſon. *Park. parad.*
485. *f.* 3.
Fl. May.   H. ♃.

2. R. foliis ſubvilloſis undulatis : ſinu baſeos dilatato,   *undula-*
petiolis ſupra planis margine acutis.   *tum.*
Rheum undulatum. *Sp. pl.* 531.
Rheum Rhabarbarum. *Syſt. veget.* 385.
Waved-leav'd Rhubarb.
Nat. of China and Siberia.
Cult. 1759, by Mr. Ph. Miller. *Mill. dict. edit.* 7. *n.* 2.
Fl. May.   H. ♃.

3. R. foliis palmatis acuminatis ſcabriuſculis : ſinu baſeos   *palma-*
dilatato, petiolis ſupra obſolete ſulcatis margine ro-   *tum.*
tundatis.
Rheum palmatum. *Sp. pl.* 531.
Officinal Rhubarb.
Nat. of China.
Cult. 1768, by Mr. Philip Miller. *Mill. dict. edit.* 8.
Fl. April and May.   H. ♃.

4. R.

compac-
tum.

4. R. foliis fublobatis obtufiffimis glaberrimis lucidis denticulatis. *Syft. veget.* 385.

Thick-leav'd Rhubarb.

Nat. of Tartary.

Introd. 1779, by Chevalier Thunberg.

Fl. June.   H. ♃.

Ribes.

5. R. foliis obtufiffimis fubverruculofis: venis fubtus fpinulofis, petiolis fupra planis margine rotundatis.

Rheum Ribes. *Sp. pl.* 532.

Warted-leav'd Rhubarb.

Nat. of the Levant.

Cult. 1724, by James Sherard, M.D. *Knowlton's mff.*

Fl.   H. ♃.

hybridum.

6. R. foliis fupra glabris fubtus pilofiufculis fublobatis acutis: finu bafeos anguftato, petiolis fupra obfolete fulcatis margine rotundatis.

Rheum hybridum. *Murray nov. comm. gotting.* 5. *p.* 50. *t.* 12. *f.* 1-3.

Baftard Rhubarb.

Nat.

Cult. 1778, by John Fothergill, M.D.

Fl. June.   H. ♃.

HEXAGYNIA.

BUTOMUS. *Gen. pl.* 507.

Cal. 0.   *Petala* 6.   *Capf.* 6, polyfpermæ.

umbella-
tus.

1. BUTOMUS. *Sp. pl.* 532. *Curtis lond.*

Flowering Rufh.

Nat. of Britain.

Fl. June and July.

H. ♃.
Claffis

Claffis X.

D E C A N D R I A

M O N O G Y N I A.

S O P H O R A. *Gen. pl.* 508.

Cal. 5-dentatus, fuperne gibbus. *Cor.* papilionacea : alis longitudine vexilli. *Legumen.*

1. S. foliis pinnatis : foliolis numerofis (17-19) lanceo-   *tetrapte-* lato-oblongis villofiufculis, leguminibus membra-   *ra.* naceo-quadrangulis, caule arboreo.
Sophora tetraptera. *Joh. Miller. ic. tab.* 1.
Winged-podded Sophora.
Nat. of New Zealand. Sir *Jofeph Banks*, Bart.
Introd. 1772.
Fl. May and June.   H. ♄ .

2. S. foliis pinnatis : foliolis numerofiffimis (33-41)   *micro-* obovatis villofiufculis, leguminibus membranaceo-   *phylla.* quadrangulis, caule arboreo.
Sophora tetraptera. *Linn. fuppl.* 230,
Small-leav'd fhrubby Sophora.
Nat. of New Zealand. Sir *Jofeph Banks*, Bart.
Introd. 1772.
Fl. May and June.   H. ♄ .

3. S. foliis pinnatis : foliolis numerofis oblongis glabris,   *flavefcens,* caule herbaceo.
Siberian Sophora,

                               Nat.

Nat. of Siberia.

Introd. 1785, by Mr. John Bell.   H. ♃.

alopecu-   4. S. foliis pinnatis : foliolis numerofis villofis oblongis,
roides.   caule herbaceo. *Syft. veget.* 391.
                  Fox-tail Sophora.
                  Nat. of the Levant.
                  Cult. 1732, by. James Sherard, M.D. *Dill. elth.*
                      136. *t.* 112. *f.* 136.
                  Fl. July and Auguft.   H. ♃.

tomentofa.   5. S. foliis pinnatis : foliolis numerofis fubrotundis to-
                      mentofis. *Sp. pl.* 533.
                  Downy Sophora.
                  Nat. of Ceylon.
                  Cult. 1739, by Mr. Philip Miller. *Mill. dict. edit.* 1.
                      Coronilla 5.
                  Fl.   S. ♄.

acciden-   6. S. foliis pinnatis : foliolis numerofis fubcordatis. *Sp.*
talis.   *pl.* 533.
                  Occidental Sophora.
                  Nat. of the Weft Indies.
                  Introd. about 1778.
                  Fl.   S. ♄.

capenfis.   7. S. foliis pinnatis : foliolis numerofis lanceolatis fub-
                      tus tomentofis, caule fruticofo. *Syft. veget.* 391.
                  Vetch-leav'd Sophora.
                  Nat. of the Cape of Good Hope.
                  Introd. 1773, by Mr. Francis Maffon.
                  Fl.   G. H. ♄.

aurea.   8. S. foliis pinnatis : foliolis numerofis oblongo-ovalibus
                      fupra glaberrimis, caule fruticofo.

                                                        Robinia

Robinia fubdecandra. *L'Herit. ſtirp. nov. tab.* 75.
Golden-flower'd Sophora.
Nat. of Africa.
Introd. 1777, by Monſ. Thouin.
Fl. July.   G. H. ♄ .

9. S. foliis pinnatis : foliolis pluribus ovatis glabris, *japonica.*
    caule arboreo. *Syſt. veget.* 391.
Shining-leav'd Sophora.
Nat. of Japan.
Introd. 1753, by Mr. James Gordon.
Fl.   H. ♄ .

10. S. foliis ternatis feſſilibus : foliolis linearibus. *Sp. pl.*   *Geniſtoi-*
    534.   *des.*
Broom-leav'd Sophora.
Nat. of the Cape of Good Hope.
Introd. 1787, by Mr. Francis Maſſon.
Fl.   G. H. ♄ .

11. S. foliis ternatis fubfeſſilibus glabris, ſtipulis enſi-   *auſtralis.*
    formibus. *Syſt. veget.* 391.
Blue Sophora.
Nat. of Carolina.
Cult. 1758, by Mr. Philip Miller.
Fl. June and July.   H. ♃ .

12. S. foliis ternatis fubfeſſilibus : foliolis obovatis gla-   *tinctoria.*
    bris, ſtipulis minutis. *Syſt. veget.* 391.
Dyers Sophora.
Nat. of Barbadoes and Virginia.
Cult. 1759. *Mill. dict. ed.* 7. *n.* 3.
Fl. July and Auguſt.   H. ♃ .

13. S. foliis ternatis petiolatis : foliolis ellipticis glabris,   *alba.*
    ſtipulis fubfubulatis brevibus. *Syſt. veget.* 391.
                                                    Crotalaria

Crotalaria alba. *Sp. pl.* 1006.
White Sophora.
Nat. of Virginia and Carolina.
Introd. 1724, by Mr. Mark Catefby.   *Mart. dec.* 4.
p. 44.
Fl. June.   H. ♃ .

lupinoi-   14. S. foliis ternatis petiolatis : foliolis ovalibus pilofis.
des.   *Sp. pl.* 534.
              Lupin-leav'd Sophora.
              Nat. of Kamtfchatka.
              Introd. 1776, by Hugh Duke of Northumberland.
              Fl.   H. ♃ .

biflora.   15. S. foliis fimplicibus obovatis fubtomentofis, pedun-
                  culis bifloris. *Sp. pl.* 534.
              Two-flower'd Sophora.
              Nat. of the Cape of Good Hope.
              Cult. 1692, by Bifhop Compton. *Pluk. phyt. t.* 185.
              f. 2.
              Fl. November——January.   G. H. ♄ .

hirfuta.   16. S. foliis fimplicibus hirfutis : fuperioribus ovatis;
                  inferioribus fubrotundis, ramis teretibus, laciniis
                  calycinis lanceolatis longitudine alarum.
              Hairy Sophora.
              Nat. of the Cape of Good Hope. Mr. *Fr. Maffon.*
              Introd. 1774.
              Fl. July and Auguft.   G. H. ♄ .

                   A N A G Y R I S. *Gen. pl.* 509.
        Vexillum alæque carina breviores in corolla papiliona-
                            cea. *Legumen.*

fœtida.   1. ANAGYRIS. *Sp. pl.* 534.

                                               Stinking

Stinking Bean Trefoil.
Nat. of Spain and Italy.
Cult. 1570, by Mr. Hugh Morgan. *Lobel. adv.* 389.
Fl. April and May. H. ♄.

C E R C I S. *Gen. pl.* 510.

Cal. 5-dentatus, inferne gibbus. *Cor.* papilionacea:
vexillo fub alis brevi. *Legumen.*

1. C. foliis cordato-orbiculatis glabris. *Sp. pl.* 534.   *Siliquaf-*
European Judas-tree.   *trum.*
Nat. of the South of Europe and the Levant.
Cult. 1596, by Mr. John Gerard. *Hort. Ger.*
Fl. May and June. H. ♄.

2. C. foliis cordatis pubefcentibus. *Sp. pl.* 535.   *canaden-*
American Judas-tree.   *fis.*
Nat. of North America.
Cult. 1730. *Hort. angl.* 73.   Siliquaftrum 3.
Fl. May and June. H. ♄.

B A U H I N I A. *Gen. pl.* 511.

Cal. 5-fidus, deciduus. *Petala* patula, oblonga, unguiculata: fuperiore magis diftante; omnia calyci
inferta. *Legumen.*

1. B. caule aculeato. *Sp. pl.* 535.   *aculeata.*
Prickly-ftalked Mountain Ebony.
Nat. of the Weft Indies.
Cult. 1752. *Mill. dict. edit.* 6. *n.* 2.
Fl. S. ♄.

2. B. foliis glabris: lobis divaricatis acutis binerviis,   *divarica-*
petalis lanceolatis.   *ta.*

Bauhinia

48 DECANDRIA MONOGYNIA. Bauhinia.

Bauhinia divaricata α. *Sp. pl.* 535. (excluſo ſynóny-
mo Plumierii.)
Dwarf Mountain Ebony.
Nat. of the Weſt Indies.
Cult. before 1742, by Robert James Lord Petre.
Fl. June——September. S. ♄.

aurita. 3. B. foliis baſi ſubtranſverſis : lobis lanceólatis porreᷓis
trinerviis, petalis lanceolatis.
Bauhinia divaricata β. *Sp. pl.* 535.
Bauhinia foliis ovato-cordatis lobis longiſſimis paral-
lelis. *Mill. ic.* 41. *tab.* 61.
Long-ear'd Mountain Ebony.
Nat. of Jamaica.
Cult. 1756, by Mr. Philip Miller. *Mill. ic. loc. cit.*
Fl. September. S. ♄.

porreᷓa. 4. B. foliis cordatis : lobis porreᷓis acutis trinerviis, pe
talis lanceolatis. *Swartz prodr.* 66.
Bauhinia non aculeata folio ampliori et bicorni. *Plum.*
gen. 23. *ic.* 32. *t.* 44. *f.* 2.
Smooth broad-leav'd Mountain Ebony.
Nat. of the Weſt Indies.
Cult. 1739, by Mr. Philip Miller. *Rand. chel. n.* 2.
Fl. July. S. ♄.

variega-
ta. 5. B. calycibus monophyllis rumpentibus, petalis ſeſſili-
bus ovatis, foliorum lobis ovatis obtuſis.
Bauhinia variegata. *Sp. pl.* 535.
Variegated Mountain Ebony.
Nat. of the Eaſt Indies.
Introd. 1690, by Mr. Bentick. *Br. Muſ. Sloan. mſſ.*
3370.
Fl. S. ♄.

6. B.

6. B. foliis cordatis fubtus pubefcentibus : lobis ovatis   *candida.*
   obtufis, calycibus fuperne attenuatis elongatis.
   Downy-leav'd Mountain Ebony.
   Nat. of the Eaft Indies.   *Claude Ruffell,* Efq.
   Introd. 1777, by Patrick Ruffel, M. D.
   Fl. May and June.   S. ♄.

7. B. floribus triandris, foliorum lobis ovalibus obtufis.   *purpurea.*
   Bauhinia purpurea.   *Sp. pl.* 536.
   Purple Mountain Ebony.
   Nat. of the Eaft Indies.
   Introd. 1778.
   Fl.   S. ♄.

H Y M E N Æ A.   *Gen. pl.* 512,

Cal. 5-partitus.   *Petala* 5, fubæqualia.   *Stylus* intor-
   tus.   *Legumen* repletum pulpa farinacea.

1. HYMENÆA.   *Sp. pl.* 537.   *Courba-*
   Locuft Tree.   *ril.*
   Nat. of the Weft Indies.
   Cult. 1739.   *Mill. dict. vol.* 2.   Courbaril.
   Fl.   S. ♄.

P A R K I N S O N I A.   *Gen. pl.* 513.

Cal. 5-fidus.   *Petala* 5, ovata : infimo reniformi.
   Stylus nullus.   *Legumen* moniliforme.

1. PARKINSONIA.   *Sp. pl.* 536.   *aculeata.*
   Prickly Parkinfonia.
   Nat. of the Weft Indies.
   Cult. 1739, by Mr. Philip Miller.   *Rand. chel.*
   Fl.   S. ♄.

CASSIA. *Gen. pl.* 514.

Cal. 5-phyllus. *Petala* 5. *Antheræ* supremæ 3 steriles; infimæ 3 rostratæ. *Legumen.*

diphylla.   1. C. foliis conjugatis, stipulis cordato-lanceolatis. *Syst. veget.* 393.
Two-leav'd Cassia.
Nat. of the West Indies.
Introd. 1781, by Mr. Francis Masson.
Fl.   S. ♂.

Absus.   2. C. foliis bijugis subovatis, glandulis duabus subulatis inter infima. *Sp. pl.* 537.
Four-leav'd Cassia.
Nat. of Egypt and the East Indies.
Introd. 1777, by Patrick Russell, M. D.
Fl. June and July.   S. ⊙.

Tora.   3. C. foliis trijugis obovatis: exterioribus majoribus, glandula subulata inter inferioria. *Syst. veget.* 393.
Oval-leav'd Cassia, or Wild Senna.
Nat. of the East Indies.
Cult. 1693, by Mr. Jacob Bobart. *Br. Muf. Sloan. mff.* 3343.
Fl. August.   S. ⊙.

bicapsularis.   4. C. foliis trijugis obovatis glabris: interioribus rotundioribus minoribus: glandula interjecta globosa. *Sp. pl.* 538.
Six-leav'd Cassia.
Nat. of the West Indies and Madeira.
Cult. 1739, by Mr. Philip Miller. *Rand. chel. n.* 3.
Fl. May and June.   S. ♄.

5. C.

5. C. foliis quinquejugis ovato-lanceolatis margine fca- *occidenta-*
bris : exterioribus majoribus, glandula bafeos pe- *lis.*
tiolorum. *Sp. pl.* 539.
Occidental Caffia.
Nat. of the Weft Indies.
Cult. 1759, by Mr. Ph. Miller. *Mill. dict. edit.* 7. *n.* 1.
Fl. S. ♄ .

6. C. foliis quinquejugis ovatis acuminatis glabris, pe- *Fiftula.*
tiolis eglandulatis. *Sp. pl.* 540.
Purging Caffia.
Nat. of the Eaft Indies.
Cult. 1731, by Mr. Ph. Miller. *Mill. dict. edit.* 1. *n.* 9.
Fl. June and July. S. ♄ .

7. C. foliis quinquejugis oblongis acutiufculis glabris, *patula.*
glandula bafeos petiolorum, ramis lævibus.
Shining Caffia.
Nat. of the Weft Indies. Mr. *Gilbert Alexander.*
Introd. 1778.
Fl. Auguft and September. S. ♄ .
Obs. *Folia* interdum fejuga. Differt a Caffia occi-
dentali et planifiliqua, foliis apice non attenuatis.

8. C. foliis fejugis fubovatis, petiolis eglandulatis. *Syft.* *Senna.*
veget. 393.
Egyptian Caffia or Senna.
Nat. of Egypt.
Cult. 1640. *Park. theat.* 225.
Fl. July and Auguft. S. ⊙ .

9. C. foliis fejugis oblongiufculis glabris: inferioribus *biflora.*
minoribus, glandula fubulata inter infima, pedicel-
lis fubbifloris. *Sp. pl.* 540.
Two-flower'd Caffia.

E 2 *Nat.*

Nat. of the Weſt Indies.
Cult. 1766, by Mr. James Gordon.
Fl. moſt part of the Winter.   S. ♄.

multi-
glandulo-
ſa.

10. C. foliis ſejugis ovali-oblongis obtuſis piloſis : ex-
terioribus majoribus, glandulis ſubulatis omnium
parium, ſiliquis linearibus.
Caffia multiglandulofa. *Jacqu. ic. colleƈt.* 1. *p.* 42.
Glandulous Caffia.
Nat. ——. Cultivated in the Iſland of Teneriffe.
Introd. 1779, by Mr. Francis Maſſon.
Fl. moſt part of the Summer.   G. H. ♄.

liguſtri-
na.

11. C. foliis ſeptemjugis lanceolatis : extimis minoribus,
glandula baſeos petiolorum. *Sp. pl.* 541.
Privet-leav'd Caffia.
Nat. of the Bahama Iſlands.
Introd. 1726, by Mr. Mark Cateſby. *Mart. dec.* 3.
p. 21.
Fl. July.   S. ♂.

ſtipula-
cea.

12. C. foliis ſuboƈtojugis ovato-lanceolatis, glandula
inter inferiora, ſtipulis ovatis maximis.
Pſeudo-Acacia foliis mucronatis, flore luteo. *Feu-
illée it.* 3. *p.* 56. *tab.* 42.
Large-ſtipul'd Caffia.
Nat. of Chili.
Introd. 1786, by Monſ. Thouin.
Fl.   S. ♄.

alata.

13. C. foliis oƈtojugis ovali-oblongis : exterioribus minoribus, petiolis eglandulatis, ſtipulis patulis. *Syſt.
veget.* 394.
Broad-leav'd Caffia.
Nat. of the Weſt Indies.

                                        Cult.

Cult. 1731, by Mr. Ph. Miller. *Mill. dict. edit.* 1. *n.* 3.
Fl. S. ♄.

14. C. foliis octojugis ovato-oblongis æqualibus, glan- *mari-*
dula bafeos petiolorum. *Sp. pl.* 541. *landica.*
Maryland Caffia.
Nat. of North America.
Introd. 1723, by Peter Collinfon, Efq. *Mart. dec.* 3.
p. 23.
Fl. Auguft——October. H. ♃.

15. C. foliis novemjugis ovali-oblongis glabris obtufi- *frondofa.*
ufculis, glandula cylindracea inter inferiora, pe-
tiolo bafi eglandulofo.
Shrubby fmooth-leav'd Caffia.
Nat. of the Weft Indies.
Introd. about 1769.
Fl. March and April. S. ♄.

16. C. foliis duodecimjugis obtufis mucronatis, glan- *auricula-*
dulis fubulatis pluribus, ftipulis reniformibus bar- *ta.*
batis. *Sp. pl.* 542.
Ear'd Caffia.
Nat. of the Eaft Indies.
Introd. 1777, by Daniel Charles Solander, LL.D.
Fl. S. ♄.

17. C. foliis duodecimjugis oblongis obtufis glabris, *javanica.*
glandula nulla. *Sp. pl.* 542.
Java Caffia.
Nat. of the Eaft Indies.
Introd. 1779.
Fl. S. ♄.

18. C. foliis multijugis, glandula petioli pedicellata, fti- *Chamæ-*
pulis enfiformibus. *Sp. pl.* 542. *crifta.*

E 3 Dwarf

Dwarf Caffia.

Nat. of the Weſt Indies.

Cult. 1699, by the Dutcheſs of Beaufort.   *Brit. Muſ.*
    Sloan. *mſſ.* 525 and 3349.

Fl. June and July.   S. ⊙.

P O I N C I A N A.   *Gen. pl.* 515.

Cal. 5-phyllus. *Petala* 5: ſummo majore.   *Stam.*
    longa : omnia fœcunda.   *Legumen.*

pulcher-   1. P. aculeïs geminis.   *Sp. pl.* 544.
rima.   Barbadoes Flower-fence.

       Nat. of both Indies.

       Cult. 1691, in Chelſea Garden.   *Br. Muſ. Sloan.*
          mſſ. 3343.

       Fl. June——September.   S, ♄.

elata.   2. P. caule inermi.   *Sp. pl.* 544.

       Smooth Flower-fence.

       Nat. of India.

       Introd. 1778, by Sir Joſeph Banks, Bart.

       Fl.   S. ♄.

C Æ S A L P I N I A.   *Gen. pl.* 516.

Cal. 5-fidus : lacinia infima majori. *Petala* 5 : infimo
            pulchriore.   *Legumen.*

braſi-   1. C. caule foliiſque inermibus.   *Sp. pl.* 544. (excluſo
lienſis.   ſynonymo Cateſbei.)

       Smooth Braſiletto.

       Nat. of Jamaica.

       Cult. 1739, by Mr. Philip Miller.   *Rand. chel. p.* 214.
          n. 13.

       Fl.   S. ♄.

                                        2. C.

2. C. caule aculeato, foliolis obcordatis ſubrotundis. *Sp.*   *veſicaria.*
   pl. 545.
Broad-leav'd prickly Braſiletto.
Nat. of Jamaica.
Introd. 1770.
Fl.   S. ♄.

3. C. caule aculeato, foliolis oblongis inæquilaterali-   *Sappan.*
   bus emarginatis. *Sp. pl.* 545.
Narrow-leav'd prickly Braſiletto.
Nat. of the Eaſt Indies.
Introd. 1773, by Sir Joſeph Banks, **Bart.**
Fl.   S. ♄.

GUILANDINA. *Gen. pl.* 517.

Cal. 1-phyllus, hypocrateriformis. *Petala* calycis collo
   inſerta, ſubæqualia. *Legumen.*

1. G. aculeata, pinnis ovatis, foliolis aculeis ſolitariis.   *Bonduc.*
   Sp. pl. 545.
Yellow Bonduc, or Nicker-tree.
Nat. of both Indies.
Introd. 1690, by Mr. Bentick.   *Br. Muſ. Sloan. mſſ.*
   3370.
Fl.   S. ♄.

2. G. aculeata, pinnis oblongo-ovatis, foliolis aculeis   *Bonducel-*
   geminis. *Sp. pl.* 545.   *la.*
Gray Bonduc.
Nat. of both Indies.
Cult. before 1640, by Mr. George Willmer.   *Park.*
   theat. 1551. *f.* 3.
Fl.   S. ♄.

Moringa. 3. G. inermis, foliis subbipinnatis: foliolis inferioribus
ternatis. *Sp. pl.* 546.
Smooth Bonduc.
Nat. of the East Indies.
Cult. 1759, by Mr. Ph. Miller. *Mill. dict. edit.* 7. *n.* 4.
Fl. S. ♄.

dioica. 4. G. inermis, foliis bipinnatis: basi apiceque simplici-
ter pinnatis. *Sp. pl.* 546.
Hardy Bonduc.
Nat. of Canada.
Cult. 1748, by Archibald Duke of Argyle.
Fl. H. ♄.

S C H O T I A. *Jacqu. collect.*
Cal. 5-fidus. *Petala* 5, calyci inserta, lateribus invicem
incumbentia, clausa. *Legumen* pedicellatum.

speciosa. 1. SCHOTIA. *Jacqu. ic. collect.* 1. *p.* 93.
Theodora speciosa. *Medic. monogr. p.* 16. *tab.* 1.
Guajacum afrum. *Sp. pl.* 547.
Lentiscus-leav'd Schotia.
Nat. of the Cape of Good Hope: near Mossel-bay.
Cult. 1759. *Mill. dict. edit.* 7. Guajacum 3.
Fl. G. H. ♄.

G U A J A C U M. *Gen. pl.* 518.
Cal. 5-fidus, inæqualis. *Petala* 5, calyci inserta,
Caps. angulata, 3-s. 5-locularis.

officinale. 1. G. foliolis bijugis obtusis. *Sp. pl.* 546.
Officinal Guajacum, or Lignum Vitæ.
Nat. of the West Indies.
Cult. 1699, by the Dutchess of Beaufort. *Br. Mus.*
Sloan. mss. 525 and 3349.
Fl. S. ♄.

DICTAM-

DICTAMNUS. *Gen. pl. 522.*

Cal. 5-phyllus. *Petala* 5, patula. *Filamenta* punctis glandulosis adspersa. *Caps.* 5, coalitæ.

1. DICTAMNUS. *Sp. pl. 548. Jacqu. austr. 5. p. 13. albus.* *t.* 428.

α flore niveo.
  White Fraxinella.
β flore rubro.
  Red Fraxinella.
Nat. of Germany, France, and Italy.
Cult. 1596, by Mr. John Gerard. *Hort. Ger.*
Fl. May——July.   H. ♃.

RUTA. *Gen. pl. 523.*

Cal. 5-partitus. *Petala* concava. *Receptac.* punctis melliferis decem cinctum. *Caps.* lobata. *Quinta pars numeri in quibusdam excluditur.*

1. R. foliis decompositis, floribus lateralibus quadrifidis. *graveo-*
  Syst. veget. 397.   *lens.*
  Common Rue.
  Nat. of the South of Europe.
  Cult. 1562. *Turn. herb.* part 2. *fol.* 122 *verso.*
  Fl. June——September.   H. ♄ .

2. R. foliis supra decompositis : laciniis linearibus, petalis   *montana.*
  imberbibus.
  Ruta montana. *Loefl. it.* 140. *Cluf. hist.* 2. *p.* 136. *fig.*
  Ruta legitima. *Allion. pedem.* 1. *p.* 280. *Jacqu. ic.*
  collect. 1. *p.* 74.
  Ruta sylvestris minor. *Bauh. pin.* 336. *Cam. epit.*
  495.
  Mountain Rue.

                                      Nat.

Nat. of the South of Europe.
Cult. 1739, by Mr. Philip Miller. *Rand. chel. n.* 4.
Fl. Auguſt and September.   H. ♄.

chalepen-   3. R. foliis ſupradecompoſitis, petalis ciliatis. *Linn.*
ſis.   *mant.* 69.
α Ruta chalepenſis latifolia, florum petalis villis ſcaten-
     tibus. *Tourn. inſt.* 257.
Broad-leav'd African Rue.
β Ruta chalepenſis tenuifolia, florum petalis villis ſcaten-
     tibus. *Moriſ. hiſt.* 2. *p.* 508. *ſ.* 5. *t.* 35. *ſ.* 8.
Narrow-leav'd African Rue.
Nat. of Africa.
Cult. 1722, in Chelſea Garden. *R. S. n.* 40.
Fl. June——September.   G. H. ♄.

pinnata.   4. R. foliis pinnatis: foliolis lanceolatis baſi attenuatis
               ſerrato-crenatis, petalis integerrimis.
Ruta pinnata. *Linn. ſuppl.* 232.
Winged-leav'd Rue.
Nat. of the Canary Iſlands. Mr. *Francis Maſſon.*
Introd. 1780.
Fl. March.   G. H. ♄.

HÆMATOXYLUM. *Gen. pl.* 525.

Cal. 5-partitus. *Petala* 5. *Capſ.* lanceolata, 1-locu-
     laris, 2-valvis: valvis navicularibus.

Campe-   1. HÆMATOXYLUM. *Sp. pl.* 549.
chianum.   Logwood.
Nat. of South America.
Cult. 1739, by Mr. Philip Miller. *Mill. dict. vol.* 2.
     Addenda. Campechia.
Fl.   S. ♄.

MURRAYA.

MURRAYA. *Linn. mant.* 554.

Cal. 5-partitus. *Cor.* campanulata: *Nectario* germen
cingente. *Bacca* monosperma.

1. MURRAYA. *Linn. mant.* 563.   *exotica.*
Afh-leav'd Murraya.
Nat. of India.
Introd. 1771, by Benjamin Torin, Efq.
Fl. Auguft and September.   G. H. ♄.

SWIETENIA. *Gen. pl.* 521.

Cal. 5-fidus. *Petala* 5. *Nectarium* cylindricum, ore
Antheras gerens. *Capf.* 5-locularis, lignofa, bafi
dehifcens. *Sem.* imbricata, alata.

1. SWIETENIA. *Sp. pl.* 548.   *Mahago-*
Mahogany Tree.   *ni.*
Nat. of the Weft Indies.
Cult. 1739, by Mr. Philip Miller. *Mill. dict. vol.* 2.
edit. 1. Addenda. Arbor 4.
Fl.   S. ♄.

MELIA. *Gen. pl.* 527.

Cal. 5-dentatus. *Petala* 5. *Nectarium* cylindrace-
um, ore Antheras gerens. *Drupa* nucleo quinque-
loculari.

1. M. foliis bipinnatis. *Sp. pl.* 550.   *Azeda-*
α Arbor fraxini folio, flore cæruleo. *Bauh. pin.* 415.   *rach.*
Common Bead-tree.
β Azedarach fempervirens & florens. *Tournef. inft.*   femper-
616.   virens.
Evergreen Bead-tree.
Nat. α. of Syria; β. of the Eaft Indies.
             Cult.

Cult. 1656, by Mr. John Tradefcant, Jun. *Muf. Trad.* 88.

Fl. June——Auguft.　　　　　　　　G. H. ♄.

ZYGOPHYLLUM. *Gen. pl.* 530.

Cal. 5-phyllus. *Petala* 5. *Nectarium* 10-phyllum, germen tegens. *Capfula* 5-locularis.

cordifoli-um.

1. Z. foliis fimplicibus oppofitis feffilibus fubrotundis.
Zygophyllum cordifolium. *Linn. fuppl.* 232.
Heart-leav'd Bean-caper.
Nat. of the Cape of Good Hope. Mr. *Fr. Maffon.*
Introd. 1774.
Fl. October.　　　　　　　　　　G. H. ♄.

Fabago.

2. Z. foliis conjugatis petiolatis : foliolis obovatis, caule herbaceo. *Syft. veget.* 400.
Common Bean-caper.
Nat. of Syria.
Cult. 1596, by Mr. John Gerard. *Hort. Ger.*
Fl. July——September.　　　　　　H. ♃.

macula-tum.

3. Z. foliis conjugatis petiolatis : foliolis lineari-lanceo-latis.
Spotted-flower'd Bean-caper.
Nat. of the Cape of Good Hope.
Introd. 1782, by George Wench, Efq.
Fl. November.　　　　　　　　　G. H. ♄.
Obs. *Petala* lutea, macula cordata rubra ad bafin om-nium, et fupra hanc linea tranfverfa rubra in tribus fuperioribus.

album.

4. Z. foliis conjugatis petiolatis : foliolis clavatis carnofis arachnoideo-incanis. *Syft. veget.* 400.

White

White Bean-caper.
Nat. of the Canary Iſlands.
Introd. 1779, by Mr. Francis Maſſon.
Fl. D. S. ♃.

5. Z. foliis conjugatis ſubpetiolatis: foliolis obovatis,   *Morgſa-*
   caule fruticoſo. *Syſt. veget.* 400.   *na.*
   Four-leav'd Bean-caper.
   Nat. of the Cape of Good Hope.
   Cult. 1732, by James Sherard, M. D. *Dill. elth.* 142.
   t. 116. *f.* 141.
   Fl. moſt part of the Summer.   G. H. ♃.

6. Z. foliis conjugatis feſſilibus: foliolis lanceolato-ova- *feſſilifoli-*
   libus margine ſcabris, caule fruticoſo. *Syſt. veget.* *um.*
   400.
α Zygophyllum capſulis globoſo-depreſſis. *Sp. pl. ed.* 1.
   p. 385.
   Round-capſul'd Seſſile-leav'd Bean-caper.
β Zygophyllum *fulvum* capſulis ovatis acutis. *Sp. pl.*
   ed. 1. *p.* 386.
   Oval-capſul'd feſſile-leav'd Bean-caper.
   Nat. of the Cape of Good Hope.
   Cult. 1713, by the Dutcheſs Dowager of Beaufort.
   Philoſoph. tranſ. n. 337. *p.* 63. *n.* 106.
   Fl. July and Auguſt.   G. H. ♃.

QUASSIA. *Gen. pl.* 529.

Cal. 5-phyllus. *Petala* 5. *Nectarium* 5-phyllum.
   Pericarpia 5, diſtantia, 1-ſperma.

1. Q. floribus monoicis, foliis abrupte pinnatis: foliolis *Simaru-*
   alternis ſubpetiolatis, petiolo nudo, floribus pani- *ba.*
   culatis. *Linn. ſuppl.* 234.
   Winged-leav'd Quaſſia.

                     Nat.

Nat. of the Weft Indies.
Introd. 1787, by Mr. Alexander Anderfon.
Fl.   S. ♃.

FAGONIA. *Gen. pl.* 531.

Cal. 5-phyllus. *Petala* 5, cordata. *Capf.* 5-locularis,
10-valvis: loculis 1-fpermis.

cretica.   1. F. fpinofa, foliolis lanceolatis planis lævibus. *Sp. pl.*
553.
Cretian Fagonia.
Nat. of the Ifland of Candia.
Cult. 1739, by Mr. Philip Miller. *Rand. chel.*
Fl. July and Auguft.   G. H. O.

TRIBULUS. *Gen. pl.* 532.

Cal. 5-partitus. *Petala* 5, patentia. *Stylus* o. *Capf.* 5,
gibbæ, fpinofæ, polyfpermæ.

maximus.   1. T. foliolis fubquadrijugis: exterioribus majoribus,
pericarpiis decafpermis muticis. *Syft. veget.* 401.
Great Caltrops.
Nat. of Jamaica.
Cult. 1739, by Mr. Philip Miller. *Rand. chel. n.* 2.
Fl. June and July.   S. ☉.

terreftris. 2. T. foliis fubfexjugis fubæqualibus, feminibus qua-
dricornibus. *Syft. veget.* 402.
Small Caltrops.
Nat. of the South of Europe.
Cult. 1596, by Mr. John Gerard. *Hort. Ger.*
Fl. June and July.   H. ☉.

LIMONIA.

LIMONIA. *Gen. pl.* 534.

Cal. 5-partitus. *Petala* 5. *Bacca* 3-locularis. *Sem.* folitaria.

1. L. foliis fimplicibus, fpinis folitariis. *Linn. mant.* 237.

monophyl-la.

Simple-leav'd Limonia.
Nat. of the Eaft Indies.
Introd. 1777, by Daniel Charles Solander, LL. D.
Fl.   S. ♄.

DIONÆA. *Linn. mant.* 151.

Cal. 5-phyllus. *Petala* 5. *Capf.* unilocularis, gibba, polyfperma.

1. DIONÆA. *Linn. mant.* 238.

Mufcipu-la.

Venus's Fly-trap.
Nat. of Carolina.
Introd. 1768, by Mr. William Young.
Fl. July and Auguft.   G. H. ♃.

JUSSIEUA. *Gen. pl.* 538.

Cal: 4-f. 5-partitus, fuperus. *Petala* 4 f. 5. *Capf.* 4-f. 5-locularis, oblonga, angulis dehifcens. *Sem.* numerofa, minuta.

1. J. erecta glabra, floribus tetrapetalis octandris feffili-bus. *Sp. pl.* 556.

erecta.

Red-ftalked Juffieua.
Nat. of America.
Cult. 1739, by Mr. Ph. Miller. *Rand. chel.* Onagra 6.
Fl. July——September.   S. ☉.

DAIS. *Gen. pl.* 540.

Involucr. 4-phyllum. *Cor.* 4-f. 5-fida. *Bacca* 1-fperma.

1. D. floribus quinquefidis decandris. *Syf. veget.* 403. Dais

cotinifo-lia.

Dais laurifolia. *Jacqu. ic. collect.* 1. *p.* 146.
Cotinus-leav'd Dais.
Nat. of the Cape of Good Hope.
Introd. 1776, by Mr. James Gordon.
Fl.   G. H. ♄.

KALMIA. *Gen. pl.* 545.

Cal. 5-partitus. *Cor.* hypocrateriformis: limbo fubtus
quinquecorni. *Capf.* 5-locularis.

latifolia.   1. K. foliis ovato-ellipticis ternis fparfifque, corymbis
terminalibus.
Kalmia latifolia. *Sp. pl.* 560.
Broad-leav'd Kalmia.
Nat. of North America.
Introd. 1734, by Peter Collinfon, Efq. *Coll. mff.*
Fl. May——July.   H. ♄.

angufti-   2. K. foliis lanceolatis, corymbis lateralibus. *Sp. pl.*
folia.   561.
rubra.   α floribus rubris.
Red-flower'd narrow-leav'd Kalmia.
carnea.   β floribus carneis.
Pale-flower'd narrow-leav'd Kalmia.
Nat. of North America.
Introd. 1736, by Peter Collinfon, Efq. *Coll. mff.*
Fl. May——July.   H. ♄.

glauca.   3. K. foliis oppofitis oblongis lævigatis fubtus glaucis
margine revolutis, corymbis terminalibus, ramulis
ancipitibus.   TAB. 8.
Kalmia glauca. *L'Herit. ftirp. nov. tom.* 2. *tab.* 9.
Glaucous Kalmia.
Nat. of Newfoundland. Sir *Jofeph Banks*, Bart.
Introd. 1767.
Fl. April and May.   H. ♄.

                                         LEDUM.

Tab. 8. Vol. 2 Page 6 4.

Kalmia glauca.

Ehret del.

M.S. Kenzie. sc.

L E D U M. *Gen. pl.* 546.

Cal. 5-fidus. *Cor.* plana, 5-partita. *Capf.* 5-locularis, bafi dehifcens.

1. L. foliis linearibus margine revolutis fubtus tomentofis.   *paluſtre.*
Ledum paluſtre. *Sp. pl.* 561.
α bipedale erectum.   erectum.
Marſh Ledum.
β fpithamæum decumbens.   decum-
Dwarf Ledum.   bens.
Nat. α. of the North of Europe; β. of Hudfon's-bay.
Introd. 1762, by Mr. John Buſh.
Fl. April and May.   H. ♄.

2. L. foliis oblongis margine revolutis fubtus tomentofis, *latifoli-*
floribus fubpentandris.   *um.*
Ledum groenlandicum. *Fl. dan.* 567. *Gunn. norv.*
1067. *Retzii fcand.* 436.
Labrador Tea, or Broad-leav'd Ledum.
Nat. of Greenland, Hudfon's-bay, Labrador, New-
foundland, and Nova Scotia.
Introd. 1763, by Mr. Bennet.
Fl. April and May.   H. ♄.

3. L. foliis ovato-oblongis planis glabris.   *buxifoli-*
Ledum buxifolium. *Bergius in act. petrop.* 1777. *um.*
part. 1. *p.* 213. *tab.* 3. *f.* 2.
Ledum ferpillifolium. *L'Herit. ſtirp. nov. tom.* 2.
tab. 10.
Box-leav'd Ledum.
Nat. of Carolina and New Jerfey.
Introd. 1736, by Peter Collinfon, Efq. *Coll. mſſ.*
Fl. April and May.   H. ♄.

RHODORA. *Gen. pl.* 547.

Cal. 5-dentatus.   *Cor.* 3-petala.   *Stam.* declinata.
Capf. 5-locularis.

canaden-   1. RHODORA. *Sp. pl.* 561. *L'Herit. ftirp. nov. tab.* 68.
fis.   Canadian Rhodora.
               Nat. of North America.
               Introd. 1767, by Sir Jofeph Banks, Bart.
               Fl. April.   H. ♄ .

RHODODENDRON. *Gen. pl.* 548.

Cal. 5-partitus.   *Cor.* fubinfundibulif.   *Stam.* declinata.   *Capf.* 5-locularis.

ferrugi-   1. R. foliis glabris fubtus leprofis, corollis infundibuli-
neum.   formibus. *Sp. pl.* 562. *Jacqu. auftr.* 3. *p.* 31.
               t. 255.
               Rufty-leav'd Rhododendron.
               Nat. of the Alps of Switzerland and Auftria.
               Cult. 1739, by Mr. Philip Miller. *Mill. dict. vol.* 2.
               Chamærhododendron 1.
               Fl. May——July.   H. ♄ .

dauri-   2. R. foliis glabris punctatis nudis, corollis rotatis. *Syft.*
cum.   *veget.* 405. *Pallas roff.* 1. *p.* 47. *t.* 32.
               Dotted-leav'd Rhododendron.
               Nat. of Siberia.
               Introd. 1780, by Anthony Chamier, Efq.
               Fl.   H. ♄ .

hirfutum. 3. R. foliis ciliatis nudis, corollis infundibuliformibus.
               Sp. pl. 562. *Jacqu. auftr.* 1. *p.* 61. *t.* 98.
               Hairy Rhododendron.
               Nat. of the Alps of Switzerland and Auftria.

                                                              Cult.

Cult. 1739. *Mill. dict. vol.* 2. Chamærhododendron 2.
Fl. May and June.   H. ♄.

4. R. foliis nitidis lanceolatis utrinque glabris, racemis   *ponticum.*
terminalibus. *Sp. pl.* 562. *Jacqu. ic. Pallas roff.* 1.
p. 43. *tab.* 29.
Purple Rhododendron.
Nat. of the Levant and Gibraltar.
Introd. 1763.
Fl. May and June.   H. ♄.

5. R. foliis nitidis ovalibus obtufis venofis margine   *maxi-*
acuto reflexo, pedunculis unifloris. *Sp. pl.* 563.   *mum.*
Broad-leav'd Rhododendron.
Nat. of North America.
Introd. 1736, by Peter Collinfon, Efq. *Coll. mff.*
Fl. June——Auguft.   H. ♄.

A N D R O M E D A. *Gen. pl.* 549.

Cal. 5-partitus. *Cor.* ovata : ore 5-fido. *Capf.*
5-locularis.

1. A. pedunculis aggregatis rameis, corollis ovato-cylin-   *mariana.*
dricis, foliis oblongo-ovatis integerrimis deciduis.
Andromeda mariana. *Sp. pl.* 564.
α foliis ovalibus.   ovalis.
Oval-leav'd Andromeda.
β foliis oblongis.   oblonga.
Oblong-leav'd Andromeda.
Nat. of North America.
Introd. 1736, by Peter Collinfon, Efq. *Coll. mff.*
Fl. June and July.   H. ♄.

2. A. pedunculis aggregatis axillaribus, corollis fubglo-   *ferrugi-*
bofis, foliis ellipticis integerrimis fubtus fquamofo-   *nea.*
fcariofis.
                    F 2   Rufty

Rusty Andromeda.
Nat. of North America.
Cult. 1776, by Mr. James Gordon.
Fl. July and August.　　　　　　　　　　H. ♄.

Polifolia.　3. A. pedunculis aggregatis, corollis ovatis, foliis alter-
　　　　　nis lanceolatis revolutis.　*Sp. pl.* 564.

latifolia.　α foliis oblongis, corollis ovatis incarnatis, laciniis caly-
　　　　　cinis patentibus ovatis albis: interdum apice rubi-
　　　　　cundis.

　　　　　Andromeda polifolia latifolia.　*L'Herit. stirp. nov.*
　　　　　tom. 2. *tab.* 11.

　　　　　Broad-leav'd Marsh Andromeda.

media.　β foliis lanceolatis, corollis oblongo-ovatis rubicundis,
　　　　　laciniis calycinis magis erectis.

　　　　　Common Marsh Andromeda, or Wild Rosemary.

angusti-　γ foliis lanceolato-linearibus, laciniis calycinis oblongis
folia.　　　rubris.

　　　　　Narrow-leav'd Marsh Andromeda.

　　　　　Nat. α. of North America; β. of Britain; and γ. of
　　　　　Newfoundland and Labrador.

　　　　　Fl. May——September.　　　　　　　　H. ♄.

Daboecia.　4. A. racemis secundis, floribus quadrifidis ovatis, foliis
　　　　　alternis lanceolatis revolutis.　*Syst. veget.* 406.

　　　　　Erica Daboecii.　*Sp. pl.* 509.

　　　　　Irish Heath, or trailing Andromeda.

　　　　　Nat. of Ireland.

　　　　　Fl. June——September.　　　　　　　　H. ♄.

droseroi-　5. A. racemis secundis, foliis linearibus pilosis viscidis.
des.　　　　　*Linn. mant.* 239.

　　　　　Clammy Andromeda.

　　　　　Nat. of the Cape of Good Hope.

　　　　　Introd. 1787, by Mr. Francis Masson.

　　　　　Fl. July.　　　　　　　　　　　　　G. H. ♄.

　　　　　　　　　　　　　　　　　　　　　6. A.

6. A. racemis terminalibus paniculatis, corollis fubrotun- *panicula-*
dis, foliis ovatis integriufculis. *L'Herit. ftirp. nov.* *ta.*
tom. 2. *tab.* 12.
Andromeda paniculata. *Sp. pl.* 564. (exclufo fyno-
nymo Catefbei.)
Panicled Andromeda.
Nat. of North America.
Cult. 1748, by Archibald Duke of Argyle.
Fl. May and June. H. ♄.

7. A. paniculis terminalibus, corollis fubpubefcentibus, *arborea.*
foliis ellipticis acuminatis denticulatis. *L'Herit.*
ftirp. nov. tom. 2.
Andromeda arborea. *Sp. pl.* 565.
Tree Andromeda.
Nat. of Carolina and Virginia.
Cult. 1752, by Mr. Ph. Miller. *Mill. dict. edit.* 6. *n.* 2.
Fl. July——September. H. ♄.

8. A. racemis terminalibus fimplicibus bracteatis, corol- *racemofa.*
lis cylindricis, foliis oblongo-lanceolatis ferratis.
L'Herit. ftirp. nov. tom. 2. *tab.* 13.
Andromeda racemofa. *Sp. pl.* 564.
Branching Andromeda.
Nat. of Penfylvania and Maryland.
Introd. 1736, by Peter Collinfon, Efq. *Coll. mff.*
Fl. July. H. ♄.

9. A. racemis axillaribus fimplicibus, corollis oblongis, *axillaris.*
foliis ovatis acutis ferrulatis. *L'Herit. ftirp. nov.*
tom. 2. *tab.* 14.
Notch'd-leav'd Andromeda.
Nat. of Carolina. Mr. *John Cree.*
Introd. 1765.
Fl. May——Auguft. H. ♄.

coriacea. 10. A. racemis axillaribus fimplicibus, foliis ovatis inte-
gerrimis nitidiffimis, ramulis triquetris. *L'Herit.*
ftirp. nov. tom. 2. *tab.* 15.
Thick-leav'd Andromeda.
Nat. of North America.   Mr. *John Cree.*
Introd. 1765.
Fl. July and Auguft.   H. ♄.

acumina- 11. A. racemis axillaribus fimplicibus, foliis ovato-lan-
ta.   ceolatis acuminatis ferratis. *L'Herit. ftirp. nov.*
tom. 2. *tab.* 16.
Andromeda lucida. *Jacqu. ic. collect.* 1. *p.* 95.
Acute-leav'd Andromeda.
Nat. of North America.   Mr. *John Cree.*
Introd. 1765.
Fl. Auguft.   H. ♄.

calycula- 12. A. pedunculis folitariis axillaribus fecundis, bracteis
ta.   binis, foliis ovalibus fquamofo-punctatis obfolete
ferrulatis.
Andromeda calyculata. *Sp. pl.* 565.
ventrico- α corollis globofis, foliis oblongo-lanceolatis.
fa.   Chamædaphne. *Buxbaum in comm. petrop.* 1, *p.* 241.
t. 8. *f.* 1.
Globe-flower'd box-leav'd Andromeda.
latifolia. β corollis oblongo-cylindraceis, foliis oblongo-ovalibus
obtufis.
Broad box-leav'd Andromeda.
angufti- γ corollis oblongo-ovalibus, foliis oblongo-lanceolatis.
folia.   Andromeda. *Gmel. fib.* 4. *p.* 119. *n.* 4.
Narrow box-leav'd Andromeda.
Nat. α. of Ruffia; β. of Newfoundland; and γ. of
North America and Siberia.
Cult. 1748, by Archibald Duke of Argyle.
Fl. February——April.   H. ♄.

                                                 E P I G Æ A.

E P I G Æ A. *Gen. pl.* 550.

Cal. exterior 3-phyllus ; *interior* 5-partitus. *Cor.* hypocrateriformis. *Capf.* 5-locularis.

1. Epigæa. *Sp. pl.* 565.   *repens.*
   Creeping Epigæa.
   Nat. of Virginia and Canada.
   Introd. 1736, by Peter Collinfon, Efq. *Coll. mff.*
   Fl. July——September.   H. ♄.

GAULTHERIA. *Gen. pl.* 551.

Cal. exterior 2-phyllus ; *interior* 5-fidus. *Cor.* ovata.
   Nectarium mucronibus 10. *Capf.* 5-locularis,
   veftita calyce interiore baccato.

1. Gaultheria. *Sp. pl.* 565.   *procum-*
   Trailing Gaultheria.   *bens.*
   Nat. of North America.
   Cult. 1765, by Mr. Philip Miller. *Mill. dict. edit.* 8.
   Fl. July.   H. ♄.

A R B U T U S. *Gen. pl.* 552.

Cal. 5-partitus. *Cor.* ovata : ore bafi pellucida. *Bacca*
   5-locularis.

1. A. caule arboreo, foliis oblongo-lanceolatis, paniculis   *Unedo.*
   glabris nutantibus.
   Arbutus Unedo. *Sp. pl.* 566.
α flore fimplici, corollis albidis.   *alba.*
   Common white-flower'd Strawberry-tree.
β flore fimplici, corollis rubicundis.   *rubra.*
   Red-flower'd Strawberry-tree.
γ flore pleno.   *plena.*
   Double-flower'd Strawberry-tree.

Nat.

Nat. of Ireland.
Fl. September——December.   H. ♄.

Andrach-   2. A. caule arboreo, foliis ovalibus integris ferratifque,
ne.   paniculis pubefcentibus erectis.
            Arbutus Andrachne. *Sp. pl.* 566.
            Oriental Strawberry-tree.
            Nat. of the Levant.
            Cult. 1724, by James Sherard, M. D. *Knowlton's mff.*
            Fl. March and April.   H. ♄.

alpina.   3. A. caulibus procumbentibus, foliis rugofis ferratis.
            Sp. pl. 566.
            Alpine Arbutus.
            Nat. of Britain.
            Fl. April and May,   H. ♄.

Uva   4. A. caulibus procumbentibus, foliis integerrimis. *Sp. pl.*
urfi.   566.
            Trailing Arbutus, or Bear-berry,
            Nat. of Britain.
            Fl. April and May.   H. ♄.

thymifo- 5. A. caulibus procumbentibus, foliis ovalibus acutis ob-
lia.   folete ferratis fubtus ftrigofis, floribus axillaribus oc-
            tandris.
            Vaccinium hifpidulum. *Sp. pl.* 500. (exclufis fyno-
                nymis Plukenetii et Raji.)
            Thyme-leav'd Arbutus.
            Nat. of North America.
            Introd. about 1776, by John Fothergill, M. D.
            Fl.   H. ♄.

CLETHRA,

C L E T H R A. *Gen. pl.* 553.

Cal. 5-partitus. *Pet.* 5. *Stigma* 3-fidum. *Capf.* 3-locularis, 3-valvis.

1. C. fruticofa, foliis obovato-lanceolatis, racemis fpi- *alnifolia.*
ciformibus, calycibus acutis, bracteis perfiftentibus.
Clethra alnifolia. *Sp. pl.* 566.

α foliis utrinque nudis. *denuda-*
Smooth Alder-leav'd Clethra. *ta.*

β foliis fubtus tomentofis. *pubef-*
Woolly Alder-leav'd Clethra. *cens.*
Nat. of North America.
Introd. 1736, by Peter Collinfon, Efq. *Coll. mff.*
Fl. Auguft——October. H. ♄.

2. C. fruticofa, foliis lanceolatis utrinque nudis, floribus *panicula-*
paniculatis. *ta.*
Panicled Clethra.
Nat. of North America. Mr. *William Young.*
Introd. 1770.
Fl. Auguft——October. H. ♄.

3. C. arborea, foliis oblongo-lanceolatis utrinque glabris, *arborea.*
racemis fpiciformibus, calycibus obtufis.
Tree Clethra.
Nat. of Madeira. Mr. *Francis Maffon.*
Introd. 1784.
Fl. G. H. ♄.

P Y R O L A. *Gen. pl.* 554.

Cal. 5-partitus. *Petala* 5. *Capf.* 5-locularis, angulis dehifcens.

1. P. ftaminibus adfcendentibus, piftillo declinato. *Sp. rotundi-*
pl. 567. *folia.*

Round-

Round-leav'd Winter-green.
Nat. of Britain.
Fl. June and July.   H. ♃.

minor.   2. P. floribus racemofis difperfis, ftaminibus piftillifque
         rectis. *Sp. pl.* 567.
         Leffer Winter-green.
         Nat. of Britain.
         Fl. June and July.   H. ♃.

fecunda.   3. P. racemo unilaterali. *Sp. pl.* 567.
         Notch'd-leav'd Winter-green.
         Nat. of Britain.
         Fl. June.   H. ♃.

umbella-   4. P. pedunculis fubumbellatis. *Sp. pl.* 567.
ta.   Umbel'd Winter-green.
         Nat. of North America.
         Introd. 1762, by Mr. John Bartram.
         Fl.   H. ♃.

maculata.   5. P. pedunculis bifloris. *Sp. pl.* 567.
         Spotted-leav'd Winter-green.
         Nat. of North America.
         Cult. 1759, by Mr. Ph. Miller. *Mill. dict. edit.* 7. *n.* 4.
         Fl. June.   H. ♃.

uniflora.   6. P. fcapo unifloro. *Sp. pl.* 568.
         One-flower'd Winter-green.
         Nat. of the North of Europe.
         Cult. 1748, by Mr. Ph. Miller. *Mill. dict. edit.* 5. *n.* 2.
         Fl. June and July.   H. ♃.

STYRAX.

S T Y R A Y   *Gen. pl.* 595.

Cal. inferus. *Cor.* infundibu.iformis. *Drupa* 2-fperma.

1. S. foliis ovatis fubtus villofis, racemis fimplicibus fo-   *officinale.*
lio brevioribus.
Styrax officinale. *Sp. pl.* 635.
Officinal Storax.
Nat. of Italy and the Levant.
Cult. 1597, by Mr. John Gerard. *Ger. herb.* 1342.
Fl. July.   H. ♄.

2. S. foliis obovatis fubtus villofis, pedunculis inferiori-   *grandifo-*
bus axillaribus folitariis unifloris.   *lium.*
Great-leav'd Storax.
Nat. of South Carolina.   Mr. *John Cree.*
Introd. 1765.
Fl. July.   H. ♄.

3. S. foliis oblongis utrinque glabris, pedunculis axilla-   *læviga-*
ribus unifloris folitariis binifve.   *tum.*
Styrax octandrum. *L'Herit. ftirp. nov. tom.* 2. *tab.* 17.
Smooth Storax.
Nat. of South Carolina.   Mr. *John Cree.*
Introd. 1765.
Fl. June and July.   H. ♄.

D I G Y N I A.

R O Y E N A.   *Gen. pl.* 555.
Cal. urceolatus.   *Cor.* 1-petala, limbo revoluta.   *Capf.*
1-locularis, 4-valvis.

1. R. foliis ovatis fcabriufculis.   *Sp. pl.* 568.   *lucida.*
Shining-leav'd African Bladder-nut.
                                              Nat.

Nat. of the Cape of Good Hope.
Introd. 1690, by Mr. Bentick.   *Br. Muf. Sloan. mff.*
3370.
Fl. May and June.   G. H. ♄.

villofa.   2. R. foliis cordatis oblongis fubtus tomentofis.   *Syft.*
veget. 410.
Heart-leav'd Royena.
Nat. of the Cape of Good Hope.
Introd. 1774, by Mr. William Malcolm.
Fl.   G. H. ♄.

glabra.   3. R. foliis lanceolatis glabris.   *Sp. pl.* 568.
Myrtle-leav'd African Bladder-nut.
Nat. of the Cape of Good Hope.
Cult. 1731, by Mr. Philip Miller.   *Mill. dict. edit.* 1.
Vitis Idæa 4.
Fl. September.   G. H. ♄.

hirfuta.   4. R. foliis lanceolatis hirfutis.   *Sp. pl.* 568.
Hairy-leav'd African Bladder-nut.
Nat. of the Cape of Good Hope.
Introd. 1767, by Mr. William Malcolm.
Fl.   G. H. ♄.

polyan-   5. R. foliis ellipticis, floribus polyandris polygamis.
dra.   Royena polyandra.   *Linn. fuppl.* 240.
Oval-leav'd Royena.
Nat. of the Cape of Good Hope.   Mr. *Fr. Maffon.*
Introd. 1774.
Fl.   G. H. ♄.

HYDRANGEA. *Gen. pl.* 557.
Capf. 2-locularis, 2-roftris, circumfcifla.

arboref-   1. HYDRANGEA.   *Sp. pl.* 568.
cens.

Shrubby

Shrubby Hydrangea.
Nat. of Virginia.
Introd. 1736, by Peter Collinfon, Efq. *Coll. mff.*
Fl. July and Auguft. H. ♄.

TRIANTHEMA. *Linn. mant.* 22.
Cal. fub apice mucronatus. *Cor.* o. *Stam.* 5 f. 10.
 Germen retufum. *Capf.* circumfciffa.

1. T. floribus pentandris monogynis. *Syft. veget.* 410. *monogy-*
 Trianthema Portulacaftrum. *Sp. pl.* 325. *na.*
 Purflain-leav'd Trianthema.
 Nat. of Jamaica.
 Cult. 1739, by Mr. Philip Miller. *Rand. chel.* Por-
 tulaca 4.
 Fl. July and Auguft. S. o.

2. T. floribus fubdecandris digynis. *Linn. mant.* 70. *decandra.*
 Trailing Trianthema.
 Nat. of India.
 Introd. 1779, by John Fothergill, M.D.
 Fl. July and Auguft. S. o.

CHRYSOSPLENIUM. *Gen. pl.* 558.
Cal. 4-f. 5-fidus, coloratus. *Cor.* o. *Capf.* 2-roftris,
 i-locularis, polyfperma.

1. C. foliis alternis. *Sp. pl.* 569. *alternifo-*
 Alternate-leav'd Golden Saxifrage. *lium.*
 Nat. of Britain.
 Fl. April. H. ♃.

2. C. foliis oppofitis. *Sp. pl.* 569. *Curtis lond.* *oppofitifo-*
 Oppofite-leav'd Golden Saxifrage. *lium.*
 Nat. of Britain.
 Fl. April. H. ♃.

SAXI-

SAXIFRAGA. *Gen. pl.* 559.

Cal. 5-partitus. *Cor.* 5-petala. *Capf.* 2-roftris, 1-locularis, polyfperma.

* Foliis *indivifis*, caule *fubnudo*.

Cotyledon. 1. S. foliis radicatis aggregatis lingulatis cartilagineoferratis, caule paniculato. *Sp. pl.* 570.

α Cotyledon media, foliis oblongis ferratis. *Bauh, pin.* 285.
Long-leav'd pyramidal Saxifrage.

β Cotyledon minor, foliis fubrotundis ferratis. *Bauh. pin.* 285. *prodr.* 133.
Round-leav'd pyramidal Saxifrage.
Nat. of the Alps of Switzerland and the Pyrenees.
Cult. 1629. *Park. parad.* 233. *f.* 5.
Fl. May and June.   H. ♃.

penfylva-   2. S. foliis lanceolatis denticulatis, caule nudo panicula-
nica.   to, floribus fubcapitatis. *Sp. pl.* 571.
Penfylvanian Saxifrage.
Nat. of North America.
Cult. 1732, by James Sherard, M. D. *Dill. elth.* 337.
t. 253. *f.* 328.
Fl. May and June.   H. ♃.

bryoides.   3. S. foliis ciliatis inflexis imbricatis, caule nudiufculo
multifloro. *Syft. veget.* 412.
Bryum Saxifrage.
Nat. of the Alps of Switzerland.
Introd. 1775, by the Doctors Pitcairn and Fothergill.
Fl.   H. ♃.

ftellaris.   4. S. foliis ferratis, caule nudo ramofo, petalis acuminatis. *Sp. pl.* 572.
                                                            Starry

Starry Saxifrage, or Kidney-wort.
Nat. of Britain.
Fl. June and July. H. ♃.

5. S. foliis ovalibus retufis obfolete ferratis petiolatis, *craffifo-*
caule nudo, panicula conglomerata. *Sp. pl.* 573. *lia.*
α foliis ovalibus.
Thick oval-leav'd Saxifrage.
β foliis cordatis fubrotundis.
Thick heart-leav'd Saxifrage.
Nat. of Siberia.
Introd. 1765, by Daniel Charles Solander, LL.D.
Fl. March and April. H. ♃.

6. S. foliis obovatis crenatis fubfeffilibus, caule nudo, flo- *nivalis.*
ribus congeftis. *Sp. pl.* 573.
Snowy Saxifrage, or Sengreen.
Nat. of Britain.
Fl. June. H. ♃.

7. S. foliis fubrotundis dentatis pilofis, ftolonibus rep- *farmen-*
tantibus, petalis duobus elongatis. *tofa.*
Saxifraga farmentofa. *Linn. fuppl.* 240. *Schreber*
monogr. Dioneæ, p. 16. *tab.* 2, 3. *Thunb. japon.* 182.
Saxifraga ftolonifera. *Meerb. ic.* 23. *Jacqu. ic. mifcell.*
2. *p.* 327.
Saxifraga ligulata. *Murray in comment. gotting.* 1781.
p. 26. *t.* 1.
China Saxifrage.
Nat. of China and Japan.
Introd. 1771, by Benjamin Torin, Efq.
Fl. June and July. H. ♃.

8. S. foliis obovatis fubretufis cartilagineo-crenatis, caule *umbrofa.*
nudo paniculato. *Sp. pl.* 574.

London-

London-pride.
Nat. of England.
Fl. April and May.   H. ♃.

cuneifo-   9. S. foliis cuneiformibus obtufiffimis repandis, caule
lia.   nudo paniculato. *Sp. pl.* 574.
               Wedge-leav'd Saxifrage.
               Nat. of the Alps of Switzerland.
               Introd. 1768, by Profeſſor de Sauſſure.
               Fl. May.   H. ♃.

Geum.   10. S. foliis reniformibus dentatis, caule nudo paniculato.
                  Sp. pl. 574.
               Kidney-leav'd Saxifrage.
               Nat. of the Alps of Europe.
               Cult. 1629. *Park. parad.* 231.
               Fl. June and July.   H. ♃.

               * * Foliis *indivifis*, caule *foliofo.*

oppofitifo- 11. S. foliis caulinis ovatis oppofitis imbricatis : ſummis
lia.   ciliatis. *Sp. pl.* 575.
               Purple-flower'd Saxifrage.
               Nat. of Britain.
               Fl. March and April.   H. ♃.

afpera.   12. S. foliis caulinis lanceolatis alternis ciliatis, caulibus
                  procumbentibus. *Syft. veget.* 413. *Jacqu. auftr.* 5.
                  p. 44. *t. app.* 31.
               Rough Saxifrage.
               Nat. of the Alps of Switzerland.
               Cult. 1748, by Mr. Ph. Miller. *Mill. dict. edit.* 5.
                  n. 14.
               Fl. Auguft.   H. ♃.

13. S.

13. S. foliis caulinis lanceolatis alternis nudis inermi- *Hirculus.*
    bus, caule erecto. *Sp. pl.* 576.
Marſh Saxifrage.
Nat. of England.
Fl. Auguſt.   H. ♃.

14. S. foliis caulinis lineari-ſubulatis ſparſis nudis iner- *aizoides.*
    mibus, caulibus decumbentibus. *Sp. pl.* 576.
Yellow Mountaiñ Saxifrage.
Nat. of England.
Fl. Auguſt.   H. ♃.

15. S. foliis caulinis reniformibus dentatis petiolatis, *rotundi-*
    caule paniculato. *Sp. pl.* 576.   *folia.*
Round-leav'd Saxifrage.
Nat. of Auſtria and Switzerland.
Cult. 1597, by Mr. John Gerard. *Ger. herb.* 644.
    f. 1.
Fl. May and June.   H. ♃.

        * * * Foliis *lobatis,* caule *erecto.*

16. S. foliis caulinis reniformibus lobatis, caule ramoſo, *granula-*
    radice granulata. *Sp. pl.* 576. *Curtis lond.*   *ta.*
 α floribus ſimplicibus.
Single white Saxifrage, or Sengreen.
 β floribus plenis.
Double white Saxifrage.
Nat. of Britain.
Fl. May.   H. ♃.

17. S. foliis radicalibus reniformibus quinquelobis mul- *geranioi-*
    tiſidis; caulinis linearibus, caule ſubnudo ramoſo. *des.*
    Sp. pl. 578.
Cranebill-leav'd Saxifrage.
Nat. of the Pyrenees.

Introd. 1770, by Monſ. Richard.
Fl. April and May.   H. ♃.

ajugifo-
lia.
18. S. foliis radicalibus palmato-quinquepartitis ; caulinis linearibus indiviſis, caulibus adſcendentibus
multifloris. *Sp. pl.* 578.
Ajuga-leav'd Saxifrage.
Nat. of the South of France.
Introd. 1770, by Monſ. Richard.
Fl. June and July.   H. ♃.

tridactyl-
ites.
19. S. foliis caulinis cuneiformibus trifidis alternis, caule
erecto ramoſo. *Sp. pl.* 578. *Curtis lond.*
Rue-leav'd Saxifrage, or Whitlow-graſs.
Nat. of Britain.
Fl. April.   H. O.

petræa.
20. S. foliis caulinis palmato-tripartitis : laciniis ſubtri-
fidis, caule ramoſiſſimo laxo. *Sp. pl.* 578.
Rock-Saxifrage.
Nat. of Norway and the Pyrenees.
Cult. 1748, by Mr. Ph. Miller. *Mill. dict. edit.* 5. *n.* 11.
Fl. April and May.   H. ♃.

cæſpitoſa.
21. S. foliis radicalibus aggregatis linearibus integris
trifidiſque, caule erecto ſubnudo ſubbifloro. *Sp.*
pl. 578.
Matted Saxifrage.
Nat. of England.
Fl. Auguſt.   H. ♃.

**** Foliis *lobatis*, caulibus *procumbentibus.*
hedera-
cea.
22. S. foliis caulinis ovatis lobatis, caule filiformi flacci-
do. *Sp. pl.* 579.
Ivy-leav'd Saxifrage.

Nat.

Nat. of the Levant.
Introd. 1788, by John Sibthorp, M. D.
Fl. July.   G. H. ☉.

23. S. foliis caulinis linearibus integris trifidifve, ftolonibus   *hypnoides.*
procumbentibus, caule erecto nudiufculo. *Sp. pl.* 579.
Mofs Saxifrage.
Nat. of Britain.
Fl. April and May.   H. ♃.

T I A R E L L A. *Gen. pl.* 560.

Cal. 5-partitus. *Cor.* 5-petala, calyci inferta: petalis
integris. *Capf.* 1-locularis, 2-valvis: valvula alte-
ra majore.

1. T. foliis cordatis. *Sp. pl.* 580.   *cordifolia.*
Heart-leav'd Tiarella.
Nat. of North America.
Cult. 1759, by Mr. Ph. Miller. *Mill. dict. edit.* 7. *n.* 1.
Fl. April and May.   H. ♃.

M I T E L L A. *Gen. pl.* 561.

Cal. 5-fidus. *Cor.* 5-petala, calyci inferta: petalis
pinnatifidis. *Capf.* 1-locularis, 2-valvis: valvulis
æqualibus.

1. M. fcapo diphyllo. *Sp. pl.* 580.   *diphylla.*
Two-leav'd Mitella.
Nat. of North America.
Cult. 1731, by Mr. Ph. Miller. *Mill. dict. edit.* 1. *n.* 2.
Fl. April and May.   H. ♃.

S C L E R A N T H U S. *Gen. pl.* 562.

Cal. 1-phyllus. *Cor.* nulla. *Sem.* 2, calyce inclufa.

1. S. calycibus fructus patulis. *Sp. pl.* 580.   *annuus.*
            G 2   Annual

Annual Knawel.
Nat. of Britain.
Fl. Auguft.   H. ⊙.

perennis.   2. S. calycibus fructus claufis. *Sp. pl.* 580.
Perennial Knawel.
Nat. of England.
Fl. Auguft.   H. ♃.

GYPSOPHILA. *Gen. pl.* 563.

Cal. 1-phyllus, campanulatus, angulatus. *Petala* 5,
ovata, feffilia. *Capf.* globofa, 1-locularis.

repens.   1. G. fol. lanceolatis, ftaminibus corolla emarginata
brevioribus. *Syft. veget.* 415. *Jacqu. auftr.* 5. *p.* 4.
t. 407.
Creeping Gypfophila.
Nat. of Siberia, Auftria, and Switzerland.
Introd. 1774, by Monf. Richard.
Fl. July——September.   H. ♃.

proftrata.   2. G. foliis lanceolatis lævibus, caulibus diffufis, piftillis
corolla campanulata longioribus. *Sp. pl.* 581.
Trailing Gypfophila.
Nat.
Cult. 1759, by Mr. Ph. Miller. *Mill. dict. edit.* 7. *n.* 3.
Fl. June——September.   H. ♃.

*panicula-
ta.*   3. G. foliis lineari-lanceolatis : inferioribus fcabris, ftaminibus minutis, ftylis corolla longioribus.
Gypfophila paniculata. *Sp. pl.* 583. *Jacqu. auftr.* 5.
p. 26. *tab. app.* 1.
Panicled Gypfophila.
Nat. of Siberia and Hungary.

                                                Cult.

Cult. 1759, by Mr. Ph. Miller. *Mill. dict. edit.* 7. *n.* 5.
Fl. June and July.   H. ♃.

4. G. foliis ovato-lanceolatis lævibus baſi cordatis am-   *viſcoſa.*
    plexicaulibus, internodiis ramorum medio viſcoſis,
    petalis retuſis.
    Gypſophila viſcoſa. *Murray in comment. gotting.* 1783.
    p. 9. *tab.* 3.
    Clammy Gypſophila.
    Nat. of the Levant.
    Introd. 1773, by Mr. William Forſyth.
    Fl. June and July.   H. ⊙.

5. G. foliis lanceolatis ſubtrinerviis, caulibus rectis. *Syſt.*   *altiſſima.*
    veget. 416.
    Upright Gypſophila.
    Nat. of Siberia.
    Introd. 1777, by Abbé Nolin.
    Fl. July.   H. ♃.

6. G. foliis linearibus carnoſis : axillaribus confertis te-   *Struthi-*
    retibus. *Syſt. veget.* 416.   *um.*
    Shrubby Gypſophila.
    Nat. of Spain.
    Introd. 1774, by Monſ. Richard.
    Fl. July and Auguſt.   H. ♃.

7. G. foliis lanceolato-linearibus obſolete triquetris læ-   *faſtigia-*
    vibus obtuſis ſecundis. *Sp. pl.* 582.   *ta.*
    Triangular-leav'd Gypſophila.
    Nat. of Sweden and Germany.
    Cult. 1759, by Mr. Ph. Miller. *Mill. dict. edit.* 7. *n.* 2.
    Fl. June and July.   H. ♃.

8. G. foliis ovato-lanceolatis ſemiamplexicaulibus. *Sp.*   *perfolia-*
    pl. 583.   *ta.*

                         G 3   *a* Spergula

α Spergula multiflora, foliis inferioribus Saponariæ, ſuperioribus Behen ſimilibus. *Dill. elth.* 368. *t.* 276.
f. 357.
Perfoliate Gypſophila.

β Gypſophila *tomentoſa,* foliis lanceolatis ſubtrinerviis
tomentoſis, caule pubeſcente. *Sp. pl.* 583.
Woolly Gypſophila.
Nat. of Spain and the Levant.
Cult. 1732, by James Sherard, M.D. *Dill. elth. loc.
cit.*
Fl. July and Auguſt.   H. ♃.

muralis. 9. G. foliis linearibus planis, calycibus aphyllis, caule
dichotomo, petalis crenatis. *Sp. pl.* 583.
Wall Gypſophila.
Nat. of Sweden, Germany, and Switzerland.
Introd. 1773, by the Rev. John Lightfoot, M.A.
Fl. June——October.   H. ♂.

ſaxifra- 10. G. foliis linearibus, calycibus angulatis : ſquamis
ga.   quatuor, corollis emarginatis. *Syſt. veget.* 416.
Small Gypſophila.
Nat. of Auſtria, France, and Switzerland.
Introd. 1774, by Monſ. Richard.
Fl. July and Auguſt.   H. ♃.

SAPONARIA. *Gen. pl.* 564.
Cal. 1-phyllus, nudus. *Petala* 5, unguiculata. *Capſ.*
oblonga, 1-locularis.

officinalis. 1. S. calycibus cylindricis, foliis ovato-lanceolatis. *Sp.
pl.* 584. *Curtis lond.*
α Saponaria major lævis. *Bauh. pin.* 206.
Common Soapwort.
β Saponaria flore pleno. *Corn. canad.* 209.
    Double-

Double-flower'd Soapwort.

γ Saponaria concava Anglica. *Bauh. pin.* 206. *hybrida.*
Hollow-leav'd Soapwort.
Nat. of England.
Fl. July——October. H. ♃.

2. S. calycibus pyramidatis quinquangularibus, foliis *Vaccaria.*
ovatis acuminatis feffilibus. *Sp. pl.* 585.
Perfoliate Soapwort.
Nat. of Germany and France.
Cult. 1596, by Mr. John Gerard. *Hort. Ger.*
Fl. July and Auguft. H. ☉.

3. S. calycibus cylindricis pubefcentibus, ramis divari- *porrigens.*
catiffimis, fructibus pendulis. *Linn. mant.* 239.
Jacqu. hort. 2. *p.* 49. *t.* 109.
Hairy Soapwort.
Nat. of the Levant.
Introd. 1777, by Abbé Nolin.
Fl. July and Auguft. H. ☉.

4. S. calycibus cylindricis villofis, caulibus dichotomis *Ocymoi-*
procumbentibus. *Sp. pl.* 585. *Jacqu. auftr.* 5. *des.*
p. 39. *tab. app.* 23.
Bafil-leav'd Soapwort.
Nat. of Italy, France, and Switzerland.
Introd. 1768, by Profeffor de Sauffure.
Fl. May——July. H. ♃.

5. S. calycibus cylindricis villofis, caule dichotomo erec- *orientalis.*
to patulo. *Sp. pl.* 585.
Small Annual Soapwort.
Nat. of the Levant.
Cult. 1732, by James Sherard, M.D. *Dill. elth.*
205. *t.* 167. *f.* 204.
Fl. June——Auguft. H. ☉.

G 4 DIAN-

DIANTHUS. *Gen. pl.* 565.

Cal. cylindricus, 1-phyllus: bafi fquamis 4. *Petala* 5, unguiculata. *Capf.* cylindrica, 1-locularis.

* *Flores aggregati.*

barbatus.   1. D. floribus aggregatis fafciculatis: fquamis calycinis ovato-fubulatis tubum æquantibus, foliis lanceolatis. *Sp. pl.* 586.

α Caryophyllus barbatus hortenfis latifolius. *Bauh. pin.* 208.

Common Sweet-William Pink.

β Caryophyllus barbatus, flore multiplici. *Bauh. pin.* 208.

Double Sweet-William Pink.

Nat.

Cult. 1596, by Mr. John Gerard. *Hort. Ger.*

Fl. June and July.   H. ♃.

carthufi-
anorum.   2. D. floribus fubaggregatis: fquamis calycinis ovatis ariftatis tubum fubæquantibus, foliis linearibus trinerviis. *Sp. pl.* 586.

Carthufian Pink.

Nat. of Italy and Germany.

Introd. 1771, by Monf. Richard.

Fl. July and Auguft.   H. ♃.

ferrugi-
neus.   3. D. floribus aggregatis, petalis bifidis: laciniis tridentatis. *Syft. veget.* 417.

Rufty Pink.

Nat. of Italy.

Cult. 1756, by Mr. Ph. Miller. *Mill. ic.* 54. *t.* 81. *f.* 1.

Fl. July and Auguft.   H. ♂.

Armeria.   4. D. floribus aggregatis fafciculatis: fquamis calycinis lanceolatis villofis tubum æquantibus. *Sp. pl.* 586.

                                  Deptford

Deptford Pink.
Nat. of England.
Fl. July.   H. ○.

5. D. floribus aggregatis capitatis : fquamis calycinis   *prolifer.*
ovatis obtufis muticis tubum fuperantibus. *Sp.*
pl. 587.
Proliferous Pink.
Nat. of England.
Fl. Auguft.   H. ⊙.

** * * *Flores folitarii, plures in eodem caule.*,
6. D. floribus folitariis : fquamis calycinis fubovatis   *Caryo-*
breviffimis, corollis crenatis. *Sp. pl.* 587.   *phyllus.*
α Caryophyllus hortenfis fimplex, flore majore. *Bauh.*
pin. 208.
Clove Pink.
β Caryophyllus maximus ruber & variegatus. *Bauh.*
pin. 207.
Common Carnation.
Nat. of England.
Fl. June——Auguft.   H. ♃.

7. D. floribus folitariis : fquamis calycinis lanceolatis   *deltoides.*
binis, corollis crenatis. *Sp.' pl.* 588.
Maiden Pink.
Nat. of Britain.
Fl. June and July.   H. ♃.

8. D. floribus fubfolitariis : fquamis calycinis lanceo-   *glaucus.*
latis quaternis brevibus, corollis crenatis. *Sp. pl.*
588.
Mountain Pink.
Nat. of Britain.
Fl. July.   H. ♃.
                                               9. D.

albens.   9. D. floribus folitariis : fquamis calycinis lanceolatis quaternis brevibus, corollis emarginatis.
Cape Pink.
Nat. of the Cape of Good Hope.   Mr. *Fr. Maſſon.*
Introd. 1787.
Fl. Auguſt.   G. H. ♃.
Obs. Petala fupra alba, fubtus e viridi albentia, apice utrinque violacea.   Differt a D. glauco petalis vix crenatis et abfentia circuli purpurei.

chinenſis.   10. D. floribus folitariis : fquamis calycinis fubulatis patulis tubum æquantibus, corollis crenatis. *Syſt. veget.* 418. *Curt. magaz.* 25.
China Pink.
Nat. of China.
Cult. before 1719. *Mill. ic.* 54.
Fl. July——September.   H. ♂.

ſuperbus.   11. D. floribus paniculatis : fquamis calycinis brevibus acuminatis, corollis multifido-capillaribus, caule erecto. *Sp. pl.* 589.
Superb Pink.
Nat. of France, Germany, and Denmark.
Cult. 1629. *Park. parad.* 316. *n.* 4.
Fl. July——September.   H. ♂.

plumarius.   12. D. floribus folitariis : fquamis calycinis fubovatis breviſſimis, corollis multifidis fauce pubefcentibus. *Sp. pl.* 589.
Feathered Pink.
Nat. of Europe and North America.
Cult. 1629. *Park. parad.* 316. *n.* 2. 3.
Fl. June——Auguſt.   H. ♃.

*** *Caule*

*** *Caule unifloro herbaceo.*

13. D. caulibus fubunifloris : fquamis calycinis ovatis   *arenari-*
obtufis, corollis multifidis, foliis linearibus. *Sp. pl.* *us.*
589.
Sand Pink.
Nat. of Britain.
Fl. May and June.   H. ♃.

14. D. caule fubunifloro, corollis crenatis, fquamis caly- *virgine-*
cinis breviffimis, foliis fubulatis. *Sp. pl.* 590. *us.*
Jacqu. auftr. 5. *p.* 34. *tab. app.* 15.
UprightPink.
Nat. of the South of France and Auftria.
Cult. 1732, by James Sherard, M.D. *Dill. elth.* 401.
t. 298. *f.* 385.
Fl. June and July.   H. ♃.

**** *Frutefcentes.*

15. D. caule fuffruticofo, foliis lineari-fubulatis, petalis *pungens.*
integris. *Linn. mant.* 240.
Pungent Pink.
Nat. of Spain.
Introd. 1781, by Monf. Thouin.
Fl. Auguft——Oĉtober.   H. ♃.

TRIGYNIA.

C U C U B A L U S. *Gen. pl.* 566.

Cal. inflatus. *Petala* **5**, unguiculata, abfque corona
ad faucem. *Capf.* 3-locularis.

1. C. calycibus campanulatis, petalis diftantibus, peri- *baccife-*
carpiis coloratis, ramis divaricatis. *Sp. pl.* 591. *rus.*
Berry-bearing Campion.
                               Nat.

Nat. of England.
Fl. July.   H. ♃.

Behen.   2. C. calycibus fubglobofis glabris reticulato-venofis,
             capfulis trilocularibus, corollis fubnudis. *Sp. pl.*
             591.
           α Lychnis fylveftris, quæ Been album vulgo. *Bauh.*
             pin. 205.
             Common Bladder Campion.
           β Lychnis maritima repens. *Bauh. pin.* 205.
             Sea Bladder Campion.
             Nat. of Britain.
             Fl. May——September.   H. ♃.

vifcofus. 3. C. floribus lateralibus undique decumbentibus, caule
             indivifo, foliis bafi reflexis. *Sp. pl.* 592.
             Clammy Campion.
             Nat. of the Levant.
             Cult. 1759. *Mill. dict. edit.* 7. *n.* 8.
             Fl. July.   H. ♂.

ftellatus. 4. C. foliis quaternis. *Sp. pl.* 592.
             Four-leav'd Campion.
             Nat. of Virginia and Canada.
             Cult. 1768, by Mr. Ph. Miller. *Mill. dict. edit.* 8.
             Fl. June——Auguft.   H. ♃.

italicus. 5. C. petalis femibifidis, calycibus clavatis, panicula di-
             chotoma erecta, genitalibus declinatis, caule erecto.
             Syft. veget. 419.
             Italian Campion.
             Nat. of Italy.
             Cult. 1768, by Mr. Ph. Miller. *Mill. dict. edit.* 8.
             Fl. May and June.   H. ♂.

tataricus. 6. C. petalis bipartitis, floribus fecundis decumbentibus,
                                                       pedunculis

pedunculis oppofitis folitariis erectis, caule fimpli-
ciffimo. *Sp. pl.* 592.
Hyffop-leav'd Campion.
Nat. of Ruffia.
Introd. about 1772.
Fl. June——Auguft. H. ♃.

7. C. petalis emarginatis, floribus fubverticillatis : verti- *fibiricus.*
cillis umbellatis aphyllis. *Sp. pl.* 592.
Siberian Campion.
Nat. of Siberia.
Introd. 1773, by Chevalier Murray.
Fl. June——Auguft. H. ♃.

8. C. petalis bipartitis, floribus paniculatis, ftaminibus *catholi-*
longis, foliis lanceolato-ovatis. *Sp. pl.* 593. *Jacqu.* *cus.*
hort. 1. *p.* 23. *t.* 59.
Panicled Campion.
Nat. of Italy.
Cult. 1711, in Chelfea Garden. *Philofoph. tranf. n.*
332. *p.* 391. *n.* 48.
Fl. Auguft and September. H. ♃.

9. C. floribus dioicis, petalis linearibus indivifis. *Sp. pl.* *Otites.*
594.
Spanifh Campion, or Catchfly.
Nat. of England.
Fl. July and Auguft. H. ♂.

S I L E N E. *Gen. pl.* 567.
Cal. ventricofus. *Petala* 5, unguiculata : coronata
ad faucem. *Capf.* 3-locularis.

○ *Floribus lateralibus folitariis.*

1. S. hirfuta, petalis integerrimis, floribuseer ctis, fruc- *anglica.*
tibus reflexis pedunculatis alternis. *Sp. pl.* 594.
Curtis lond.
Englifh

English Catchfly.
Nat. of England.
Fl. June and July.   H. ☉.

lufitanica.   2. S. hirfuta, petalis dentatis, floribus erectis, fructibus
divaricato-reflexis alternis. *Sp. pl.* 594.
Portugal Catchfly.
Nat. of Portugal.
Cult. 1732, by James Sherard, M. D. *Dill. elth.* 420.
t. 311. *f.* 401.
Fl. June and July.   H. ☉.

quinque-   3. S. petalis integerrimis fubrotundis, fructibus erectis
vulnera.   alternis. *Sp. pl.* 595.
Variegated Catchfly, or Mountain Lychnis.
Nat. of England.
Fl. July and Auguft.   H. ○.

nocturna.   4. S. floribus fpicatis alternis fecundis feffilibus, petalis
bifidis. *Sp. pl.* 595.
Spik'd night-flowering Catchfly.
Nat. of France and Spain.
Cult. 1683, by Mr. James Sutherland. *Sutherl. hort.*
edin. 214. *n.* 1.
Fl. July.   H. ☉.

** *Floribus lateralibus confertis.*

nutans.   5. S. petalis bifidis, floribus lateralibus fecundis cernuis,
panicula nutante. *Sp. pl.* 596.
Nottingham Catchfly.
Nat. of England.
Fl. June and July.   H. ♃.

fruticofa.   6. S. petalis bifidis, caule fruticofo, foliis lato-lanceolatis, panicula trichotoma. *Sp. pl.* 597.

Shrubby

Shrubby Catchfly.
Nat. of Germany and Sicily.
Cult. 1683. *Sutherl. hort. edin.* 211. *n.* 3.
Fl. June and July.   H. ♄.

7. S. petalis bifidis, foliis radicalibus cochleariformibus *gigantea.*
obtufiffimis, floribus fubverticillatis. *Syft. veget.*
421.
Gigantic Catchfly.
Nat. of Africa.
Cult. 1768, by Mr. Philip Miller. *Mill. dict. edit.* 8.
n. 13.
Fl. June and July.   G. H. ♂.

8. S. petalis emarginatis, foliis fuborbiculatis carnofis *craffife-*
hirfutis, racemo fecundo. *Syft. veget.* 421.   *lia.*
Thick-leav'd Catchfly.
Nat. of the Cape of Good Hope.
Introd. 1774, by Mr. Francis Maffon.
Fl. July and Auguft.   G. H. ♂.

9. S. petalis femibifidis, foliis ovatis fcabriufculis acu- *viridiflo-*
tis, panicula elongata fubaphylla. *Sp. pl.* 597.   *ra.*
Green-flower'd Catchfly.
Nat. of Portugal and Spain.
Cult. 1739, by Mr. Ph. Miller. *Rand. chel.* Lych-
nis 23.
Fl. June and July.   H. ♂.

        *** *Floribus ex dichotomia caulis.*
10. S. calycibus fructus globofis acuminatis ftriis tri- *conoidea.*
ginta, foliis glabris, petalis integris. *Sp. pl.* 598.
Corn Catchfly.
Nat. of England.
Fl. June and July.   H. ⊙.
                                  11. S.

conica. 11. S. calycibus fructus conicis ftriis triginta, foliis mollibus, petalis bifidis. *Sp. pl.* 598. *Jacqu. auftr.* 3. *p.* 30. *t.* 253.
Conic Catchfly.
Nat. of France and Spain.
Cult. 1739, by Mr. Ph. Miller. *Rand. chel.* Lychnis 38.
Fl. June and July. H. ☉.

pendula. 12. S. calycibus fructiferis pendulis inflatis : angulis decem fcabris. *Sp. pl.* 599.
Pendulous Catchfly.
Nat. of Candia and Sicily.
Cult. 1732, by James Sherard, M. D. ·*Dill. elth.* 421. *t.* 312. *f.* 402.
Fl. May and June. H. ☉.

noctiflora. 13. S. calycibus decemangularibus : dentibus tubum æquantibus, caule dichotomo, petalis bifidis. *Syft. veget.* 421.
Fork'd night-flowering Catchfly.
Nat. of England.
Fl. July. H. ☉.

ornata. 14. S. calycibus fructus oblongis carinatis pilofis, petalis bifidis, foliis lanceolatis pubefcentibus vifcofis planis, caule vifcido.
Dark-colour'd Catchfly.
Nat. of the Cape of Good Hope. Mr. *Fr. Maffon. Introd.* 1775.
Fl. May——September. G. H. ♂.

undulata. 15. S. calycibus fructus clavato-cylindricis pilofis, petalis bifidis, foliis lanceolatis pubefcentibus undulatis, caule adfcendente.
Waved-leav'd Catchfly.

Nat.

Nat. of the Cape of Good Hope. Mr. *Fr. Maſſin.*
Introd. 1775.
Fl. Auguſt.   G. H. ♂ .

16. S. foliis lanceolatis ſubciliatis, pedunculis trifidis,   *antirrhi-*
   petalis emarginatis, calycibus ovatis. *Sp. pl.* 600.   *na.*
   Snap-dragon-leav'd Catchfly.
   Nat. of Virginia and Carolina.
   Cult. 1732, by James Sherard, M. D. *Dill. elth.*
   422. *t.* 313. *f.* 403.
   Fl. June and July.   H. ⊙ .

17. S. caule dichotomo paniculato, calycibus lævibus,   *inaperta.*
   petalis breviſſimis emarginatis, foliis glabris lan-
   ceolatis. *Sp. pl.* 600.
   Small-flower'd Catchfly.
   Nat. of Madeira.
   Cult. 1732, by James Sherard, M. D. *Dill. elth.*
   424. *t.* 315. *f.* 407.
   Fl. July.   H. ⊙ .

18. S. caule dichotomo paniculato, calycibus ſtriatis, pe-   *portenſis.*
   talis bifidis, foliis linearibus. *Sp. pl.* 600.
   Oporto Catchfly.
   Nat. of Portugal.
   Cult. 1759, by Mr. Philip Miller.
   Fl. July and Auguſt.   H. ⊙ .

19. S. erecta lævis, calycibus erectis decangulis fulcatis,   *cretica.*
   petalis bifidis. *Sp. pl.* 601.
   Daiſy-leav'd Catchfly.
   Nat. of Candia.
   Cult. 1732, by James Sherard, M. D. *Dill. elth.*
   422. *t.* 314. *f.* 404. 405.
   Fl. June——Auguſt.   H. ⊙ .

VOL. II.   H   2C. S.

Armeria.   20. S. floribus fasciculatis fastigiatis, foliis superioribus cordatis glabris, petalis integris. *Syst. veget.* 422.

α floribus rubris.

Lobel's red Catchfly.

β floribus albis.

Lobel's white Catchfly.

Nat. of England.

Fl. July and August.   H. ☉.

orchidea.   21. S. petalis bilobis : laminis basi utrinque processu subulato auctis, foliis lævibus : inferioribus subrotundo-spathulatis ; petiolis ciliatis.

Silene orchidea. *Linn. suppl.* 241. *Syst. veget.* 422.

Silene Atocion. *Jacqu. hort.* 3. *p.* 19. *t.* 32. *Syst. veget.* 421.

Orchis-flower'd Catchfly.

Nat. of the Levant.

Introd. 1781, by P. M. A. Broussonet, M. D.

Fl. May and June.   H. ☉.

alpestris.   22. S. petalis quadridentatis, caule dichotomo, capsulis ovato-oblongis, foliis lineari-lanceolatis glabris erectis, pedunculis viscidis.

Silene alpestris. *Jacqu. austr.* 1. *p.* 60. *t.* 96.

Lychnis alpestris. *Linn. suppl.* 244.

Lychnis quadrifida. *Scop. carn.* 1. *p.* 307.

Lychnis viscosa alba angustifolia major. *Bauh. pin.* 205.

Lychnis sylvestris decima. *Cluf. hist.* 1. *p.* 291.

Caryophyllus minimus humilis alter exoticus flore candido amœno. *Lob. ic.* 445.

Austrian Catchfly.

Nat. of Austria.

Introd. 1774, by Joseph Nicholas de Jacquin, M. D.

Fl. May——July.   H. ♃.

23. S.

23. S. floribus erectis, petalis emarginatis, calycibus te-   *rupestris.*
retibus, foliis lanceolatis. *Sp.*   602.
Rock Catchfly.
Nat. of Sweden and Switzerland.
Introd. 1771, by Monf. Richard.
Fl. June——Auguft.   H. ♂.

24. S. caulibus fubunifloris, pedunculis longitudine cau-   *Saxifra-*
lis, floribus hermaphroditis femineifque, petalis bi-   *ga.*
fidis. *Syft. veget.* 422.
Saxifrage Catchfly.
Nat. of France and Italy.
Introd. 1772, by Monf. Richard.
Fl. June——Auguft.   H. ♃.

25. S. caulibus fubunifloris decumbentibus, foliis lan-   *vallefia.*
ceolatis tomentofis longitudine calycis. *Sp. pl.*
603.
Woolly-leav'd Catchfly.
Nat. of the Alps of Switzerland.
Cult. 1768, by Mr. Philip Miller. *Mill. dict. edit.* 8.
Fl. June——Auguft.   H. ♃.

26. S. acaulis depreffa, petalis emarginatis. *Syft. veget.*   *acaulis.*
422.
Dwarf Catchfly.
Nat. of Britain.
Fl. June and July.   H. ♃.

STELLARIA. *Gen. pl.* 568.

Cal. 5-phyllus, patens. *Petala* 5, bipartita. *Capf.*
1-locularis, polyfperma.

1. S. foliis cordatis petiolatis, panicula pedunculis ra-   *nemorum.*
molis. *Sp. pl.* 603.
Broad-leav'd Stichwort.

Nat.

Nat. of Britain.
Fl. April——June.   H. ♃.

dichoto-   2. S. foliis ovatis feffilibus, caule dichotomo, floribus
ma.   folitariis, pedunculis fructiferis reflexis. *Sp. pl.* 603.
            Forked Stichwort.
            Nat. of Siberia.
            Introd. about 1774.
            Fl. July.   H. ☉.

Holoſtea.   3. S. foliis lanceolatis ferrulatis, petalis bifidis. *Sp. pl.*
               603. *Curtis lond.*
            Great Stichwort.
            Nat. of Britain.
            Fl. April——June.   H. ♃.

graminea.   4. S. foliis linearibus integerrimis, floribus paniculatis.
               Sp. pl. 604.
            Grafs-leav'd Stichwort.
            Nat. of Britain.
            Fl. April and June.   H. ♃.

            A R E N A R I A.   *Gen. pl.* 569.
       Cal. 5-phyllus, patens. *Petala* 5, integra. *Capſ.* 1-lo-
                          cularis, polyſperma.

peploides. 1. A. foliis ovatis acutis carnofis. *Sp. pl.* 605.
            Sea Sandwort.
            Nat. of Britain.
            Fl. May and June.   H. ♃.

tetraque-   2. A. foliis ovatis carinatis recurvis quadrifariam imbri-
tra.   catis. *Sp. pl.* 605.
            Square Sandwort.
            Nat. of the Pyrenean Mountains.
            Introd. 1776, by Caſimir Gomez Ortega, M.D.
            Fl. Auguſt,   H. ♃.
                                                        3. A.

3. A. foliis ovatis acutis petiolatis nervofis. *Sp. pl.* 605.   *trinervia.*
   Curtis lond.
   Plantain-leav'd Sandwort.
   Nat. of Britain.
   Fl. May.   H. ☉.

4. A. foliis ovatis nervofis ciliatis acutis. *Syft. veget.*   *ciliata.*
   423.
   Ciliated Sandwort.
   Nat. of Iceland.
   Introd. 1773, by Sir Jofeph Banks, Bart.
   Fl. March——Auguft.   H. ♂.

5. A. foliis ovatis lucidis fubcarnofis, caule repente, pe-   *balearica.*
   dunculis unifloris. *Syft. veget.* 423.   *L'Herit.*
   ftirp. nov. p. 29. *tab.* 15.
   Balearic Sandwort.
   Nat. of the Balearic Iflands.
   Introd. 1787, by Monf. Cels.
   Fl. June——Auguft.   H. ♃.

6. A. foliis fubovatis acutis feffilibus, corollis calyce bre-   *ferpyllifo-*
   vioribus. *Sp. pl.* 606. *Curtis lond.*   *lia.*
   Thyme-leav'd Sandwort.
   Nat. of Britain.
   Fl. June.   H. ☉.

7. A. foliis filiformibus, ftipulis membranaceis vaginan-   *rubra.*
   tibus. *Sp. pl.* 606.
α Alfine fpergulæ facie minor : five Spergula minor flore   *campef-*
   fubcæruleo. *Bauh. pin.* 251.   *tris.*
   Red Sandwort, Purple Chickweed, or Spurry.
β Arenaria foliis linearibus longitudine internodiorum.   *marina.*
   Hort. cliff. 173.
   Sea Sandwort, or Spurry.

Nat. of Britain.

Fl. June and July.   H. ⊙.

media.   8. A. foliis linearibus carnofis, ftipulis membranaceis.
          Sp. pl. 606.
          Downy Sandwort.
          Nat. of England.
          Fl. July.   H. ⊙.

verna.   9. A. foliis fubulatis, caulibus paniculatis, calycibus acu-
          minatis ftriatis. *Linn. mant.* 72. *Jacqu. auftr.*
          5. *p.* 2. *t.* 404.
          Spring Sandwort.
          Nat. of England.
          Fl. May——Auguft.   H. ♃.

tenuifolia. 10. A. foliis fubulatis, caule paniculato, capfulis erectis,
            petalis calyce brevioribus lanceolatis. *Syft. veget.*
            424.
            Fine-leav'd Sandwort.
            Nat. of England.
            Fl. June and July.   H. ♃.

laricifo- 11. A. foliis fetaceis, caule fuperne nudiufculo, calycibus
lia.   fubhirfutis. *Sp. pl.* 607. *Jacqu. auftr.* 3. *p.* 39.
              t. 272.
              Larch-leav'd Sandwort.
              Nat. of Britain.
              Fl. Auguft.   H. ♃.

ftriata.   12. A. foliis linearibus erectis adpreffis, calycibus oblon-
           gis ftriatis. *Syft. veget.* 424.
           Striated Sandwort.
           Nat. of Switzerland.
           Introd. 1768, by Profeffor de Sauffure.
           Fl. June——Auguft.   H. ♃.

                                                        13. A.

13. A. foliis fubulatis, caule erecto ftricto, floribus fafci-   *fafcicula-*
culatis, petalis breviffimis. *Syft. veget.* 424. *Jacqu.*   *ta.*
auftr. 2. *p.* 49. *t.* 182.
Cluſter-flowering Sandwort.
Nat. of Auſtria and the South of France.
Introd. 1787, by Mr. Zier.
Fl. Auguſt.   H. ☉.

C H E R L E R I A. *Gen. pl.* 570.

Cal. 5-phyllus. *Nectaria* 5, bifida, petaloidea. *Antheræ* alternæ fteriles. *Capf.* 1-locularis, 3-valvis,
trifperma.

1. CHERLERIA. *Sp. pl.* 608. *Jacqu. auftr.* 3. *p.* 46.   *Sedoides.*
t. 284.
Stone-crop Cherleria.
Nat. of Scotland.
Fl. July and Auguſt.   H. ♃.

G A R I D E L L A. *Gen. pl.* 571.

Cal. 5-phyllus, petaloideus. *Nectaria* 5, bilabiata,
bifida. *Capf.* 3, connexæ, polyfpermæ.

1. GARIDELLA. *Sp. pl.* 608.   *Nigellaf-*
Fennel-leav'd Garidella.   *trum.*
Nat. of the South of France.
Cult. 1748. *Mill. dict. edit.* 5.
Fl. June and July.   H. ☉.

M A L P I G H I A. *Gen. pl.* 572.

Cal. 5-phyllus, baſi extus poris mellif ris. *Petala* 5, fubrotunda, unguiculata. *Bacca* 1-locularis, 3-fperma.

1. M. folis ovatis integerrimis glabris, pedunculis um-   *glabra.*
bellatis. *Sp. pl.* 609.

H 4   Smooth-

Smooth-leav'd Barbadoes Cherry.
Nat. of the Weſt Indies.
Cult. 1757, by Mr. Ph. Miller.   *Mill. ic.* 121. *t.* 181,
f. 2.
Fl. December——March.   S. ♄,

punicifo-   2. M. foliis ovatis integerrimis glabris, pedunculis uni-
lia.   floris. *Sp. pl.* 609.
            Pomegranate-leav'd Barbadoes Cherry.
            Nat. of the Weſt Indies.
            Introd. 1690, by Mr. Bentick.   *Br. Muſ. Sloan. mſſ.*
            3370.
            Fl.   S. ♄.

nitida.   3. M. foliis lanceolatis integerrimis glabris, ſpicis-late-
                   ralibus. *Sp. pl.* 609.
            Shining-leav'd Barbadoes Cherry.
            Nat. of the Weſt Indies.
            Introd. before 1733, by William Houſtoun, M. D,
                   Mill. dict. edit. 8.
            Fl. at various Seaſons.   S. ♄.

urens.   4. M. foliis oblongo-ovatis : ſetis ſubtus decumbentibus
                   rigidis, pedunculis unifloris aggregatis. *Syſt. veget,*
                   426.
            Stinging Barbadoes Cherry.
            Nat. of South America.
            Introd. about 1753.   *Mill. ic.* 121. *t.* 181. *f.* 1.
            Fl. July——October.   S. ♄,

anguſtifo- 5. M. foliis lineari-lanceolatis : ſetis utrinque decum-
lia.   bentibus rigidis, pedunculis umbellatis. *Sp. pl.*
                   610.
            Narrow-leav'd Barbadoes Cherry,
            Nat. of the Weſt Indies.
                                                    Cult,

Cult. 1752, by Mr. Ph. Miller. *Mill. dict. edit.* 6. *n.* 3.
Fl.   S. ♄.

6. M. foliis oblongis obtufis pubefcentibus, racemis axil-   *canefcens.*
    laribus compofitis.
    Downy-leav'd Barbadoes Cherry.
    Nat. of the Weft Indies.
    Cult. before 1742, by Robert James Lord Petre.
    Fl.   S. ♄.

BANISTERIA. *Gen. pl.* 573.

Cal. 5-partitus, bafi extus poris melliferis. *Pet.* fubro-
    tunda, unguiculata. *Sem.* 3, membranaceo-alata.

1. B. foliis ovatis, fpicis lateralibus, feminibus erectis. *purpurea.*
    Sp. pl. 611.
    Purple Banifteria.
    Nat. of the Weft Indies.
    Cult. 1759, by Mr. Ph. Miller. *Mill. dict. edit.* 7. *n.* 7.
    Fl.   S. ♄.

PENTAGYNIA.

SPONDIAS. *Gen. pl.* 577.

Cal. 5-dentatus. *Cor.* 5-petala. *Drupa* nucleo 5-
         loculari.

1. S. petiolis teretibus, foliolis nitidis acuminatis. *Syft.*   *Myroba-*
    veget. 428.   *lanus.*
    Spondias lutea. *Sp. pl.* 613.
    Yellow Hog Plum.
    Nat. of the Weft Indies.
    Cult. 1768, by Mr. Ph. Miller. *Mill. dict. edit.* 8.
    Fl.   S. ♄.

COTYLE-

COTYLEDON. *Gen. pl.* 578.

Cal. 5-fidus.   *Cor.* 1-petala.   *Squamæ* nectariferæ 5, ad basin germinis.   *Capf.* 5.

orbicula-
ta.
1. C. foliis ovato-spathulatis obtusis cum acumine lævibus, floribus paniculatis.
Cotyledon orbiculata. *Sp. pl.* 614.
α foliis ovato-spathulatis, caudice erecto.
Oval-leav'd Navelwort.
β foliis oblongo-spathulatis, caudice erecto.
Oblong-leav'd Navelwort.
γ foliis ovato-spathulatis, caudice ramosissimo divaricato.
Branching Navelwort.
δ foliis orbiculato-spathulatis, caudice erecto.
Round-leav'd Navelwort.
Nat. of the Cape of Good Hope.
Introd. 1690, by Mr. Bentick. *Br. Muf. Sloan. mff.*
3370.
Fl. July——September.   D. S. ♄.

fafcicula-
ris.
2. C. foliis cuneiformibus fafciculatis terminalibus, caudice incraffato : ramis carnofis fubconicis.
Cotyledon frutefcens, folio oblongo, viridi, floribus ramofis, pendulis, reflexis. *Burm. afr.* 41. *t.* 18.
Clufter-leav'd Navelwort.
Nat. of the Cape of Good Hope.
Cult. 1759, by Mr. Philip Miller.
Fl. July—— September.   D. S. ♄.

fpuria.
3. C. foliis fpathulatis obtufis cum acumine nudis.
Cotyledon fpuria. *Sp. pl.* 614. (exclufis fynonymis Burmanni, præter fequens.)
Cotyledon acaulon, foliis ad radicem feffilibus, oblongis,

longis, anguftis, floribus pendulis, reflexis. *Burm.*
afr. 43. *t.* 19. *f.* 1.
Narrow-leav'd Navelwort.
Nat. of the Cape of Good Hope.
Cult. 1731, by Mr. Ph. Miller. *Mill. dict. edit.* 1. *n.* 3.
Fl. July and Auguft. D. S. ♄.

4. C. foliis fuborbiculatis furfure punctatis fubtus con- *hemi-*
vexis, floribus fubfeffilibus. *fphærica.*
Cotyledon hemifphærica. *Sp. pl.* 614.
Thick-leav'd Navelwort.
Nat. of the Cape of Good Hope.
Cult. 1732, by James Sherard, M.D. *Dill. clth.* 112.
t. 95. *f.* 111.
Fl. June and July. D. S. ♄.

5. C. foliis ovalibus crenatis, caule fpicato. *Sp. pl.* 614. *ferrata.*
Notch'd-leav'd Navelwort.
Nat. of Candia and Siberia.
Cult. 1732, by James Sherard, M.D. *Dill. elth.* 113.
t. 95. *f.* 112.
Fl. June and July. G. H. ♄.

6. C. foliis peltatis crenatis, caule fubfimplici, floribus *Umbili-*
pendulis, bracteis integris. *Hudf. angl.* 194. *cus.*
Cotyledon Umbilicus β. tuberofa. *Sp. pl.* 615.
Common Navelwort, or Wall Pennywort.
Nat. of Britain.
Fl. June and July. H. ♃.

7. C. foliis peltatis crenatis, caule fubfimplici, floribus *lutea,*
erectis, bracteis dentatis. *Hudf. angl.* 194.
Cotyledon Umbilicus α. repens. *Sp. pl.* 615.
Yellow Navelwort.
Nat. of England.
Fl. June and July. H. ♃.
8. C.

laciniata. 8. C. foliis trifidis, floribus quadrifidis. *Sp. pl.* 615.
Cut-leav'd Navelwort.
Nat. of the Eaft Indies.
Cult. 1731, by Mr. Ph. Miller. *Mill. dict. edit.* 1. *n.* 5.
Fl. July and Auguft.   D. S. ♄.

S E D U M. *Gen. pl.* 579.

Cal. 5-fidus. *Cor.* 5-petala. *Squamæ* nectariferæ 5,
ad bafin germinis. *Capf.* 5.

* *Planifolia.*

Telephi- 1. S. foliis planiufculis ferratis, corymbo foliofo, caule
um.   erecto. *Sp. pl.* 616. *Curtis lond.*
album.   α Telephium vulgare. *Bauh. pin.* 287.
Small common white Orpine.
maxi-   β Telephium hifpanicum. *Cluf. hift.* 2. *p.* 66.
mum.   Great white Orpine.
purpure- γ Telephium purpureum majus. *Bauh. pin.* 287.
um.   Great purple Orpine.
          δ Telephium purpureum minus. *Bauh. pin.* 287.
Small purple Orpine.
Nat. α. and γ. of Britain; β. and δ. of Portugal.
Fl. July and Auguft.   H. ♃.

Anacamp- 2. S. foliis cuneiformibus bafi attenuatis fubfeffilibus,
feros.   caulibus decumbentibus, floribus corymbofis.
Sedum Anacampferos. *Sp. pl.* 616.
Evergreen Orpine.
Nat. of the South of France.
Cult. 1596, by Mr. John Gerard. *Hort. Ger.*
Fl. July and Auguft.   H. ♃.

divarica- 3. S. foliis cuneiformi-rhombeis emarginatis petiolatis,
tum.   caulibus ramofis, paniculis terminalibus divaricatis.
Spreading Stone-crop.
                                          Nat.

Nat. of Madeira. Mr. *Francis Maſſon.*
Introd. 1777.
Fl. June and July.   G. H. ♄.

4. S. foliis lanceolatis ſerratis planis, caule erecto, cyma   *Aizoon.*
    feſſili terminali. *Sp. pl.* 617.
    Yellow Stone-crop.
    Nat. of Siberia.
    Cult. 1759. *Mill. dict. edit.* 7. *n.* 13.
    Fl. July——September.   H. ♃.

5. S. foliis cuneiformibus concavis ſubdentatis aggrega-   *hybridum.*
    tis, ramis repentibus, cyma terminali. *Sp. pl.* 617.
    Germander-leav'd Stone-crop.
    Nat. of Siberia.
    Cult. 1766, by Peter Collinſon, Eſq. *Coll. mſſ.*
    Fl. May and June.   H. ♃.

6. S. foliis planis cordatis dentatis petiolatis, corymbis   *populifo-*
    terminalibus.   *lium.*
    Sedum populifolium. *Pallas it.* 3. *p.* 730. *tab. O.*
    fig. 2. *Linn. ſuppl.* 242.
    Poplar-leav'd Stone-crop.
    Nat. of Siberia.
    Introd. 1780, by Peter Simon Pallas, M.D.
    Fl. July and Auguſt.   H. ♄.

7. S. foliis planiuſculis angulatis, floribus lateralibus feſ-   *ſtellatum.*
    filibus ſolitariis. *Sp. pl.* 617.
    Starry Stone-crop.
    Nat. of France and Italy.
    Cult. 1739, in Chelſea Garden. *Rand. chel. n.* 10.
    Fl. June and July.   H. ☉.

8. S.

Cepæa. 8. S. foliis planis, caule ramofo, floribus paniculatis. *Sp. pl.* 617.
Purflane-leav'd Stone-crop.
Nat. of France and Switzerland.
Cult. 1640. *Park. theat.* 728. *f.* 6.
Fl. July and Auguft. H. ☉.

** *Teretifolia.*

dafyphyl-lum. 9. S. foliis oppofitis ovatis obtufis carnofis, caule infirmo, floribus fparfis. *Sp. pl.* 618. *Curtis lond.*
Jacqu. hort. 2. *p.* 71. *t.* 153.
Round-leav'd Stone-crop.
Nat. of England.
Fl. July. H. ☉.

reflexum. 10. S. foliis fubulatis fparfis bafi folutis : inferioribus recurvatis. *Sp. pl.* 618.
α Sedum minus luteum folio acuto. *Bauh. pin.* 283.
Great reflex-leav'd Stone-crop, or Prick-madam.
β Sedum minus luteum ramulis reflexis. *Bauh. pin.* 283.
Small reflex'd-leav'd Stone-crop.
Nat. of Britain.
Fl. June and July. H. ♃.

virens. 11. S. foliis fubulatis fparfis bafi folutis, floribus cymofis, petalis calyce lanceolato fefquilongioribus.
Green Stone-crop.
Nat. of Portugal.
Introd. 1774, by Meffrs. Kennedy and Lee.
Fl. June and July. H. ♃.

rupeftre. 12. S. foliis fubulatis fparfis bafi folutis glaucis, floribus cymofis, petalis calyce duplo longioribus.
Sedum

Sedum rupeftre. *Sp. pl.* 618.
Rock Stone-crop.
Nat. of England.
Fl. Auguft.   H. ♃.

13. S. foliis linearibus tereti-depreffis fparfis, cyma pa-   *hifpani-*
    tula, floribus he apetalis.   *cum.*
    Sedum hifpanicum. *Sp. pl.* 618.
    Spanifh Stone-crop.
    Nat. of Spain.
    Cult. 1732, by James Sherard, M. D. *Dill. elth.*
    342. *t.* 256. *f.* 332.
    Fl. July.   H. ♃.

14. S. foliis oblongis obtufis teretiufculis feffilibus pa-   *album.*
    tentibus, cyma ramofa. *Sp. pl.* 619. *Curtis lond.*
    White Stone-crop.
    Nat. of England.
    Fl. June and July.   H. ♃.

15. S. foliis fubovatis adnato-feffilibus gibbis erectiuf-   *acre.*
    culis alternis, cyma trifida. *Sp. pl.* 619. *Curtis*
    lond.
    Wall Stone-crop, or Pepper.
    Nat. of England.
    Fl. June.   H. ♃.

16. S. foliis fubovatis adnato-feffilibus gibbis erectiufeu-   *fexangu-*
    lis fexfariam imbricatis. *Sp. pl.* 620. *Curtis lond.*   *lare.*
    Six-angled Stone-crop.
    Nat. of England.
    Fl. July.   H. ♃.

17. S. foliis fubovatis adnato-feffilibus gibbis alternis,   *anglicum.*
    cyma ramofa bifida. *Hudf. angl.* 196.
                                         Englifh

Englifh Stone-crop.
Nat. of England.
Fl. July and Auguft.   H. ♃.

annuum. 18. S. caule erecto folitario annuo, foliis ovatis feffilibus
gibbis alternis, cyma recurva. *Sp. pl.* 620.
Annual Stone-crop.
Nat. of the North of Europe.
Cult. 1768.   *Mill. dict. edit.* 8.
Fl. Auguft.   H. ☉.

villofum. 19. S. caule erecto, foliis planiufculis pedunculifque fub-
pilofis. *Sp. pl.* 620.
Hairy Stone-crop.
Nat. of Britain.
Fl. June.   H. ☉.

nudum. 20. S. foliis fparfis oblongo-cylindraceis obtufis, caulibus
fruticofis ramofiffimis : ramis tortuofis, cymis ter-
minalibus.
Naked-branch'd Stone-crop.
Nat. of Madeira,   Mr. *Francis Maffon.*
Introd. 1777.
Fl. July and Auguft.   G. H. ♄.

P E N T H O R U M. *Gen. pl.* 580.

Cal. 5-fidus. *Petala* nulla vel quinque. *Capf.* 5-cufpi-
data, 5-locularis.

Sedoides. 1. PENTHORUM. *Sp. pl.* 620.
American Penthorum.
Nat. of Virginia.
Cult. 1768, by Mr. Philip Miller. *Mill. dict. edit.* 8.
Fl. July and Auguft.   H. ♃.

O X A L I S.

O X A L I S. *Gen. pl.* 582.

Cal. 5-phyllus. *Petala* unguibus connexa. *Capf.* angulis dehifcens, 5-gona.

1. O. foliis ovatis indivifis. *Thunb. oxalis, n.* 1. *Syft. veget.* 432.
Simple-leav'd Wood Sorrel.
Nat. of the Cape of Good Hope.
Introd. 1774, by Mr. Francis Maffon.
Fl.   G. H. ♃.

monophylla.

2. O. fcapis unifloris, foliis ternatis : foliolis obcordatis pilofis. *Thunb. oxalis, n.* 5. *Syft. veget.* 432. *Curtis lond.*
Common Wood Sorrel.
Nat. of Britain.
Fl. April.   H. ♃.

Acetofella.

3. O. fcapis unifloris, foliis ternatis : foliolis fubrotundis ciliatis. *Thunb. oxalis, n.* 8. *Syft. veget.* 432.
Purple Wood Sorrel.
Nat. of the Cape of Good Hope.
Cult. 1690, in the Royal Garden at Hampton-court. *Catal. mff.*
Fl. February and March.   G. H. ♃.

purpurea.

4. O. fcapis umbelliferis, foliis ternatis glabris, calycibus callofis. *Thunb. oxalis, n.* 10. *Syft. veget.* 433. *Jacqu. hort.* 2. *p.* 84. *t.* 180.
Violet-colour'd Wood Sorrel.
Nat. of North America.
Introd. 1772, by Samuel Martin, M. D.
Fl. May and June.   G. H. ♃.

violacea.

caprina. 5. O. ſcapis umbelliferis, foliis ternatis glabris, floribus
erectis. *Thunb. oxalis, n. 11. Syſt. veget.* 433.
Oxalis Pes capræ. *Sp. pl.* 622.
Goat's-foot Wood Sorrel.
Nat. of the Cape of Good Hope.
Cult. 1757, by Mr. Philip Miller. *Mill. ic.* 130.
t. 195. *f.* 1.
Fl. March——June.   G. H. ♃.

incarna- 6. O. caule bulbifero, pedunculis unifloris, foliis terna-
ta.   tis : foliolis obcordatis glabris. *Thunb. oxalis, n.* 15.
Syſt. veget. 433.
Fleſh-colour'd Wood Sorrel.
Nat. of the Cape of Good Hope.
Cult. 1739, by Mr. Ph. Miller. *Rand. chel.* Oxys 5.
Fl. April——June.   G. H. ♃.

birta. 7. O. caule erecto hirto, pedunculis unifloris, foliis ter-
natis ſubſeſſilibus : foliolis oblongis emarginatis
glabris. *Thunb. oxalis, n.* 18. *Syſt. veget.* 433.
Hairy Wood Sorrel.
Nat. of the Cape of Good Hope.
Introd. 1787, by Mr. Francis Maſſon.
Fl. October.   G. H. ♃.

verſicolor. 8. O. caule erecto hirto, pedunculis unifloris, foliis ter-
natis : foliolis linearibus calloſis. *Thunb. oxalis,*
n. 19. *Syſt. veget.* 434.
Striped-flower'd Wood Sorrel.
Nat. of the Cape of Good Hope.
Introd. 1774, by Mr. Francis Maſſon.
Fl. January and February.   G. H. ♃.

cornicu- 9. O. caule decumbente herbaceo, pedunculis umbelli-
lata.   feris. *Thunb. oxalis, n.* 20. *Syſt. veget.* 434.
Yellow

Yellow.Wood Sorrel.

Nat. of the South of Europe.

Cult. 1656, by Mr. John Tradefcant, Jun. *Trad. muf.* 149.

Fl. May——July.   H. ☉.

10. O. caule erecto herbaceo, pedunculis umbelliferis. *Thunb. oxalis, n.* 21. *Syft. veget.* 434.   *ftricta.*

American Wood Sorrel.

Nat. of North America.

Cult. 1658, in Oxford Garden. *Hort. oxon. edit.* 2. *p.* 135.

Fl. June——October.   H. ♃.

11. O. foliis multifidis glabris. *Thunb. oxalis, n.* 24. *Syft. veget.* 434.   *flava.*

Narrow-leav'd Wood Sorrel.

Nat. of the Cape of Good Hope.

Introd. 1775, by Mr. Francis Maffon.

Fl.   G. H. ♃.

A G R O S T E M M A. *Gen. pl.* 583.

Cal. 1-phyllus, coriaceus. *Petala* 5, unguiculata: *Limbo* obtufo, indivifo. *Capf.* 1-locularis.

1. A. hirfuta, calycibus corollam æquantibus, petalis integris nudis. *Sp. pl.* 624. *Curtis land.*   *Githago.*

Corn Rofe Campion, or Cockle.

Nat. of Britain.

Fl. June.   H. ☉.

2. A. tomentofa, foliis ovato-lanceolatis, petalis emarginatis coronatis ferratis. *Syft. veget.* 435. *Curtis magaz.* 24.   *Coronaria.*

α floribus fimplicibus rubris.

I 2   Common

Common red Rose Campion.

β floribus simplicibus albis.

White Rose Campion.

γ floribus plenis.

Double Rose Campion.

Nat. of Italy.

Cult. 1596, by Mr. John Gerard.   *Hort. Ger.*

Fl. June———September.   H. ♂.

Flos jo-   3. A. tomentosa, petalis emarginatis.   *Sp. pl.* 625.
vis.   Umbel'd Rose Campion.

Nat. of Switzerland and Germany.

Cult. 1726, by Mr. Philip Miller.   *R. S. n.* 235.

Fl. July.   H. ♃.

Coeli rosa. 4. A. glabra, foliis lineari-lanceolatis, petalis emargina-
              tis coronatis.   *Sp. pl.* 624.

Smooth-leav'd Rose Campion.

Nat. of Sicily and the Levant.

Cult. 1739, by Mr. Philip Miller.   *Rand. chel.* Lychnis 29.

Fl. July and August.   H. ⊙.

L Y C H N I S.   *Gen. pl.* 584.

Cal. 1-phyllus, oblongus, lævis.   *Petala* 5, unguiculata : *Limbo* subbifido.   *Capf.* 5-locularis.

chalcedo- 1. L. floribus fasciculatis fastigiatis.   *Sp. pl.* 625.
nica.   α Lychnis hirfuta, flore coccineo, major. *Bauh. pin.* 203.
             Single-flower'd Scarlet Lychnis.

β Lychnis hirfuta, flore candido, major. *Tourn. inst.* 334.
Single-flower'd Blush Lychnis.

γ Lychnis chalcedonica, flore pleno miniato seu auran-
tiaco.   *Boerh. ludgb.* 1. *p.* 211.

Double-flower'd Scarlet Lychnis.

Nat.

Nat. of Ruffia.
Cult. 1596, by Mr. John Gerard. *Hort. Ger.*
Fl. June and July. H. ♃.

2. L. petalis quadrifidis, fructu fubrotundo. *Sp. pl.* 625. *Flos cu-*
Curtis *lond.* *culi.*
α floribus fimplicibus rubris.
Red-flower'd Meadow Lychnis, or Cuckow-flower.
β floribus fimplicibus albis.
White-flower'd Meadow Lychnis.
γ Lychnis caule erecto, calycibus ftriatis acutis, petalis
diffectis. *Mill. ic.* 114. *t.* 170. *f.* 2.
Double red-flower'd Meadow Lychnis.
Nat. of Britain.
Fl. June——September. H. ♃.

3. L. glabra, floribus axillaribus terminalibufque folita- *coronata.*
riis, petalis laciniatis. *Thunb. japon.* 187. *Syft.*
veget. 435.
Lychnis grandiflora. *Jacqu. ic. collect.* 1. *p.* 149.
L'Herit. ftirp. nov. tom. 2. *tab.* 18.
Chinefe Lychnis.
Nat. of China and Japan.
Introd. 1774, by John Fothergill, M.D.
Fl. June and July. G. H. ♃.

4. L. petalis fubintegris. *Syft. veget.* 435. *Vifcaria.*
α floribus fimplicibus.
Single-flower'd Vifcous Catchfly.
β floribus plenis.
Double-flower'd Vifcous Catchfly.
Nat. of Britain.
Fl. May——July. H. ♃.

5. L. petalis bifidis, floribus tetragynis. *Syft. veget.* 435. *alpina.*
Alpine Lychnis.

I 3 *Nat.*

Nat. of the Alps in Europe and Siberia.
Cult. 1768, by Mr. Philip Miller. *Mill. dict. edit.* 8.
Fl. April and May.   H. ♂.

læta.

6. L. petalis bifidis, floribus folitariis, foliis lineari-lan-
    ceolatis glabris, calycibus decemcarinatis.
Small Portugal Campion.
Nat. of Portugal. *Edward Gray,* M. D.
Introd. 1778.
Fl. July.   H. ☉.

dioica.

7. L. floribus dioicis. *Sp. pl.* 626. *Curtis lond.*
α Lychnis fylveftris f. aquatica purpurea fimplex. *Bauh.*
    pin. 204.
Single-flower'd red Lychnis.
β Lychnis fylveftris alba fimplex. *Bauh. pin.* 204.
Single-flower'd white Lychnis.
γ Lychnis purpurea multiplex. *Bauh. pin.* 204.
. Double-flower'd red Lychnis.
♂ Lychnis alba multiplex. *Bauh. pin.* 204.
Double-flower'd white Lychnis.
Nat. of Britain.
Fl. May——Auguft.   H. ♃.

C E R A S T I U M. *Gen. pl.* 585.
Cal. 5-phyllus. *Petala* bifida. *Capf.* unilocularis,
    apice dehifcens.

* *Capfulis oblongis.*

perfolia-
tum.

1. C. foliis connatis. *Sp. pl.* 627.
Perfoliate Cerastium.
Nat. of Greece.
Cult. 1731, by Mr. Philip Miller. *Mill. dict. edit.* 1.
    Myofotis 3.
Fl. June.   H. ☉.

                                                         2. C.

2. C. foliis ovatis, petalis calyci æqualibus, caulibus dif-   *vulga-*
fuſis. *Sp. pl.* 627. *Curtis lond.*   *tum.*
Common Ceraſtium, or Mouſe-ear-chickweed.
Nat. of Britain.
Fl. June.   H. ☉.

3. C. erectum villoſo-viſcoſum. *Sp. pl.* 627. *Curtis lond.*   *viſcoſum.*
Clammy Ceraſtium.
Nat. of Britain.
Fl. April and May.   H. ☉.

4. C. floribus pentandris, petalis emarginatis. *Sp. pl.*   *ſemide-*
627. *Curtis lond.*   *candrum.*
Leaſt Ceraſtium, or Mouſe-ear-chickweed.
Nat. of Britain.
Fl. April.   H. ☉.

5. C. foliis lineari-lanceolatis obtuſis glabris, corollis ca-   *arvenſe.*
lyce majoribus. *Sp. pl.* 628.
Corn Ceraſtium, or Mouſe-ear-chickweed.
Nat. of Britain.
Fl. May —— July.   H. ♃.

6. C. foliis lanceolatis, caule dichotomo ramoſiſſimo,   *dichoto-*
capſulis erectis. *Sp. pl.* 628.   *mum.*
Forked Ceraſtium.
Nat. of Spain.
Cult. 1731, by Mr. Philip Miller. *Mill. dict. edit.* 1.
Myoſotis 1.
Fl. June.   H. ☉.

** *Capſulis ſubrotundis.*

7. C. foliis lanceolatis, pedunculis ramoſis, capſulis ſub-   *repens.*
rotundis. *Sp. pl.* 628.
Creeping Ceraſtium.

Nat.

Nat. of France and Italy.
Cult. 1759, by Mr. Ph. Miller. *Mill. diÄ. edit.* 7. *n.* 1.
Fl. May——July.   H. ♃.

aquati-
cum.   8. **C.** foliis cordatis feffilibus, floribus folitariis, fruÄibus
     pendulis. *Sp. pl.* 629. *Curtis lond.*
     Water Ceraftium.
     Nat. of Britain.
     Fl. July.   H. ♃.

dioicum.   9. **C.** hirtum vifcidum, foliis lanceolatis, floribus dioicis,
     petalis calyce triplo majoribus.
     Spanifh Ceraftium.
     Nat. of Spain,
     Cult. 1766, in Oxford Garden.
     Fl. June.   H. ♃.

latifoli-
um.   10. **C.** foliis ovatis fubtomentofis, ramis fubunifloris, cap-
     fulis globofis. *Sp. pl.* 629.
     Mountain, or Broad-leav'd Ceraftium.
     Nat. of Britain.
     Fl. June and July.   H. ♃.

tomento-
fum.   11. **C.** foliis oblongis tomentofis, pedunculis ramofis, cap-
     fulis globofis. *Sp. pl.* 629.
     White Ceraftium.
     Nat. of the South of Europe.
     Cult. 1759. *Mill. diÄ. edit.* 7. *n.* 2.
     Fl. June.   H. ♃.

SPERGULA. *Gen. pl.* 586.
Cal. 5-phyllus. *Petala* 5, integra. *Capf.* ovata, 1-lo-
cularis, 5-valvis.

arvenfis.   1. **S.** foliis verticillatis, floribus decandris. *Sp. pl.* 630.
     Curtis lond.

     Corn

Corn Spurrey.
Nat. of Britain.
Fl. Auguft.   H. ☉.

2. S. foliis verticillatis, floribus pentandris. *Sp. pl.* 630.   *pentan-*
   Small Spurrey.   *dra.*
   Nat. of England.
   Fl. July.   H. ☉.

3. S. foliis oppofitis fubulatis lævibus, caulibus fimplici-   *nodofa.*
   bus. *Sp. pl.* 630. *Curtis lond.*
   Knotted Spurrey.
   Nat. of Britain.
   Fl. July.   H. ♃.

4. S. foliis oppofitis linearibus lævibus, pedunculis foli-   *Saginoi-*
   tariis longiffimis. *Sp. pl.* 631. *Curtis lond.*   *des.*
   Spergula laricina. *Hudf. angl.* 203.
   Pearlwort Spurrey.
   Nat. of England.
   Fl. June——Auguft.   H. ♃.

FORSKOHLEA. *Linn. mant.* 11.

Cal. 5-phyllus, corolla longior. *Petala* 10, fpathulata.
   Pericarp. nullum. *Sem.* 5, lana connexa.

1. F. pilofo-hifpida, foliis ellipticis muticis, laciniis caly-   *tenacifli-*
   cinis oblongo-lanceolatis acutis.   *ma.*
   Forfkohlea tenaciffima. *Linn. mant.* 72. *Jacqu.*
   hort. 1. *p.* 18. *t.* 48.
   Caidbeja adhærens. *Forfk. defcr. p.* 82.
   Clammy Forfkohlea.
   Nat. of Egypt.
   Introd. 1770, by Monf. Richard.
   Fl. July and Auguft.   G. H. ☉.
                                                   2. F.

candida.　2. F. ſcabra, foliis ellipticis undulatis muticis, laciniis
　　　　　calycinis ovatis obtuſis.
　　　　Forſkohlea candida.　*Linn. ſuppl.* 245.
　　　　Forſkohlea ſcabra.　*Retz. obſ. bot.* 3. *p.* 31. *n.* 49.
　　　　Rough Forſkohlea.
　　　　Nat. of the Cape of Good Hope.　Mr. *Fr. Maſſon.*
　　　　Introd. 1774.
　　　　Fl. June and July.　　　　　　　　　　G. H. ♃.

anguſtifo-　3. F. ſtrigoſa, foliis lanceolatis : dentibus ſpinoſo-ſeta-
lia.　　　　ceis, laciniis calycinis lanceolato-ſubulatis.
　　　　Forſkohlea anguſtifolia.　*Retz. obſ. bot.* 3. *p.* 31. *n.* 50.
　　　　　Murray in commentat. gotting. 1784. *p.* 24. *tab.* 2.
　　　　Narrow-leav'd Forſkohlea.
　　　　Nat. of the Iſland of Teneriffe.　Mr. *Francis Maſſon.*
　　　　Introd. 1779.
　　　　Fl. July and Auguſt.　　　　　　　　　G. H. ☉.

DECAGYNIA.

PHYTOLACCA. *Gen. pl.* 588.

Cal. 0.　*Petala* 5, calycina.　*Bacca* ſupera, 10-locu-
laris, 10-ſperma.

octandra.　1. P. floribus octandris octagynis.　*Sp. pl.* 631.
　　　　White-flower'd Phytolacca.
　　　　Nat. of Mexico.
　　　　Cult. 1732, by James Sherard, M. D.　*Dill. elth.* 318.
　　　　t. 239. *f.* 308.
　　　　Fl. July——November.　　　　　　　　S. ♃.

decandra.　2. P. floribus decandris decagynis.　*Sp. pl.* 631.
　　　　Branching Phytolaca, or Virginian Poke.

　　　　　　　　　　　　　　　　　　　　　　Nat.

Nat. of Virginia.
Cult. 1640. *Park. theat.* 347. *f.* 8.
Fl. Auguſt and September. S. ♃.

3. P. floribus icoſandris decagynis. *Sp. pl.* 631.   *icoſandra.*
Red Phytolaca.
Nat. of the Eaſt Indies.
Cult. 1758, by Mr. Ph. Miller. *Mill. ic.* 138. *t.* 207.
Fl. July——November. S. ♃.

4. P. floribus pentadecandris penta-octogynis. *L'Herit.*   *dodecan-*
 ſtirp. nov. tab. 69.   *dra.*
African Phytolacca.
Nat. of Africa. *James Bruce,* Eſq.
Introd. 1775.
Fl. May and June. S. ♄.

5. P. floribus dioicis. *Sp. pl.* 632. *L'Herit. ſtirp. nov.*   *dioica.*
 tab. 70.
Tree Phytolacca.
Nat. of South America.
Cult. 1768, by Mr. Philip Miller. *Mill. dict. edit.* 8.
Fl. S. ♄.

Claſſis

Claſſis XI.

DODECANDRIA

MONOGYNIA.

A S A R U M. *Gen. pl.* 589.

Cal. 3- ſ. 4-fidus, germini inſidens. *Cor.* o. *Capſ.* coriacea, coronata.

europæ-
um.

1. A. foliis reniformibus obtuſis binis. *Sp. pl.* 633.
   Common Aſarabacca.
   Nat. of England.
   Fl. May.   H. ♃.

cana-
denſe.

2. A. foliis reniformibus mucronatis. *Sp. pl.* 633.
   Canadian Aſarabacca.
   Nat. of Canada.
   Cult. 1731. *Mill. dict. edit.* 1. *n.* 2.
   Fl. April——July.   H. ♃.

virgini-
cum.

3. A. foliis cordatis obtuſis glabris petiolatis. *Sp. pl.*
      633.
   Sweet-ſcented Aſarabacca.
   Nat. of Virginia and Carolina.
   Cult. 1759, by Mr. Ph. Miller. *Mill. dict. edit.* 7. *n.* 3.
   Fl. April and May.   H. ♃.

BOCCO-

B O C C O N I A. *Gen. pl.* 591.

Cal. 2-phyllus. *Cor.* o. *Stylus* bifidus. *Bacca* exa-
rida, monofperma.

1. Bocconia. *Sp. pl.* 634.   *frutef-*
Shrubby Bocconia, or Tree-celandine.   *cens.*
Nat. of Jamaica and Mexico.
Cult. 1739, by Mr. Philip Miller.· *Rand. chel.*
Fl. January——April.   S. ♄.

H A L E S I A. *Gen. pl.* 596.

Cal. 4-dentatus, fuperus. *Cor.* 4-fida. *Nux* 4-angu-
laris, 2-fperma.

1. H. foliis lanceolato-ovatis: petiolis glandulofis. *Sp.* *tetrapte-*
pl. 636.   *ra.*
Snow-drop-tree.
Nat. of Carolina.
Introd. 1756, by John Ellis, Efq. *Philof. tranfa&.*
vol. 51. *p.* 931.
Fl. April and May.   H. ♄.

C A N E L L A. (Winterana. *Gen. pl.* 598.)

Cal. 3-lobus. *Pet.* 5. *Antheræ* 16, adnatæ nectario
urceolato. *Bacca* 3-locularis. *Sem.* 2.

1. Canella. *Syft. veget.* 443.   *alba.*
Winterania Canella. *Sp. pl.* 636. (exclufo fynonymo
Clufii.) *Linn. fuppl.* 247.
Laurel-leav'd Canella.
Nat. of the Weft Indies.
Cult. 1739, by Mr. Philip Miller. *Rand. chel. Ad-
denda, n.* 3.
Fl.   S. ♄.

TRIUM-

TRIUMFETTA. *Gen. pl.* 600.

Cor. 5-petala.   *Cal.* 5-phyllus.   *Capf.* hifpida, in
4 diffiliens.

Lappula.   1. T. foliis bafi emarginatis, floribus ecalyculatis. *Syft.*
veget. 444.
Prickly-feeded Triumfetta.
Nat. of Jamaica and Brazil.
Cult. 1739, by Mr. Philip Miller. *Rand. chel. n.* 1.
Fl. July and Auguft.   S. ♄.

Bartra-   2. T. foliis bafi integris indivifis. *Syft. veget.* 444.
mia.   Currant-leav'd Triumfetta.
Nat. of the Eaft Indies.
Cult. 1759, by Mr. Philip Miller. *Mill. dict. edit.* 7.
Bartramia.
Fl.   S. ♄.

femitrilo- 3. T. foliis femitrilobis, floribus completis. *Linn. mant.*
ba.   73. *Jacqu. hort.* 3. *p.* 41. *t.* 76.
Mallow-leav'd Triumfetta.
Nat. of the Weft Indies.
Introd. 1773, by John Earl of Bute.
Fl. July.   S. ♄.

PEGANUM. *Gen. pl.* 601.

Cor. 5-petala.   *Cal.* 5-phyllus, f. o.   *Capf.* 3-locularis,
3-valvis, polyfperma.

Harma-   1. P. foliis multifidis. *Sp. pl.* 638.
la.   Syrian Rue.
Nat. of Spain and the Levant.
Cult. 1570. *Lobel. adv.* 390.
Fl. July and Auguft.   H. ♃.

NITRA-

NITRARIA. *Gen. pl.* 602.

Cor. 5-petala, petalis apice fornicatis. *Cal.* 5-fidus.
Stam. 15. *Drupa* monofperma.

1. NITRARIA. *Sp. pl.* 638. *Pallas roff.* 1. *p.* 79. *tab.* 50.   *Schoberi.*
Thick-leav'd Nitraria.
Nat. of Siberia.
Introd. 1778, by Monf. Thouin.
Fl.   H. ♄.

PORTULACA. *Gen. pl.* 603.

Cor. 5-petala. *Cal.* 2-fidus. *Capf.* 1-locularis, circumfciffa, aut 3-valvis.

1. P. foliis cuneiformibus, floribus feffilibus. *Sp. pl.* 638.   *oleracea.*
Garden Purflane.
Nat. of both Indies.
Cult. 1562. *Turn. herb. part* 2. *fol.* 102 *verfo.*
Fl. June and July.   H. ☉.

2. P. foliis fubulatis alternis : axillis pilofis, floribus fef-   *pilofa.*
filibus terminalibus. *Sp. pl.* 639.
Hairy Purflane.
Nat. of the Weft Indies.
Cult. 1690, in the Royal Garden at Hampton-court.
Catal. mff.
Fl. June and July.   H. ☉.

3. P. bracteis quaternis, floribus quadrifidis, caule geni-   *quadrifi-*
culis pilofis. *Syft. veget.* 445.   *da.*
Creeping Annual Purflane.
Nat. of the Eaft Indies.
Introd. 1773, by Sir Jofeph Banks, Bart.
Fl. Auguft and September.   S. ☉.

4. P.

Anacamp- 4. P. foliis ovatis gibbis, pedunculo multifloro, caule
feros.   fruticofo. *Sp. pl.* 639.
           Round-leav'd Purflane.
           Nat. of the Cape of Good Hope.
           Cult. 1732, by James Sherard, M.D. *Dill. elth.* 375.
           t. 281. *f.* 363.
           Fl. July.   G. H. ♄.

patens.   5. P. foliis lanceolato-ovatis planis, panicula ramofa, ca-
               lycibus diphyllis. *Linn. mant.* 242.
           Panicled Purflane.
           Nat. of the Weft Indies.
           Introd. 1776, by Chevalier Murray.
           Fl. July.   S. ♄.

L Y T H R U M. *Gen. pl.* 604.

Cal. 12-fidus. *Petala* 6, calyci inferta. *Capf.* 2-lo-
            cularis, polyfperma.

Salicaria. 1. L. foliis oppofitis cordato-lanceolatis, floribus fpicatis
               dodecandris. *Sp. pl.* 640. *Curtis lond.*
            Common, or Purple Willow-herb.
            Nat. of Britain.
            Fl. July.   H. ♃.

virgatum. 2. L. foliis oppofitis lanceolatis, panicula virgata, flori-
               bus dodecandris ternis. *Syft. veget.* 446. *Jacqu.*
               auftr. 1. *p.* 8. *t.* 7.
            Fine-branched Willow-herb.
            Nat. of Auftria and Siberia.
            Introd. 1776, by Jofeph Nicholas de Jacquin, M.D.
            Fl. June——September.   H. ♃.

Hyffopi-   3. L. foliis alternis linearibus, floribus hexandris. *Sp. pl.*
folia.   642. *Jacqu. auftr.* 2. *p.* 20. *t.* 133.

                                                      Hyffop-

Hyffop-leav'd Willow-herb.
Nat. of England.
Fl. Auguft. H. ☉.

C U P H E A. *Jacquin.*

Cal. 6-dentatus, inæqualis. *Pet.* 6, inæqualia, calyci
inferta. *Capf.* 1-locularis: conceptaculo triquetro.

1. CUPHEA. *Jacqu. hort.* 2. *p.* 83. *t.* 177. *Roth beytr.* 1.   *vifcofiffi-*
   p. 124.   *ma.*
   Lythrum Cuphea. *Linn. fuppl.* 249.
   Clammy Cuphea.
   Nat. of America.
   Introd. 1776, by Jofeph Nicholas de Jacquin, M.D.
   Fl. July and Auguft.   G. H. ☉.

D I G Y N I A.

H E L I O C A R P U S. *Gen. pl.* 606.

Cal. 4-phyllus. *Petala* 4. *Styli* fimplices. *Capf.*
2-locularis, compreffa, utrinque longitudinaliter
radiata.

1. HELIOCARPUS. *Sp. pl.* 643.   *america-*
   American Heliocarpus.   *na.*
   Nat. of Vera Cruz.
   Introd. before 1733, by William Houftoun, M.D.
   Mill. dict. edit. 8.
   Fl.   S. ♄.

A G R I M O N I A. *Gen. pl.* 607.

Cal. 5-dentatus, altero obvallatus. *Petala* 5. *Sem.* 2,
             in fundo calycis.

1. A. fructibus hifpidis, foliis caulinis pinnatis: foliolis   *Eupato-*
   oblongo-ovatis, fpicis elevatis, petalis calyce duplo   *ria.*
   longioribus.

VOL. II.   K   Agrimonia

Agrimonia Eupatoria. *Sp. pl.* 643. *Curtis lond.*

α Agrimonia vulgaris. *Park. theat.* 594.

Common Agrimony.

minor. β Agrimonia foliis caulinis pinnatis, foliolis obtusis dentatis. *Mill. dict.*

Dwarf Agrimony.

Nat. of Britain; β. of Italy.

Fl. June and July.· H. ♃.

odorata. 2. A. fructibus hispidis, foliis caulinis pinnatis : foliolis oblongis: inferioribus diminutis, petalis calyce duplo longioribus.

Agrimonia odorata. *Mill. dict. Camer. hort.* 7.

Sweet-scented Agrimony.

Nat. of Italy.

Cult. 1640. *Park. theat.* 594. *n.* 2.

Fl. July. H. ♃.

repens. 3. A. fructibus hispidis, foliis caulinis pinnatis : foliolis oblongis, spicis subsessilibus, petalis calyce triplo longioribus.

Agrimonia repens. *Sp. pl.* 643.

Creeping Agrimony.

Nat. of the Levant.

Cult. 1739, by Mr. Philip Miller. *Rand. chel. n.* 4.

Fl. July——September. H. ♃.

parviflo-ra. 4. A. fructibus hispidis, foliis caulinis pinnatis : foliolis plurimis lanceolatis, petalis calyce sesquilongioribus.

Small-flower'd Agrimony.

Nat. of North America.

Cult. 1766, by Mr. James Gordon.

Fl. July. H. ♃.

5. A.

5. A. foliis caulinis ternatis, fructibus glabris. *Sp. pl.*   *Agrimo-*
   643.   *noides.*
   Three-leav'd Agrimony.
   Nat. of Italy.
   Cult. 1739, by Mr. Philip Miller.   *Rand. chel. n.* 5.
   Fl. June and July.   H. ♃.

T R I G Y N I A.

R E S E D A. *Gen. pl.* 608.

Cal. 1-phyllus, partitus. *Petala* laciniata. *Capf.* ore
   dehifcens, 1-locularis.

1. R. foliis lanceolatis integris bafi utrinque unidentatis, *Luteola.*
   calycibus quadrifidis. *Syft. veget.* 448.
   Dyers Weed, or Refeda.
   Nat. of Britain.
   Fl. June.   H. ☉.

2. R. foliis lanceolatis undulatis pilofis. *Syft. veget.* 448. *canefcens.*
   Hoary Refeda.
   Nat. of Spain.
   Cult. 1739, by Mr. Philip Miller. *Rand. chel.* Sefa-
   moides.
   Fl. May——July.   H. ♃.

3. R. foliis linearibus bafi dentatis, floribus tetragynis. *glauca.*
   Sp. pl. 644.
   Glaucous Refeda.
   Nat. of the South of Europe.
   Cult. 1748, by Mr. Ph. Miller. *Mill. dict. edit.* 5. *n.* 8.
   Fl. May——July.   G. H. ♃.

K 2   4. R.

dipetala. 4. R. foliis linearibus integerrimis, floribus tetragynis dipetalis: petalis indiviſis.
Flax-leav'd Reſeda.
Nat. of the Cape of Good Hope. Mr. *Fr. Maſſon.*
Introd. 1774.
Fl. Auguſt. G. H. ♂.

Seſamoi-des. 5. R. foliis lanceolatis integris, fruⅽtibus ſtellatis. *Sp. pl.* 644.
Spear-leav'd Reſeda.
Nat. of the South of France.
Introd. 1787, by Mr. Zier.
Fl. July and Auguſt. H. ☉.

alba. 6. R. foliis pinnatis, floribus tetragynis, calycibus ſex-partitis. *Sp. pl.* 645.
Upright white Reſeda.
Nat. of Spain and the South of France.
Cult. 1633. *Ger. emac.* 277. *f.* 2.
Fl. May——Oⅽtober. H. ☉.

undata. 7. R. foliis pinnatis undulatis, floribus trigynis tetragy-niſve. *Syſt. veget.* 448.
Waved-leav'd Reſeda.
Nat. of Spain.
Cult. 1759, by Mr. Ph. Miller. *Mill. diⅽt. edit.* 7. *n.* 5.
Fl. June——Auguſt. H. ♃.

lutea. 8. R. foliis omnibus trifidis: inferioribus pinnatis. *Sp. pl.* 645. *Jacqu. auſtr.* 4. *p.* 28. *t.* 353.
Yellow Reſeda.
Nat. of Britain.
Fl. July. H. ☉.

Phyteu-ma. 9. R. foliis integris trilobiſque, calycibus ſexpartitis maximis.

maximis. *Sp. pl.* 645. *Jacqu. auſtr.* 2. *p.* 20.
t. 132.
Trifid Reſeda.
Nat. of the South of Europe.
Cult. 1739, by Mr. Ph. Miller. *Mill. diƈt. vol.* 2. *n.* 5.
Fl. June——September.   H. ⊙.

10. R. foliis integris trilobiſque, calycibus florem æquan-   *odorata.*
    tibus. *Sp. pl.* 646. *Curtis magaz.* 29.
    Sweet Reſeda, or Mignonette.
    Nat. of Egypt.
    Cult. 1752, by Mr. Ph. Miller. *Mill. diƈt. edit.* 6. *n.* 9.
    Fl. moſt part of the Summer.   H. ⊙.

EUPHORBIA. *Gen. pl.* 609.

Cor. 4-ſ. 5-petala, calyci inſidens. *Cal.* 1-phyllus,
       ventricoſus. *Capſ.* 3-cocca.

* *Fruticoſæ, aculeatæ.*

1. E. aculeata ſubnuda triangularis articulata : ramis pa-   *antiquo-*
   tentibus. *Sp. pl.* 646.   *rum.*
   Triangular Spurge.
   Nat. of India.
   Cult. 1731, by Mr. Philip Miller. *Mill. diƈt. edit.* 1.
   Euphorbium 1.
   Fl.   D. S. ♄.

2. E. aculeata nuda ſubquadrangularis : aculeis gemi-   *canarien-*
   natis. *Sp. pl.* 646.   *ſis.*
   Canary Spurge.
   Nat. of the Canary Iſlands.
   Cult. 1697, by the Dutcheſs of Beaufort. *Br. Muſ.*
   Stoan. *mſſ.* 3357. *fol.* 21.
   Fl. March and April.   D. S. ♄.

heptago-
na.

3. E. aculeata nuda feptemangularis : fpinis folitariis fu-
    bulatis floriferis. *Sp. pl.* 647.
    Seven-angled Spurge.
    Nat. of the Cape of Good Hope.
    Cult. 1731. *Mill. dict. ed.* 1. Euphorbium 11.
    Fl. July——November.   D. S. ♄.

mammil-
laris.

4. E. aculeata nuda : angulis tuberofis fpinis interftinctis.
        Sp. pl. 6 7.
    Warty-angled Spurge.
    Nat. of the Cape of Good Hope.
    Cult. 1759. *Mill. dict. ed.* 7. *n.* 8.
    Fl. July and Auguft.   D. S. ♄.

cereifor-
mis.

5. E. aculeata nuda multangularis, fpinis folitariis fubu-
        latis. *Sp. pl.* 647.
    Naked Spurge.
    Nat. of the Cape of Good Hope.
    Cult 1731. *Mill. dict. ed.* 1. *n.* 5.
    Fl. June and July.   D. S. ♄.

officina-
rum.

6. E. aculeata nuda multangularis : aculeis geminatis.
        Sp. pl. 647.
    Officinal Spurge.
    Nat. of Africa.
    Cult. 1597, by Mr. John Gerard. *Ger. herb.* 1014. *f.* 1.
    Fl. June and July.   D. S. ♄.

neriifolia.

7. E. aculeata : angulis oblique tuberculatis. *Syft. veget.*
        449.
    Oleander-leav'd Spurge.
    Nat. of India.
    Cult. 1699, in the Royal Garden at Hampton-court.
        Morif. hift. 3. *p.* 344. *n.* 2.
    Fl. June and July.   D. S. ♄.
                                        * * *Fruticofæ,*

** *Fruticosæ, inermes.*

(Caulis nec dichotomus, nec umbelliferus.)

8. E. inermis fubglobofa multangularis.   *melofor-*
Melon Spurge.   *mis.*

Nat. of the Cape of Good Hope. Mr. *Fr. Maffon.*
Introd. 1774.

Fl. May——September.   G. H. ♄.

DESCR. *Caudex* carnofus, fubglobofus, diametro tri-
unciali, glaber, multiangulatus : *porcæ* 8, 10 vel
plures, bafi latæ, carinatæ : carinæ florigeræ, cicatricibus pedunculorum glandulifque alternatim
notatæ. *Pedunculi* cylindracei, craffitie pennæ columbinæ, articulati, villis breviffimis ad-
fperfi, plerumque primum trichotomi, dein dicho-
tomi ; raro fimplices. *Bractea* ad divifuras pedunculorum et ad bafin finguli floris oppofitæ, oblongæ, acutiufculæ, adpreffæ, lineam longæ. *Calyx*
monophyllus, campanulatus, bracteis paulo longior,
limbo quinquefido : *laciniæ* ovatæ, obtufæ, inflexæ,
concavæ. *Petala* 5, calyci inter lacinias inferta,
fubrotundo-reniformia, obtufiffima, laciniis calycis
duplo majora, patentia, carnofa, convexa, poris raris pertufa, viridia. *Filamenta* viginti et quod ex-
currit, receptaculo inferta, calyce parum longiora,
fubulata, villofa. *Antheræ* flavæ, didymæ. *Pifil-
lum* nullum.

9. E. inermis imbricata, tuberculis foliolo lineari inftruc-   *Caput*
tis, floribus fubpedunculatis, petalis palmatis.   *medufæ.*
Euphorbia Caput-medufæ. *Sp. pl.* 648.

α Tithymalus aizoides Africanus fimplici fquamato   major.
caule. *Comm. præl.* 57. *t.* 7.
Great Medufa's-head Spurge.

β Euphorbium anacanthum, angufto Polygoni folio.   minor.
Ifnard act. parif. 1720. *p.* 386. *Breyn. ic.* 29. *t.* 19.

Small Medufa's-head Spurge.

gemina-   γ Euphorbia procumbens, ramis geminatis, caule glabro
ta.   oblongo cinereo. *Burm. afr.* 18. *t.* 9. *f.* 1.

Leaft Medufa's-head Spurge.

Nat. of Africa.

Cult. 1731, by Mr. Philip Miller. Euphorbium 8.
  Mill. dict. edit. 1.

Fl. June——Auguft.   D. S. ♄.

Clava.   10. E. inermis imbricata, tuberculis foliolo lanceolato
               inftructis, floribus pedunculatis, petalis integerri-
               mis.

Euphorbia Clava. *Jacqu. ic. collect.* 1. *p.* 104.

Euphorbium acaulon erectum tuberofum, linearibus
  foliis, flore ac fructu foliaceis. *Burm. afr.* 12. *t.* 6.
  f. 1.

Tithymalus aizoides Africanus fimplici fquamato caule,
  chamænerii folio. *Commel. præl.* 58. *t.* 8.

Club Spurge.

Nat. of the Cape of Good Hope. Mr. *Fr. Maſſon.*

Introd. 1774.

Fl. January——Auguft.   G. H. ♄.

anacan-   11. E. inermis imbricata, tuberculis foliolo fubrotundo
tha.   inftructis, floribus terminalibus folitariis feſſilibus,
               petalis palmatis.

Euphorbium anacanthum, fquamofum, lobis florum
  tridentatis. *Iſnard act. pariſ.* 1720. *p.* 387. *n.* 12.
  & *p.* 392. *tab.* 11.

Euphorbium erectum, aphyllum, ramis rotundis, tuberculis quadragonis. *Burm. afr.* 16. *tab.* 7. *f.* 2.

Scaly Spurge.

Nat. of the Cape of Good Hope.

Cult. 1731. *Mill. dict. ed.* 1. Euphobium 7.

Fl. September and October.   D. S. ♄.

                                                        12. E

12. E. inermis feminuda fruticofa filiformis flaccida, fo- *maurita-*
    liis alternis. *Sp. pl.* 649.   *nica.*
    Mauritanian Spurge.
    Nat. of Africa.
    Cult. 1732, by James Sherard, M. D. *Dill. elth.* 384.
    t. 289. *f.* 373.
    Fl. June——Auguft.   D. S. ♄.

13. E. inermis fruticofa ftricta, umbellis quinquefidis *pifcato-*
    terminalibus, involucellis oblongis, foliis lineari- *ria.*
    lanceolatis lævibus.
    Smooth fpear-leav'd Spurge.
    Nat. of Madeira and the Canary Iflands. Mr. *Fr.*
    Maffon.
    Introd. 1777.
    Fl.   D. S. ♄.

14. E. inermis fruticofa ftricta, capitulo terminali, foliis *balfami-*
    lanceolatis lævibus glaucis. *fera.*
    Balfam Spurge.
    Nat. of the Canary Iflands. Mr. *Francis Maffon.*
    Introd. 1779.
    Fl.   D. S. ♄.

15. E. inermis feminuda fruticofa filiformis erecta, ramis *Tirucalli.*
    patulis determinate confertis. *Sp. pl.* 649.
    Indian Tree Spurge.
    Nat. of India.
    Cult. 1731, by Mr. Philip Miller. *Mill. dict. edit.* 1.
    Tithymalus 8.
    Fl.   D. S. ♄.

16. E. inermis fruticofa, foliis diftiche alternis ovatis. *Tithyma-*
    Sp. pl. 649. *loides.*
α Tithymalus Coraffavicus myrtifolius, flore coccineo *myrtifo-*
    mellifero. *Herm. par.* 234. *t.* 234. *lia.*
                              Myrtle-

Myrtle-leav'd Spurge.

padifolia.   β Tithymaloides Laurocerafi folio non ferrato. *Dill.*
  elth. 383. *t.* 288. *f.* 372.
  Laurel-leav'd Spurge.
  Nat. of South America.
  Cult. 1732, by Sir Charles Wager. *Dill. elth. loc.
  cit.*
  Fl. June and July.   D. S. ♄.

hetero-   17. E. inermis, foliis ferratis petiolatis difformibus ovatis
phylla.   lanceolatis panduriformibus. *Sp. pl.* 649.
  Various-leav'd Spurge.
  Nat. of the Weft Indies.
  Cult. 1690, in the Royal Garden at Hampton-court.
  Catal. mff.
  Fl. April——September.   S. ♄.

cotinifo-   18. E. foliis oppofitis fubcordatis petiolatis emarginatis
lia.   integerrimis, caule fruticofo. *Sp. pl.* 650.
  Venus's Sumach-leav'd Spurge.
  Nat. of South America.
  'Introd. 1690, by Mr. Bentick. *Br. Muf. Sloan. mff.*
  3370.
  Fl. July and Auguft.   S. ♄.

      *** *Dichotomæ.* (*Umbella bifida aut nulla.*)
hyperici-   19. E. dichotoma, foliis ferratis ovali-oblongis glabris,
folia.   corymbis terminalibus, ramis divaricatis. *Sp. pl.*
  650.
  St. John's-wort-leav'd Spurge.
  Nat. of the Weft Indies.
  Cult. 1739, by Mr. Philip Miller. *Rand. chel.* Tithymalus 33.
  Fl. June——September.   S. ☉.

                                              20. E.

20. E. dichotoma, foliis ovalibus obfolete-ferratis, pe- *proftrata.*
dunculis axillaribus fubtrifloris, caulibus diffufis
glabris.
Trailing red Spurge.
Nat. of the Weft Indies.
Cult. 1758, by Mr. Philip Miller.
Fl. July and Auguft. S. ☉.
DESCR. *Caules* herbacei, fpithamæi, procumbentes,
teretes, ramofi, rubicundi. *Folia* oppoíita, bre-
viffime petiolata, obtufiufcula, glabra, obfolete trinervia, fupra viridia, fubtus glaucefcentia, trilinearia. *Flores* axillares, breviter pedicellati, fæpe terni,
interdum folitarii. *Petala* purpurea.

21. E. dichotoma, foliis ferratis oblongis pilofis, floribus *maculata,*
axillaribus folitariis, ramis patulis. *Sp. pl.* 652.
Jacqu. hort. 2. *p.* 87. *t.* 186.
Spotted Spurge.
Nat. of America.
Cult. before 1660, by Mr. Walker. *Pluk. phyt.*
t. 65. *f.* 8.
Fl. July. H. ☉.

22. E. dichotoma, foliis fubcrenatis linearibus, floribus *hyffopifo-*
fafciculatis terminalibus, caule erecto. *Sp. pl.* 651. *lia.*
Hyffop-leav'd Spurge.
Nat. of the Weft Indies.
Introd. 1787, by Mr. Alexander Anderfon.
Fl. Auguft and September. S. ☉.

23. E. dichotoma, foliis crenulatis fubrotundis glabris, *Chamæ-*
floribus folitariis axillaribus, caulibus procumben- *fyce.*
tibus. *Sp. pl.* 652.
Crenated Annual Spurge.
Nat. of the South of Europe and Siberia.
Cult.

Cult. 1752, by Mr. Philip Miller. *Mill. dict. edit.* 6.
n. 27.
Fl. July.   H. ☉.

Peplis.   24. E. dichotoma, foliis integerrimis semicordatis, flori-
bus solitariis axillaribus, caulibus procumbenti-
bus. *Sp. pl.* 652.
Purple Spurge.
Nat. of England.
Fl. June.   H. ☉.

**** *Umbella trifida.*

Peplus.   25. E. umbella trifida : dichotoma, involucellis ovatis, fo-
liis integerrimis obovatis petiolatis. *Sp. pl.* 653.
Curtis lond.
Petty Spurge.
Nat. of Britain.
Fl. July.   H. ☉.

exigua.   26. E. umbella trifida : dichotoma, involucellis lanceo-
latis, foliis linearibus. *Sp. pl.* 654. *Curtis lond.*
Dwarf Spurge.
Nat. of Britain.
Fl. July.   H. ☉.

***** *Umbella quadrifida.*

Lathyris.   27. E. umbella quadrifida : dichotoma, foliis oppositis
integerrimis. *Sp. pl.* 655.
Caper Spurge.
Nat. of France and Italy.
Cult. 1597. *Ger. herb.* 405. *f.* 13.
Fl. May and June.   H. ♂.

Apios.   28. E. umbella quadrifida : bifida, involucellis reniformibus : primis obcordatis. *Syst. veget.* 451.
Pear-rooted Spurge.

Nat.

Nat. of the Island of Candia.
Cult. 1596, by Mr. John Gerard. *Hort. Ger.*
Fl. June and July. H. ♃.

29. E. umbella quadri-vel quinquefida: bis dichotoma, *læta.*
involucellis primis oblongis; superioribus rhom-
beo-subrotundis, foliis lineari-lanceolatis subemar-
ginatis integerrimis.
Mezerion-leav'd Spurge.
Nat.
Cult. 1758, by Mr. Philip Miller.
Fl. June and July. G. H. ♄.
DESCR. *Tota* planta glabra. *Caulis* fruticosus, teres,
lævis. *Folia* sparsa, sessilia, sesquiuncialia. *Involu-
cra* universalia foliis simillima. *Involucella* primæ
dichotomiæ ovali-oblonga, subemarginata, foliis
dimidio breviora; involucella secundæ dichotomiæ et floralia elliptico-subrotunda, emarginata.

****** *Umbella quinquefida.*

30. E. umbella subquinquefida simplici, involucellis ova- *spinosa.*
tis: primariis triphyllis, foliis oblongis integerri-
mis, caule fruticoso. *Sp. pl.* 655.
Prickly Spurge.
Nat. of the Levant.
Cult. 1752, by Mr. Philip Miller. *Mill. dict. edit.* 6.
n. 16.
Fl. May——September. H. ♄.

31. E. umbella quinquefida: bifida, involucellis subova- *dulcis.*
tis, foliis lanceolatis obtusis integerrimis. *Sp. pl.*
656. *Jacqu. austr.* 3. *p.* 8. *t.* 213.
Sweet Spurge.
Nat. of the South of Europe.
Cult. 1759, by Mr. Philip Miller.
Fl. May and June. H. ♃.

32. E.

Pithyuſa. 32. E. umbella quinquefida: bifida, involucellis ova-
tis mucronatis, foliis lanceolatis: infimis involu-
tis retrorſum imbricatis. *Sp. pl.* 656.
Juniper-leav'd Spurge.
Nat. of the South of Europe.
Introd. 1785, by Mr. John Græfer.
Fl. June and July.   G. H. ♄.

portland- 33. E. umbella quinquefida : dichotoma, involucellis
ica.   ſubcordatis concavis, foliis lineari-lanceolatis acu-
tis glabris patentibus. *Sp. pl.* 656.
Portland Spurge.
Nat. of England.
Fl. May——September.   H. ♃.

Paralias. 34. E. umbella ſubquinquefida: bifida, involucellis cor-
dato-reniformibus, foliis ſurſum imbricatis. *Sp.
pl.* 657. *Jacqu. hort.* 2. *p.* 88. *t.* 188.
Sea Spurge.
Nat. of England.
Fl. July.   H. ♃.

juncea. 35. E. umbella quinquefida: dichotoma, foliis involu-
criſque lineari-lanceolatis acutis, involucellis ovato-
oblongis acuminatis.
Linear-leav'd Spurge.
Nat. of the Iſland of Porto Santo near Madeira.
Mr. *Francis Maſſon.*
Introd. 1779.
Fl. July.   G. H. ♃.

helioſco- 36. E. umbella quinquefida : trifida: dichotoma, invo-
pia.   lucellis obovatis, foliis cuneiformibus ſerratis. *Sp.
pl.* 658. *Curtis lond.*
Sun Spurge, or Wart-wort.

                                                        Nat.

Nat. of Britain.
Fl. July.   H. ☉.

37. E. umbella quinquefida : trifida: dichotoma, invo- *ferrata.*
lucellis diphyllis reniformibus, foliis amplexicau-
libus cordatis ferratis. *Sp. pl.* 658.
Narrow notch'd-leav'd Spurge.
Nat. of the South of Europe.
Introd. 1780, by Sign. Giovanni Fabroni.
Fl.   G. H. ♃.

38. E. umbella quinquefida : fubtrifida : bifida, involu- . *verruce-*
cellis ovatis, foliis lanceolatis ferrulatis villofis, *fa.*
capfulis verrucofis. *Sp. pl.* 658.
Warty-fruited Spurge.
Nat. of England.
Fl. Auguft.   H. ♂.

39. E. umbella quinquefida : trifida, involucellis ovali- *punicea.*
bus acuminatis coloratis, capfulis glabris, foliis
lanceolato-cuneiformibus fubtus glaucis. *Swartz*
prodr. 76.
Scarlet-flower'd Spurge.
Nat. of Jamaica.
Introd. 1778, by Matthew Wallen, Efq.
Fl. January.   S. ♄.

40. E. umbella quinquefida : trifida : dichotoma, invo- *coralloi-*
lucellis ovatis, foliis lanceolatis, capfulis lanatis. *des.*
Sp. pl. 659.
Coral-ftalk'd Spurge.
Nat. of Sicily, Barbary, and the Levant.
Cult. 1739, by Mr. Philip Miller. *Mill. dict. vol.* 2.
Tithymalus 35.
Fl. June——Auguft.   H. ♃.

41. E.

pilofa. 41. E. umbella quinquefida: trifida: bifida, involucellis
ovatis, petalis integris, foliis lanceolatis fubpilofis
apice ferrulatis. *Syft. veget.* 453.
Hairy Spurge.
Nat. of Siberia.
Cult. 1758, by Mr. Philip Miller.
Fl. May——Auguft.   H. ♃.

orientalis. 42. E. umbella quinquefida: quadrifida: dichotoma, in-
volucellis fubrotundis acutis, foliis lanceolatis. *Sp.*
pl. 660.
Willow-leav'd Spurge.
Nat. of the Levant.
Cult. 1739, by Mr. Philip Miller. *Rand. chel.* Ti-
thymalus 13.
Fl. June and July.   H. ♃.

platyphyl- 43. E. umbella quinquefida: trifida: dichotoma, involu-
los.   cellis carina pilofis, foliis ferratis lanceolatis, capfulis verrucofis. *Syft. veget.* 453. *Jacqu. auftr.* 4.
p. 40. *t.* 376.
Broad notch'd-leav'd Spurge.
Nat. of England.
Fl. July.   H. ☉.

******* *Umbella multifida.*

Efula. 44. E. umbella multifida: bifida, involucellis fubcor-
datis, petalis fubbicornibus, ramis fterilibus foliis
uniformibus. *Sp. p.* 660.
Gromwell-leav'd Spurge.
Nat. of Holland and Germany.
Cult. 1570. *Lobel. adv.* 151.
Fl. May——July.   H. ♃.

45. E.

45. E. umbella multifida: dichotoma, involucellis fub-   *Cyparif-*
cordatis, ramis fterilibus foliis fetaceis; caulinis   *fias.*
lanceolatis. *Sp. pl.* 661. *Jacqu. auftr.* 5. *p.* 16.
t. 435.
Cyprefs Spurge.
Nat. of Germany and France.
Cult. 1640. *Park. theat.* 193. *f.* 3.
Fl. May——September.   H. ♃.

46. E. umbella fuboctofida: bifida, involucellis fubova-   *Myrfi-*
tis, foliis fpathulatis patentibus carnofis mucronatis   *nites.*
margine fcabris. *Sp. pl.* 661.
Glaucous Spurge.
Nat. of Italy, France, and Spain.
Cult. 1570. *Lobel. adv.* 150.
Fl. April——June.   H. ♃.

47. E. umbella multifida: fubtrifida: bifida, involucellis   *paluftris.*
ovatis, foliis lanceolatis, ramis fterilibus. *Sp. pl.* 662.
Marfh Spurge.
Nat. of Germany, France, and Sweden.
Cult. 1570. *Lobel. adv.* 151.
Fl. May——Auguft.   H. ♃.

48. E. umbella multifida: bifida, involucellis late cor-   *emargi-*
datis, foliis oblongis emarginatis glabris, caule ra-   *nata.*
mofo, capfulis fubverrucofis.
Freckled Spurge.
Nat. of Italy.
Cult. 1758, by Mr. Philip Miller.
Fl. July and Auguft.   H. ♃.

49. E. umbella multifida: bifida, involucellis ovatis,   *hyberna.*
foliis oblongis emarginatis fubtus villofiufculis,
caule fimplici, capfulis verrucofo-ramentaceis.
Euphorbia hyberna. *Sp. pl.* 662.
Vol. II.   L   Irifh

Irish Spurge.

Nat. of Ireland and England.

Fl. May and June.   H. ♃,

amygda-   50. E. umbella multifida: dichotoma, involucellis per-
loides.   foliatis orbiculatis, foliis obtusis. *Sp. pl.* 662.

Wood Spurge.

Nat. of England.

Fl. April——July.   H. ♃.

Chara-   51. E. umbella multifida: bifida, involucellis perfoliatis
sias.   emarginatis, foliis integerrimis, caule frutescente.
       Syst. veget. 454. *Jacqu. ic. collect.* 1. *p.* 57.

Red Spurge.

Nat. of England.

Fl. June.   H. ♄.

ARISTOTELIA. *L'Herit. stirp. nov.*

Cor. 5-petala. *Cal.* 5-phyllus. *Bacca* 3-locularis.
Sem. bina.

Macqui.   1. ARISTOTELIA. *L'Herit. stirp. nov. p.* 31. *tab.* 16.

Shining-leav'd Aristotelia.

Nat. of Chili.

Introd. about 1773, by Messrs. Kennedy and Lee.

Fl. April and May.   H. ♄.

PENTAGYNIA.

G L I N U S. *Gen. pl.* 610.

Cal. 5-phyllus. *Cor.* 0. *Nectaria* setis bifidis. *Capf.*
5-angularis, 5-locularis, 5-valvis, polysperma.

lotoides.   1. G. caule piloso, foliis obovatis. *Syst. veget.* 455.

Hairy Glinus.

Nat. of the South of Europe and the Levant.

Introd. 1788, by Monf. Thouin.

Fl. July.   G. H. ☉.

DODECA-

DODECAGYNIA.

SEMPERVIVUM. *Gen. pl.* 612.

Cal. 12-partitus. *Petala* 12. *Capf.* 12, polyfpermæ.

1. S. caule arborefcente lævi ramofo, foliis cuneiformibus glabriufculis ciliatis : ciliis patulis mollibus.   *arbore-um.*
Sempervivum arboreum. *Sp. pl.* 664.
Tree Houfeleek.
Nat. of Portugal and the Levant.
Cult. 1727. *Bradl. fucc.* 4. *p.* 1. *t.* 31.
Fl. December——March.   G. H. ♄.

2. S. caule frutefcente, foliis orbiculato-fpathulatis villofis, nectariis fubquadratis truncatis.   *canari-enfe.*
Sempervivum canarienfe. *Sp. pl.* 664.
Canary Houfeleek.
Nat. of the Canary Iflands.
Cult. 1699, by the Dutchefs of Beaufort. *Br. Muf.*
  Sloan. mff. 525 and 3349.
Fl. June and July.   G. H. ♄.
OBS. Petala 9. Stamina 18. Piftilla 9.

3. S. caule frutefcente, foliis cuneiformibus vifcidis ciliatis : ciliis cartilagineis adpreffis.   *glutino-fum.*
Clammy Houfeleek.
Nat. of Madeira. Mr. *Francis Maffon.*
Introd. 1777.
Fl. July and Auguft.   D. S. ♄.
OBS. Petala lutea, 8 vel 9. Stamina 16-18. Piftilla 8 vel 9.

L 2   4. S.

glandulo-
fum.

4. S. caule frutefcente, foliis orbiculato-fpathulatis mar-
gine glandulofis: glandulis globofis, nectariis cu-
neiformibus truncatis.
Glandulous-leav'd Houfeleek.
Nat. of Madeira.   Mr. *Francis Maffon.*
Introd. 1777.
Fl. March——May.   D. S. ♄.

tectorum.

5. S. foliis ciliatis, propaginibus patulis, floribus dode-
candris dodecagynis, nectariis cuneiformibus carun-
culatis.
Sempervivum tectorum.   *Sp. pl.* 664.   *Curtis lond.*
Common Houfeleek.
Nat. of Britain.
Fl. July——September.   H. ♃.

globife-
rum.

6. S. foliis ciliatis, propaginibus globofis.   *Sp. pl.* 665.
Jacqu. auftr. 5. *p.* 50. *tab. app.* 40.
Globular Houfeleek.
Nat. of Germany.
Cult. 1731, by Mr. Philip Miller.   *Mill. dict. edit.* 1.
Sedum 8.
Fl. June and July.   H. ♃.

villofum.

7. S. foliis fpathulato-cuneiformibus obtufis villofis, nec-
tariis palmatis: lacinulis fubulatis.
Hairy Houfeleek.
Nat. of Madeira.   Mr. *Francis Maffon.*
Introd. 1777.
Fl. June.   G. H. ☉.
OBS. Petala 8, flava.   Stamina 12-16. Piftilla 8.

tortuo-
fum.

8. S. foliis obovatis fubtus gibbis villofis, nectariis bilobis.
Gouty Houfeleek.

Nat.

Nat. of the Canary Iflands. Mr. *Francis Maſſon.*
Introd. 1779.
Fl. July and Auguſt.   G. H. ♄.
Obs. Petala 8, flava.   Stamina 16.   Piftilla 8.

9. S. foliis pilis intertextis, propaginibus globofis. *Sp.*   *arach-*
   pl. 665. *Jacqu. auſtr.* 5. *p.* 51. *tab. app.* 42.   *noideum.*
   Cobweb Houfeleek.
   Nat. of the Alps of Italy and Switzerland.
   Cult. 1699, by the Dutchefs of Beaufort. *Br. Muſ.*
   Sloan. mſſ. 525 and 3349.
   Fl. June and July.   H. ♃.

10. S. foliis integerrimis, propaginibus patulis. *Syſt. veget.*   *monta-*
    456. *Jacqu. auſtr.* 5. *p.* 50. *tab. app.* 41.   *num.*
    Mountain Houfeleek.
    Nat. of Switzerland.
    Cult. 1759, by Mr. Ph. Miller. *Mill. dict. edit.* 7. *n.* 3.
    Fl. June and July.   H. ♃.

11. S. foliis fparfis : inferioribus teretibus ; fuperioribus   *fediforme.*
    depreffis, *Syſt. veget.* 456. *Jacqu. hort.* 1. *p.* 35.
    t. 81.
    Stone-crop-leav'd Houfeleek.
    Nat. of the South of Europe.
    Introd. 1769, by Meſfrs. Kennedy and Lee.
    Fl. July.   H. ♃.

12. S. foliis teretibus clavatis confertis, pedunculis nudis   *monan-*
    fubunifloris, nectariis obcordatis.   *thes.*
    Clufter'd Houfeleek.
    Nat. of the Canary Iflands. Mr. *Francis Maſſon.*
    Introd. 1777.
    Fl. July.   G. H. ♃.
    Obs. Numerus partium fructificationis variat a qui-
    nario ad octonarium.

Claſſis

Classis XII.

ICOSANDRIA

MONOGYNIA.

C A C T U S. *Gen. pl.* 613.

Cal. 1-phyllus, fuperus, imbricatus. *Cor.* multiplex,
Bacca 1-locularis, polyfperma.

* Echinomelocacti, *fubrotundi.*

mammil-laris.	1. C. fubrotundus tectus tuberculis ovatis barbatis, *Sp, pl.* 666.
	α fimplex, fpinis rubicundis.
	Cactus mammillaris. *Mill. dict.*
	Red-fpined fmall Melon Thiftle.
prolifer.	β multiplex, fpinis albentibus.
	Cactus proliferus. *Mill. dict.*
	White-fpined fmall Melon Thiftle.
	Nat. of the Weft Indies.
	Cult. 1690, in the Royal Garden at Hampton-court, *Catal. mff.*
	Fl. July and Auguft. D. S. ♄.
Melocac-tus.	2. C. fubrotundus quatuordecim-angularis. *Sp. pl,* 666.
commu-nis.	α ovatus, 17-angulatus.
	Common Melon Thiftle, or Turk's Cap,
depreffus.	β fubrotundus, depreffus, 10-angulatus.
	Flat Melon Thiftle,

Nat.

Nat. of the Weſt Indies.
Cult. 1727. *Bradl. ſucc.* 4. *p.* 9. *t.* 32.
Fl. July and Auguſt. D. S. ♄.

* * Cerei *erecti* (*ſtantes per ſe.*)

3. C. erectus oblongus ſeptemangularis. *Sp. pl.* 666. *heptago-*
Seven-angled Torch-thiſtle. *nus.*
Nat. of the Weſt Indies.
Introd. 1728, by Mr. Ph. Miller. *Mill. dict. edit.* 8.
Fl. D. S. ♄.

4. C. erectus quadrangularis longus: angulis compreſſis. *tetrago-*
Sp. pl. 667. *nus.*
Four-angled Torch-thiſtle.
Nat. of South America.
Cult. 1731, by Mr. Philip Miller. *Mill. dict. edit.* 1.
Cereus 2.
Fl. July. D. S. ♄.

5. C. erectus ſexangularis longus: angulis diſtantibus. *hexago-*
Syſt. veget. 459. *nus.*
Six-angled Torch-thiſtle.
Nat. of Surinam.
Introd. 1690, by Mr. Bentick. *Br. Muſ. Sloan. mſſ.*
3370. *p.* 1. *n.* 70.
Fl. July and Auguſt. D. S. ♄.

6. C. erectus ſubquinquangularis longus articulatus. *Sp.* *pentago-*
pl. 667. *nus.*
Five-angled Torch-thiſtle.
Nat. of America.
Introd. about 1769.
Fl. July. D. S. ♄.

7. C. erectus longus octangularis: angulis compreſſis *repandus.*
undatis, ſpinis lana longioribus. *Sp. pl.* 667.
L 4 Wavy-

Wavy-angled Torch-thiftle.
Nat. of the Weft Indies.
Introd. 1728, by Mr. Ph. Miller.   *Mill. dict. edit.* 8.
Fl. Auguft.   D. S. ♄.

lanugina-   8. C. erectus longus fubnovemangularis : angulis obfo-
fus.   letis, fpinis lana brevioribus.   *Sp. pl.* 667.
              Woolly Torch-thiftle.
              Nat. of the Weft Indies.
              Introd. 1690, by Mr. Bentick.   *Br. Muf. Sloan, mff.*
                 3370. *p.* 1. *n.* 69.
              Fl.   D. S. ♄.

peruvia-   9. C. erectus longus fuboctangularis : angulis obtufis.
nus.   *Sp. pl.* 667.
              Peruvian Torch-thiftle,
              Nat. of Jamaica and Peru.
              Introd. 1728, by Mr. Ph. Miller.   *Mill. dict. edit.* 8.
              Fl.   D. S. ♄.

Royeni.   10. C. erectus articulatus novemangularis : articulis fub-
                ovatis, fpinis lanam æquantibus.   *Syft. veget.* 459.
              Nine-angled Torch-thiftle.
              Nat. of America.
              Introd. 1728, by Mr. Ph. Miller.   *Mill. dict. edit.* 8,
              Fl.   D. S. ♄.

                 *** Cerei *repentes, radiculis lateralibus,*

grandi-   11. C. repens fubquinquangularis.   *Sp. pl.* 668.
florus.   Great Night-flowering Cereus.
              Nat. of Jamaica and Vera Cruz.
              Cult. before 1700, in the Royal Garden at Hampton-
                 court.   *Pluk. mant.* 76.
              Fl. June and July.   D. S. ♄.

                                              12. C.

12. C. repens decemangularis. *Sp. pl.* 668.   *flagelli-*
Small creeping Cereus.   *formis.*
Nat. of Peru.
Introd. 1734, by Bernard de Juffieu, M.D. *Mill.*
dict. edit. 8.
Fl. March——June.   D. S. ♄.

13. C. repens triangularis. *Sp. pl.* 669.   *triangu-*
Triangular Cereus.   *laris.*
Nat. of the West Indies.
Cult. 1690, in the Royal Garden at Hampton-court.
Catal. mss.
Fl. July.   D. S. ♄.

14. C. pendulus, ramis verticillatis teretibus glabris *pendulus.*
muticis. *Swartz prodr.* 77.
Caffytha baccifera. *J. Miller illustr.*
Slender Cereus.
Nat. of the West Indies.
Introd. 1758, by Mr. Philip Miller.
Fl. September.   D. S. ♄.

* * * * Opuntiæ, *compressæ, articulis proliferis.*
15. C. articulato-prolifer laxus, articulis ovatis, fpinis *Opuntia.*
fetaceis. *Sp. pl.* 669.
Common Indian Fig.
Nat. of America and the South of Europe.
Cult. 1596, by Mr. John Gerard. *Hort. Ger.*
Fl. July and August,   D. S. ♄.

16. C. articulato-prolifer, articulis ovato-oblongis, fpinis *Ficus in-*
fetaceis. *Sp. pl.* 669.   *dica.*
White-fpined Indian Fig.
Nat. of South America,

                      Cult.

Cult. 1759. *Mill. dict. edit.* 7. Opuntia 2.

Fl. July and Auguſt.   D. S. ♄,

Tuna.   17. C. articulato-prolifer, articulis ovato-oblongis, ſpinis ſubulatis. *Sp. pl.* 669.

α Tuna major, ſpinis validis flavicantibus, flore gilvo.
Dill. elth. 396. *t.* 295. *f.* 380.
Yellow-ſpined Indian Fig.

β Tuna elatior, ſpinis validis nigricantibus. *Dill.*
elth. 395. *t.* 294. *f.* 379.
Black-ſpined Indian Fig.
Nat. of Jamaica and South America.
Cult. 1732, by James Sherard, M. D. *Dill. elth.*
loc. cit.
Fl. July and Auguſt,   D. S. ♄,

cochinil-
lifer.   18. C. articulato-prolifer, articulis ovato-oblongis ſub-
inermibus. *Sp. pl.* 670.
Cochineal Fig.
Nat. of South America.
Cult. 1732, by James Sherard, M. D. *Dill. elth.*
399. *t.* 297. *f.* 383.
Fl. September,   D. S. ♄,

curaſſa-
vicus.   19. C. articulato-prolifer, articulis cylindrico-ventrico-
fis compreſſis. *Sp. pl.* 670.
Small Indian Fig.
Nat. of Curaſſao,
Introd. 1690, by Mr. Bentick. *Br. Muſ. Sloan, mſſ.*
3370.
Fl.   D. S. ♄.

Phyllan-
thus.   20. C. prolifer enſiformi-compreſſus ferrato-repandus,
Sp. pl. 670.
Spleenwort-leav'd Indian Fig.

Nat.

Nat. of South America.
Cult. 1731, by Mr. Philip Miller. *Mill. dict. edit.* 1.
　Opuntia 11.
Fl. June. 　　　　　　　　　　　　D. S. ♄.

21. C. caule erecto compresso, ramis oppositis bifariis　*spinosissi-*
　compressis, spinis setaceis. 　　　　　　　　　　　　*mus.*
　Cactus spinosissimus. *Martyn hort. cantabr.* 88.
　Opuntia spinosissima. *Mill. dict.*
　Cluster-spined Indian Fig.
　Nat. of Jamaica.
　Introd. before 1733, by William Houstoun, M. D.
　　Mill. dict. edit. 8.
　Fl. 　　　　　　　　　　　　　D. S. ♄.

22. C. caule tereti arboreo: aculeis geminis recurvis,　*Pereskia.*
　foliis lanceolato-ovatis. *Sp. pl.* 671.
　Barbadoes Gooseberry.
　Nat. of the West Indies.
　Cult. 1696, in the Royal Garden at Hampton-court.
　　Catal. mss.
　Fl. 　　　　　　　　　　　　　D. S. ♄.

PHILADELPHUS. *Gen. pl.* 614.

Cal. 4- s. 5-partitus, superus. *Petala* 4- s. 5. *Caps.* 4- s.
　5-locularis, polysperma.

1. P. foliis subdentatis. *Sp. pl.* 671. 　　　　　　　*coronari-*
α foliis ovato-oblongis magnis, frutice orgyali. 　　　*us.*
　Common Syringa.
β foliis ovatis minoribus, frutice tripedali. 　　　　nanus.
　Philadelphus nanus. *Mill. dict.*
　Dwarf Syringa.
　Nat. of the South of Europe.

　　　　　　　　　　　　　　　Cult.

Cult. 1597, by Mr. John Gerard.   *Ger. herb.* 1213.
f. 1.

Fl. May and June.   H. ♄.

fcoparius.   2. P. foliis lanceolatis integerrimis rigidis trinerviis, floribus omnibus quinquefidis, laciniis calycinis colo-
ratis, deciduis.

linifolius.   α foliis lanceolatis.

Melaleuca fcoparia.   *Linn. fuppl.* 343.   *Forft. pl.*
efcul. p. 78. *fl. auftr. p.* 37.

Leptofpermum fcoparium.   *Forft. gen.* 36.

Tea Plant.   *Cook's voyage, vol.* 1. *p.* 100. *tab.* 22.

Narrow-leav'd Philadelphus.

myrtifo-   β foliis ovato-ellipticis.
lius.
Myrtle-leav'd Philadelphus.

Nat. of New Zealand.   Sir *Jofeph Banks*, Bart.

Introd. 1772.

Fl. June and July.   G. H. ♄.

aromati-   3. P. foliis lineari-lanceolatis enerviis integerrimis, flori-
cus.
bus omnibus quinquefidis, laciniis calycinis coloratis deciduis.

Sweet-fcented Philadelphus.

Nat. of New Zealand.   Sir *Jofeph Banks*, Bart.

Introd. 1772.

Fl. July and Auguft.   G. H. ♄.

laniger.   4. P. foliis oblongis acutis integerrimis pubefcentibus,
calycibus lanatis.

canef-   α foliis ovalibus, villis canefcentibus.
cens.
Hoary Philadelphus.

piliger.   β foliis lanceolato-oblongis pilofis fubobliquis apice reflexis.

Hairy Philadelphus.

Nat. of New South Wales.

Introd.

Introd. 1774, by Tobias Furneaux, Esq.
Fl. June and July.   G. H. ♄.

EUCALYPTUS. *L'Herit. sert. angl.*

Calyx superus, persistens, truncatus, ante anthesin tectus *operculo* hemisphærico, deciduo. *Cor.* o. *Caps.* 4-locularis, apice dehiscens, polysperma.

1. EUCALYPTUS. *L'Herit. sert. angl. tab.* 20.   *obliqua.*
Oblique-leav'd Eucalyptus.
Nat. of New South Wales.
Introd. 1774, by Tobias Furneaux, Esq.
Fl. July.   G. H. ♄.

PSIDIUM. *Gen. pl.* 615.

Cal. 5-fidus, superus. *Petala* 5. *Bacca* 1-locularis,
      polysperma.

1. P. foliis lineatis obtusiusculis, pedunculis unifloris.   *pyrisc-*
      Sp. pl. 672.   *rum.*
White Guava.
Nat. of the West Indies.
Cult. 1656, by Mr. John Tradescant, Jun. *Mus. Trad.*
      119.
Fl. June.   S. ♄.

2. P. foliis lineatis acuminatis, pedunculis trifloris. *Sp.*   *pomise-*
      pl. 672.   *rum.*
Red Guava.
Nat. of the West Indies.
Introd. 1692, by the Dutchess of Beaufort. *Br. Mus.*
      Sloan. mss. 525 and 3349.
Fl. June.   S. ♄.

EUGENIA.

EUGENIA. *Gen. pl.* 616.

Cal. 4-partitus, fuperus. *Petala.* 4. *Drupa* 1-fperma, 4-angularis.

malac- 1. E. foliis integerrimis, pedunculis ramofis lateralibus.
cenfis.   *Sp. pl.* 672.
       Broad-leav'd Eugenia.
       Nat. of the Eaft Indies.
       Cult. 1768, by Mr. Ph. Miller. *Mill. dict. edit.* 8.
       Fl.   S. ♄.

Jambos. 2. E. foliis integerrimis, pedunculis ramofis terminali-
       bus. *Sp. pl.* 672.
       Narrow-leav'd Eugenia.
       Nat. of the Eaft Indies.
       Cult. 1768, by Mr. Ph. Miller. *Mill. dict. edit.* 8.
       Fl. May——July.   S. ♄.

MYRTUS. *Gen. pl.* 617.

Cal. 5-fidus, fuperus. *Petala* 5. *Bacca* 2- f. 3-fperma.

communis. 1. M. floribus folitariis : involucro diphyllo. *Sp. pl.*
       673.
romana. α Myrtus foliis ovatis pedunculis longioribus. *Mill.*
       dict. ic. 123. *t.* 184. *f.* 1.
       Common broad-leav'd Myrtle.
tarentina. β Myrtus foliis ovatis, baccis rotundioribus. *Mill. dict.*
       Box-leav'd Myrtle.
italica. γ Myrtus foliis ovato-lanceolatis acutis, ramis erectiori-
       bus. *Mill. dict.*
       Italian, or Upright Myrtle.
bœtica. δ Myrtus foliis ovato-lanceolatis confertis. *Mill. dict.*
       Orange-leav'd Myrtle.
lufitani- ε Myrtus *acuta*, foliis lanceolato-ovatis acutis. *Mill. dict.*
ca.   Portugal Myrtle.

                                         ζ Myrtus

ζ Myrtus foliis lanceolatis acuminatis. *Mill. dict.*   belgica.
Broad-leav'd Dutch Myrtle.

η Myrtus *minima*, foliis lineari-lanceolatis acuminatis.   mucro-
Mill. dict.   nata.
Rofemary-leav'd Myrtle.
Nat. of Afia, Africa, and the South of Europe.
Cult. 1629, by Mr. John Parkinfon. *Park. parad.* 427.
Fl. July and Auguft.   G. H. ♄.

2. M. pedunculis unifloris, foliis triplinerviis fubtus to-   *tomentofa.*
mentofis.
Arbor finenfis Canellæ folio minore, trinervi, prona
parte villofo, fructu Caryophylli aromatici majoris,
villis fimiliter obducto. *Pluk. amalth.* 21. *t.* 372. *f.* 1.
Woolly-leav'd Myrtle.
Nat. of China.
Introd. about 1776, by Mrs. Norman.
Fl. June and July.   S. ♄.

3. M. pedunculis multifloris axillaribus, foliis ovatis   *Gregii.*
ellipticis acutis integerrimis fubtus pubefcentibus.
Swartz prodr. 78.
Round-leav'd Myrtle.
Nat. of Dominica.
Introd. 1776, by John Greg, Efq.
Fl.   S. ♄.

4. M. pedunculis dichotomis paniculatis tomentofis, fo-   *Chytra-*
liis geminis fubovatis terminalibus. *Sp. pl.* 675.   *culia.*
Forked Myrtle.
Nat. of Jamaica.
Introd. 1778, by Thomas Clark, M. D.
Fl.   S. ♄.

5. M. pedunculis multifloris, foliis geminis fubovatis   *Zuzygi-*
terminalibus, ramis dichotomis. *Sp. pl.* 675.   *um.*
Oval-

Oval-leav'd Myrtle.
Nat. of the Weft Indies.
Introd. 1778, by Thomas Clark, M. D.
Fl.   S. ♄.

Pimenta.   6. M. floribus trichotomo-paniculatis, foliis oblongo-
lanceolatis.
Myrtus Pimenta. *Sp. pl.* 676.
α foliis oblongo-lanceolatis acuminatis : acumine obtufo.
Long-leav'd Pimento, Jamaica Pepper, or All-fpice.
β foliis ovalibus obtufis.
Short-leav'd Pimento.
Nat. of the Weft Indies.
Cult. 1739, by Mr. Ph. Miller. *Rand. chel. n.* 14.
Fl. July.   S. ♄.

P U N I C A. *Gen. pl.* 618.

Cal. 5-fidus, fuperus. *Petala* 5. *Pomum* multiloculare,
polyfpermum.

Grana-   1. P. foliis lanceolatis, caule arboreo. *Sp. pl.* 676.
tum.   α floribus fimplicibus.
Common Pomegranate Tree.
β floribus plenis.
Double-flower'd Pomegranate Tree.
Nat. of Spain, Italy, and Barbary.
Cult. 1596, by Mr. John Gerard. *Hort. Ger.*
Fl. June——September.   H. ♄.

nana.   2. P. foliis linearibus, caule fruticofo. *Sp. pl.* 676.
Dwarf Pomegranate Tree.
Nat. of the Weft Indies.
Cult. 1731, by Mr. Ph. Miller. *Mill. dict. edit.* i. *n.* 5.
Fl. July——September.   G. H. ♄.

AMYG-

AMYGDALUS. *Gen. pl.* 619.

Cal. 5-fidus, inferus. *Pet.* 5. *Drupa* nuce poris
  perforata.

1. A. foliorum ferraturis omnibus acutis, floribus feffili- *Perfica.*
    bus folitariis. *Sp. pl.* 676.
α fructibus lanuginofis.
Common Peach Tree.
β fructibus glabris.   Nectari-
Common Nectarine Tree.   na.
γ flore pleno.   plena.
Dou le-flower'd Peach Tree.
Nat.
Cult. 1562. *Turn. herb. part* 2. *fol.* 48 *verfo.*
Fl. April and May.   H. 1.

2. A. foliis ferraturis infimis glandulofis, floribus foffili- *communis.*
    bus geminis. *Sp. pl.* 677.
α Amygdalus fativa. *Bauh. pin.* 441.
Sweet Almond Tree.
β Amygdalus amara. *Tourn. inft.* 627.
Bitter Almond Tree.
Nat. of Barbary.
Cult. 1570. *Lobel. adv.* 423.
Fl. March and April.   H. ♄.

3. A. foliis venofo-rugofis. *Linn. mant.* 74.   *pumila.*
Double-flower'd Dwarf Almond.
Nat.
Cult. 1731, by Mr. Philip Miller. *Mill. dict. edit.* 1.
    Perfica 3.
Fl. May and June.   H. ♄.

nana.   4. A. foliis bafi attenuatis. *Sp. pl.* 677. *Pallas roff.* 1.
      p. 12. *t.* 6.
      Common Dwarf Almond.
      Nat. of Ruffia.
      Cult. 1683, by Mr. James Sutherland. *Sutherl. hort.*
      edin. 22. *n.* 4.
      Fl. March and April.   H. ♄.

orientalis. 5. A. foliis lanceolatis integerrimis argenteis peren-
      nantibus, petiolo breviore. *Mill. dict.*
      Amygdalus orientalis, foliis argenteis fplendentibus.
      Du Hamel arb. 48.
      Silvery-leav'd Almond.
      Nat. of the Levant.
      Cult. 1759, by Mr. Ph. Miller. *Mill. dict. edit.* 7. *n.* 4.
      Fl.   H. ♄.

P R U N U S. *Gen. pl.* 620.

    Cal. 5-fidus, inferus. *Petala* 5. *Drupæ* nux futuris
prominulis.

Padus. 1. P. floribus racemofis : racemis pendulis, foliis deci-
      duis fubrugofis bafi biglandulofis.
      Prunus Padus. *Sp. pl.* 677.
      Common Bird-cherry Tree.
      Nat. of Britain.
      Fl. April and May.   H. ♄.

rubra. 2. P. floribus racemofis : racemis erectis, foliis deciduis
      lævibus bafi biglandulofis.
      Cerafus racemofa fylveftris, fructu non eduli rubro.
      Hort. angl. 18. *n.* 2.
      Cornifh Bird-cherry Tree.
      Nat.

Cult.

Cult. 1724. *Furber's catal.*
Fl. May and June.   H. ♄.

3. P. floribus racemofis, foliis deciduis bafi antice glandulofis. *Sp. pl.* 677.   *virginia-na.*
Common American Bird-cherry Tree.
Nat. of North America.
Cult. 1629. *Park. parad.* 597. *f.* 6.
Fl. May and June.   H. ♄.

4. P. floribus racemofis, foliis fempervirentibus oblongo-lanceolatis ferratis eglandulofis.   *carolinia-na.*
Prunus Padus carolina. *Du Roi hort. harbecc.* 2. *p.* 198.
Padus caroliniana. *Mill. dict.*
Evergreen Bird-cherry, Tree.
Nat. of South Carolina.
Cult. 1759, by Mr. Philip Miller. *Mill. dict. edit.* 7. Padus 6.
Fl. May.   H. ♄.

5. P. floribus racemofis: racemis lateralibus, foliis perennantibus eglandulofis oblongis acuminatis integris utrinque glabris. *Swartz prodr.* 80.   *occiden-talis.*
Weft India Laurel.
Nat. of Jamaica.
Introd. 1784, by Matthew Wallen, Efq.
Fl.   S. ♄.

6. P. floribus racemofis, foliis fempervirentibus ovato-lanceolatis ferratis eglandulofis.   *lufitani-ca.*
Prunus lufitanica. *Sp. pl.* 678.
Portugal Laurel.
Nat. of Portugal and Madeira.
Cult. 1722, by Mr. Thomas Fairchild. *Knowlton's mff.*
Fl. June. ¹   H. ♄.

Lauro-   7. P. floribus racemofis, foliis fempervirentibus dorfo bi-
Cerafus.   glandulofis. *Sp. pl.* 678.
         Common Laurel.
         Nat. of the Levant.
         Cult. before 1629, by Mr. James Cole. *Park. parad.*
         401.
         Fl. April.   H. ♄.

Mahaleb.   8. P. floribus corymbofis terminalibus, foliis ovatis.
             Syft. veget. 463. *Jacqu. auftr.* 3. *p.* 15. *t.* 227.
         Perfum'd Cherry Tree.
         Nat. of Auftria and Switzerland.
         Cult. 1714, by the Dutchefs of Beaufort. *Br. Muf.*
         H. S. 139. *fol*5.
         Fl. April and May.   H. ♄.

Armenia-   9. P. floribus feffilibus, foliis fubcordatis.   *Sp. pl.* 679.
ca.   Common Apricot Tree.
         Nat.
         Cult. 1562.   *Turn. herb. part* 2. *fol.* 48 *verfo.*
         Fl. February.   H. ♄.

pumila.   10. P. floribus fubumbellatis, foliis angufto-lanceolatis.
             Syft. veget. 463.
         Dwarf Canadian Cherry Tree.
         Nat. of North America.
         Cult. 1756, by Mr. Ph. Miller. *Mill. ic.* 60. *t.* 89. *f.* 2.
         Fl. May.   H. ♄.

Cerafus. 11. P. umbellis fubpedunculatis, foliis ovato-lanceolatis
             glabris conduplicatis.   *Syft. veget.* 463.
         α Cerafa fativa, rotunda, rubra & acida. *Bauh. pin.*
             449.
         Cultivated Cherry Tree.
         β Cerafus hortenfis pleno flore. *Bauh. pin.* 450. *Mill.*
             ic. 59. *t.* 89. *f.* 1.

                                                  Double-

Double-flowering Cherry Tree.
Nat. of England.
Fl. April. H. ♄ .

12. P. umbellis feffilibus, foliis ovato-lanceolatis con- *avium.*
    duplicatis fubtus pubefcentibus. *Sp. pl.* 680.
    Small-fruited Cherry Tree.
    Nat. of Britain.
    Fl. April. H. ♄ .

13. P. umbellis fubfeffilibus aggregatis multifloris tan- *penfylva-*
    dem paniculæformibus, foliis oblongo-lanceolatis *nica.*
    acuminatis glabris bafi glandulofis.
    Prunus penfylvanica. *Linn. fuppl.* 252.
    Upright Cherry Tree.
    Nat. of North America.
    Introd. 1773, by Meffrs. Kennedy and Lee.
    Fl. May. H. ♄ .

14. P. umbellis feffilibus folitariis paucifloris, foliis deci- *nigra.*
    duis ovatis acuminatis, petiolis biglandulofis.
    Black Cherry Tree.
    Nat. of Canada.
    Introd. 1773, by Meffrs. Kennedy and Lee.
    Fl. April and May. H. ♄ .

15. P. pedunculis fubfolitariis, foliis lanceolato-ovatis *domeftica.*
    convolutis, ramis muticis. *Sp. pl.* 680.
    Common Plum Tree.
    Nat. of England.
    Fl. April. H. ♄ .

16. P. pedunculis geminis, foliis ovatis fubtus villofis *infititia.*
    convolutis, ramis fpinefcentibus. *Sp. pl.* 680.
    Common Bullace Plum Tree.

M 3 *Nat.*

Nat. of Britain.
Fl. April.   H. ♄.

fpinofa.   17. P. pedunculis folitariis, foliis lanceolatis glabris, ra-
mis fpinofis. *Sp. pl.* 681.
Sloe Tree.
Nat. of Britain.
Fl. March and April.   H. ♄.

PLINIA. *Gen. pl.* 671. conf. *mant.* 244.

Cal. 5- f. 4-partitus.   *Petala* 5 f. 4.   *Drupa* fupera,
fulcata.

peduncu-   1. P. floribus pedunculatis polyandris. *Linn. fuppl.* 253.
lata.   Plinia rubra. *Linn. mant.* 243.
Myrtus brafiliana. *Sp. pl.* 674.
Eugenia uniflora. *Sp. pl.* 673.
Pedunculated Plinia.
Nat. of Brafil. (Cultivated in Madeira and the Weft
Indies.)
Cult. 1759, by Mr. Philip Miller.
Fl. January and February.   S. ♄.

CHRYSOBALANUS. *Gen. pl.* 621.

Cal. 5-fidus.   *Petala* 5.   *Drupæ* nux 5-fulcata, 5-valvis.

Icaco.   1. CHRYSOBALANUS. *Sp. pl.* 681.
Cocoa Plum Tree.
Nat. of the Weft Indies.
Cult. 1752, by Mr. Ph. Miller. *Mill. dict. edit.* 6.
Fl.   S. ♄.

DIGYNIA.

DIGYNIA.

CRATÆGUS. *Gen. pl.* 622.

Cal. 5-fidus. *Petala* 5. *Bacca* infera, 2-fperma.

1. C. inermis, foliis ovatis incifis ferratis fubtus tomen-   *Aria.*
    tofis.
   Cratægus Aria. *Sp. pl.* 681.
   α foliis incifo-ferratis.   indivifa.
   Common white Beam Tree.
   β Cratægus inermis, foliis ellipticis ferratis tranfverfali- fuecica.
    ter finuatis fubtus villofis. *Fl. lapp.* 199.
   Swedifh white Beam Tree.
   Nat. of Britain; ß. of Sweden.
   Fl. May and June.   H. ♄.

2. C. inermis, foliis glabris feptemangulis: lobis infimis   *tormina-*
    divaricatis, calycibus villofis.   *lis.*
   Cratægus torminalis. *Sp. pl.* 681. *Jacqu. auftr.* 5.
    p. 21. *t.* 443.
   Wild Haw Tree, or Service.
   Nat. of England.
   Fl. May and June.   H. ♄.

3. C. fpinofa, foliis cordato-ovatis incifo-angulatis gla-   *coccinea.*
    bris, petiolis calycibufque glandulofis, floribus pen-
    tagynis.
   Cratægus coccinea. *Sp. pl.* 632.
   Great American Hawthorn.
   Nat. of Virginia and Canada.
   Cult. 1696, by Mr. Jacob Bobart. *Br. Muf. Sloan.*
    mff. 33+3.
   Fl. April and May.   H. ♄.

4. C.

cordata.   4. C. fpinofa, foliis cordato-ovatis incifo-angulatis glabris, petiolis calycibufque eglandulofis, floribus pentagynis.

Mefpilus *cordata*, foliis cordato-ovatis accuminatis
acute ferratis, ramis fpinofis. *Mill. dict. ic.* 119.
tab. 179.

Mefpilus Phœnopyrum. *Ehrhart in Linn. fuppl.* 254.
Mœnch hort. weiffenft. 61.

Maple-leav'd Hawthorn.

Nat. of North America.

Introd. 1738, by Mr. Ph. Miller. *Mill. ic. loc. cit.*

Fl. May.   H. ♄.

pyrifolia.   5. C. fpinofa inermifve, foliis ovato-ellipticis incifo-
ferratis fubplicatis fubhirtis, calycibus villofiufculis:
foliolis lineari-lanceolatis ferratis, floribus trigynis.

Cratægus Leucophleos. *Mœnch hort. weiffenft.* 31.
tab. 2.

Pear-leav'd Hawthorn.

Nat. of North America.

Introd. 1765, by Meffrs. Kennedy and Lee.

Fl. June.   H. ♄.

elliptica.   6. C. fpinofa, foliis ellipticis inæqualiter ferratis glabris,
petiolis calycibufque glandulofis, baccis globofis
pentafpermis.

Oval-leav'd Hawthorn.

Nat. of North America.

Introd. 1765, by Meffrs. Kennedy and Lee.

Fl. May.   H. ♄.

glandulo-   7. C. fpinofa, foliis obovato-cuneiformibus angulatis
fa.   glabris nitidis, petiolis ftipulis calycibufque glandulofis, baccis ovalibus pentafpermis.

Cratægus glandulofus. *Mœnch hort. weiffenft.* 31.

                                                Hollow-

Hollow-leav'd Hawthorn.
Nat. of North America.
Cult. 1750, by Archibald Duke of Argyle.
Fl. May and June.   H. ♄.
Obs. In hac fpecie fpinæ validæ, in fequentibus dua-
bus tenues, apice fere incurvo, et in C. flava fæpe
foliolis parvis inſtruǎæ.

8. C. fpinofa, foliis obovato-cuneiformibus angulatis gla-   *flava.*
bris nitidis, petiolis ſtipulis calycibufque glandulo-
fis, baccis turbinatis tetrafpermis.
Mefpilus flexifpina. *Mœnch hort. weiſſenſt.* 62. *tab.* 4.
(exclufo fynonymo du Roi, quod fequentis.)
Yellow Pear-berried Hawthorn.
Nat. of North America.
Cult. 1758, by Mr. Philip Miller.
Fl. May.   H. ♄.

9. C. fpinofa, foliis cuneiformi-ovatis incifis ferratis, fo-   *parvifo-*
liolis calycinis lanceolatis incifis longitudine fruc-   *lia.*
tus, floribus pentagynis.
Cratægus uniflora.   *du Roi hort. harbecc.* 1. *p.* 184.
Mefpilus foliis lanceolato-ovatis ferratis fubtus villo-
fis, floribus folitariis, calycibus foliaceis, fpinis lon-
giſſimis tenuioribus.   *Mill. dicǎ. edit.* 7. *n.* 17.
Goofeberry-leav'd Hawthorn.
Nat. of Virginia.
Cult. before 1713, by Biſhop Compton.   *Mill. dicǎ.*
loc. cit.
Fl. May and June.   H. ♄.

10. C. fpinofa inermifve, foliis obovato-cuneiformibus   *punǎata.*
glabris ferratis, calycibus fubvillofis : foliolis fubulatis integris.
                                    Cratægus

Cratægus punctata. *Jacqu. hort.* 1. *p.* 10. *t.* 28.
Syft. veget. 465.

rubra.   α fructibus rubris.
           Great-fruited Hawthorn.

aurea.   β fructibus flavis.
           Great yellow-fruited Hawthorn.
           Nat. of North America.
           Cult. 1746, by Archibald Duke of Argyle.
           Fl. May.   H. ♄.

Crus   11. C. fpinofa, foliis obovato-cuneiformibus fubfeffilibus
Galli.   nitidis coriaceis, foliolis calycinis lanceolatis fub-
           ferratis, floribus digynis.
           Cratægus Crus Galli. *Sp. pl.* 682.

fplen-   α foliis obovato-cuneiformibus.
deus.   Common Cockfpur Hawthorn.

pyracan-   β foliis oblongo-lanceolatis fubcuneiformibus.
thifolia.   Pyracantha-leav'd Cockfpur Hawthorn.

falicifo-   γ foliis lanceolatis.
lia.   Willow-leav'd Cockfpur Hawthorn.
           Nat. of North America.
           Cult. 1691, by the Hon. Charles Howard. *Pluk.*
           phyt. t. 46. *f.* 1.
           Fl. May and June.   H. ♄.

Oxya-   12. C. foliis obtufis fubtrifidis ferratis.   *Sp. pl.* 683.
cantha.   α Mefpilus apii folio fylveftris fpinofa five Oxyacantha.
           Bauh. pin. 454.

vulgaris.   Common white Hawthorn.

major.   β Mefpilus apii folio fylveftris fpinofa, folio et fructu
           majore. *Raj. fyn.* 454.
           Great-fruited common Hawthorn.

præcox.   γ floribus præcocibus.
           Glaftonbury Hawthorn.

                                        δ floribus

δ floribus plenis.   plena.
Double-flower'd Hawthorn.

ε fructibus flavis.   aurea.
Yellow-berried common Hawthorn.
Nat. of Britain.
Fl. α, β, δ and ε, May; γ. January and February.
$$\text{H. } ♄ .$$

13. C. foliis obtusis subtrifidis subdentatis. *Sp. pl.* 683. *Azarolus.*
Parsley-leav'd Hawthorn.
Nat. of the South of Europe.
Cult. 1656, by Mr. John Tradescant. *Muf. Trad.*
141.
Fl. May.   H. ♄ .

T R I G Y N I A.

S O R B U S. *Gen. pl.* 623.
Cal. 5-fidus. *Petala* 5. *Bacca* infera, 3-fperma.

1. S. foliis pinnatis utrinque glabris. *Sp. pl.* 683.   aucupa-
Mountain Ash, or Service Tree.   ria.
Nat. of Britain.
Fl. May.   H. ♄ .

2. S. foliis femipinnatis fubtus tomentofis. *Sp. pl.* 684. *hybrida.*
Baftard Service Tree.
Nat. of Sweden and Norway.
Introd. 1779, by Chevalier Thunberg.
Fl. May and June.   H. ♄ .

3. S. foliis pinnatis fubtus villofis. *Sp. pl.* 684. *Jacqu.* *domeftica.*
auftr. 5. *p.* 23. *t.* 447.
True Service, or Sorb Tree.
$$\text{Nat.}$$

Nat. of England.

Fl. Máy.   H. ♄.

S E S U V I U M.   *Gen. pl.* 624.

Cal. 5-partitus, coloratus.   *Petala* nulla.   *Capf.* ovata,
trilocularis, circumfciffa, polyfperma.

Portula-   1. SESUVIUM.   *Sp. pl.* 684.
caftrum.   Purflane-leav'd Sefuvium.
                  Nat. of the Weft Indies.
                  Cult. 1692, in the Royal Garden at Hampton-court.
                     Pluk. phyt. t. 216. *f.* 1.
                  Fl.   S. ♃.

P E N T A G Y N I A.

M E S P I L U S.   *Gen. pl.* 625.

Cal. 5-fidus.   *Pet.* 5.   *Bacca* infera, 5-fperma.

Pyracan-   1. M. fpinofa, foliis lanceolato-ovatis crenatis, calycibus
tha.   fructus obtufis.   *Sp. pl.* 685.   *Pallas roff.* 1. *p.* 29.
                  t. 13. *f.* 2.
                  Evergreen Thorn, or Mefpilus.
                  Nat. of the South of Europe.
                  Cult. 1629.   *Park. parad.* 605. *f.* 2.
                  Fl. May.   H. ♄.

germani-   2. M. inermis, foliis lanceolatis fubtus tomentofis, flori-
ca.   • bus feffilibus folitariis.   *Sp. pl.* 684.   *Pallas roff.* 1.
                     p. 29. *t.* 13. *f.* 1.
ftricta.   α foliis duplicato-ferratis.
                  Narrow-leav'd Dutch Medlar.

                                                               β foliis

β foliis fubintegris.   diffufa.
Broad-leav'd Dutch Medlar.
Nat. of England.
Fl. June and July.   H. ♄.

3. M. inermis, foliis lanceolatis crenatis fubtus tomen- *arbutifo-*
   tofis. *Sp. pl.* 685.   *lia.*
α baccis rubris.   rubra.
Red-fruited Arbutus-leav'd Mefpilus.
β baccis nigris.   nigra.
Black-fruited Arbutus-leav'd Mefpilus.
γ baccis albis.   alba.
White-fruited Arbutus-leav'd Mefpilus.
Nat. of Virginia.
Cult. 1700, by Lord Clarendon. *Br. Muf. Sloan. mff.*
  3343.
Fl. May.   H. ♄.

4. M. inermis, foliis ovalibus ferratis fubtus hirfutis. *Amelan-*
   Syft. veget. 466. *Jacqu. auftr.* 3. *p.* 55. *t.* 300.   *chier.*
Alpine Mefpilus.
Nat. of the South of Europe.
Cult. 1730. *Hort. angl.* 49. *n.* 12.
Fl. May.   H. ♄.

5. M. inermis, foliis ovalibus acute ferratis glabris, flo- *Chamæ-*
   ribus corymbofo-capitatis. *Syft. veget.* 466.   *Mefpilus.*
Baftard Quince, or Mefpilus.
Nat. of the Pyrenees and the Alps of Auftria.
Cult. 1683, by Mr. James Sutherland. *Sutherl. hort.*
  edin. 80. *n.* 2.
Fl.   *a*   H. ♄.

6. M. inermis, foliis ovato-oblongis glabris ferratis acu- *canaden-*
   tiufculis. *Syft. veget.* 466.   *fis.*
                     Snowy

Snowy Mefpilus.
Nat. of Canada and Virginia.
Cult. 1746, by Archibald Duke of Argyle.
Fl April and May.   H. ♄.

Cotoneaf-ter.   7. M. inermis, foliis ovatis integerrimis acutiufculis fub-
tus tomentofis, germinibus glabris, baccis di- vel
trifpermis.
Mefpilus Cotoneafter. *Sp. pl.* 686. *Pallas roff.* 1.
p. 30. *t.* 14.
Dwarf Mefpilus.
Nat. of Europe and Siberia.
Cult. 1656, by Mr. John Tradefcant, Jun. *Trad.
muf.* 111.
Fl. April and May.   H. ♄.

tomentofa.   8. M. inermis, foliis ovatis integerrimis obtufis fubtus
tomentofis, germinibus lanatis, baccis pentafper-
mis.
Mefpilus orientalis. *Mill. dict.* (exclufo fynonymo.)
Quince-leav'd Mefpilus.
Nat.
Cult. 1759, by Mr. Ph. Miller. *Mill. dict. edit.* 7. *n.* 22.
Fl. April and May.   H. ♄.

P Y R U S. *Gen. pl.* 626.

Cal. 5-fidus. *Petala* 5. *Pomum* inferum 5-loculare,
polyfpermum.

communis.   1. P. foliis ovatis ferratis, pedunculis corymbofis.
Pyrus communis. *Sp. pl.* 686.
Common Pear Tree.
Nat. of England.
Fl. April.   H. ♄.

2. P.

2. P. foliis ferratis fubtus tomentofis, floribus corymbo-   *Pollve-*
    fis. *Linn. mant.* 244.   *ria.*
    Woolly-leav'd Pear Tree.
    Nat. of Germany.
    Introd. 1786, by Mr. John Græfer.
    Fl.   H. ♄.

3. P. umbellis feffilibus, foliis ovato-oblongis acuminatis   *Malus.*
    ferratis glabris, unguibus calyce brevioribus, ftylis
    glabris.
    Pyrus Malus. *Sp. pl.* 686.
α Malus fativa. *Raj. fyn.* 451.
    Common Apple Tree.
β Cratægus cerafi folio, floribus magnis. *Mill. ic.* 180.
    t. 269. (exclufo fynonymo Ammanni.)
    Siberian Crab Tree.
    Nat. of Britain.
    Fl. April and May.   H. ♄.

4. P. umbellis feffilibus, foliis ovali-oblongis ferratis læ-   *fpectabi-*
    vibus, unguibus calyce longioribus, ftylis bafi la-   *lis.*
    natis.
    Chinefe Apple Tree.
    Nat. of China.
    Cult. 1780, by John Fothergill, M. D.
    Fl. May.   H. ♄.

5. P. foliis æqualiter ferrulatis, pedunculis confertis, po-   *baccata.*
    mis baccatis, calycibus deciduis.
    Pyrùs baccata. *Linn. mant.* 75. (exclufo fynonymo
    Milleri.) *Pallas roff.* 1. *p.* 23. *t.* 10.
    Small-fruited Crab Tree.
    Nat. of Siberia.
    Introd. 1784, by Mr. John Græfer.
    Fl. April.   H. ♄.
                         6. P.

coronaria. 6. P. foliis cordatis incifo-ferratis angulatis glabris, pe-
dunculis corymbofis.
Pyrus coronaria.   *Sp. pl.* 687.
Sweet-fcented Crab Tree.
Nat. of Virginia.
Cult. 1724.   *Furber's calal.*
Fl. May.   H. ♄.

anguftifo- 7. P. foliis lanceolato-oblongis nitidis dentato-ferratis
lia.   bafi attenuatis integris, pedunculis corymbofis.
Narrow-leav'd Crab Tree.
Nat. of North America.
Cult. 1750, by Mr. Chriftopher Gray.
Fl. May.   H. ♄.

Cydonia. 8. P. foliis integerrimis, floribus folitariis.   *Sp. pl.* 687.
Jacqu. auftr. 4. *p.* 22. *t.* 342.
Common Quince Tree.
Nat. of Auftria.
Cult. 1597.   *Ger. herb.* 1264.
Fl. May and June.   H. ♄.

falicifolia. 9. P. foliis lineari-lanceolatis canis fubtus albo-tomento-
fis, floribus axillaribus folitariis fubfeffilibus.   *Linn.*
fuppl. 255.   *Pallas roff.* 1. *p.* 20. *t.* 9.
Willow-leav'd Crab Tree.
Nat. of the Levant.
Introd. 1780, by Peter Simon Pallas, M.D.
Fl.   H. ♄.

TETRAGONIA. *Gen. pl.* 627

Cal. 3-5-partitus.   *Cor.* 0.   *Drupa* infera, nuce 3-8-
loculari.

fruticofa. 1. T. fruticofa, foliis linearibus, fructibus alatis.

                                                  Tetragonia

Tetragonia fruticofa. *Sp. pl.* 687.
Shrubby Tetragonia.
Nat. of the Cape of Good Hope.
Cult. 1712, by Bifhop Compton. *Philof. tranfact.*
n. 333. *p.* 422. *n.* 78.
Fl. July——September.   G. H. ♄.

2. T. fruticofa pruinofa, foliis obovatis, fructibus alatis.   *decum-*
    Tetragonia *decumbens,* foliis ovatis integerrimis, caule   *bens.*
    fruticofo decumbente. *Mill. dict. ic.* 176. *t.* 263.
    f. 1.
Trailing Tetragonia.
Nat. of the Cape of Good Hope.
Cult. 1758, by Mr. Ph. Miller. *Mill. ic. loc. cit.*
Fl. July——September.   G. H. ♄.

3. T. herbacea lævis, foliis ovatis petiolatis, fructibus   *herbacea.*
    alatis.
Tetragonia herbacea. *Sp. pl.* 687.
Herbaceous Tetragonia.
Nat. of the Cape of Good Hope.
Cult. 1759, by Mr. Ph. Miller. *Mill. dict. ed.* 7. *n.* 3.
Fl. June and July.   G. H. ♃.

4. T. herbacea, foliis rhombeo-ovatis, fructibus echinatis.   *echinata.*
Hedge-hog Tetragonia.
Nat. of the Cape of Good Hope. Mr. *Fr. Maffon.*
Introd. 1774.
Fl. May——Auguft.   G. H. ♂.
DESCR. *Caulis* herbaceus, prope radicem divifus in
    Ramos diffufos, petiolis decurrentibus angulatos, vix
    pedales. *Folia* fucculenta, patentia, uncialia : *Pe-*
    tioli foliis dimidio breviores. *Pedunculi* axillares,
    folitarii, filiformes, papulis nitidis tecti, purpurei,
    breviffimi. *Flores* penduli, papulis cryftallinis prui-

nofi. *Calyx* tri- vel tetraphyllus, intus e viridi-
flavicans. *Stamina* fæpius tria, raro quatuor. *Germen* fubtus planum, triquetrum : anguli proceffibus
conicis pluribus echinati. *Styli* tres. *Nux* trilocularis.

expanfa.   5. T. herbacea, foliis ovato-rhombeis, fructibus quadri-
cornibus.
Tetragonia expanfa. *Murray in commentat. gotting.* 6.
(1783) *p.* 13. *t.* 5. *Scop. infubr.* 1. *p.* 32. *t.* 14.
Demidovia tetragonoides. *Pallas hort. Demidov.*
p. 150. *tab.* 1.
Horn'd Tetragonia.
Nat. of New Zealand. Sir *Jof. Banks,* Bart.
Introd. 1772.
Fl. Auguft and September.   G. H. ♂.

cryftalli-   6. T. herbacea pruinofa, foliis ovatis feffilibus, fructibus
na.   inermibus.
Tetragonia cryftallina. *L'Herit. ftirp. nov. p.* 81.
tab. 39.
Diamond Tetragonia.
Nat. of Peru.
Introd. 1788, by Monf. Thouin.
Fl. June.   S. ☉.

MESEMBRYANTHEMUM. *Gen. pl.* 628.

Cal. 5-fidus. *Petala* numerofa, linearia. *Capf.* carnofa, infera, polyfperma.

* *Albis corollis.*
nodiflo-   1. M. foliis alternis teretiufculis obtufis bafi ciliatis. *Sp.*
rum.   *pl.* 687.
Egyptian Fig Marygold.

Nat.

Nat. of Italy and Egypt.
Cult. 1748, by Mr. Philip Miller. *Mill. dict. edit.* 5.
Ficoides 42.
Fl. Auguft. G. H. ⊙.

2. M. foliis oppofitis connatis femiteretibus, ftipulis mem-   *ciliatum.*
branaceis reflexis laceris ciliiformibus.
Ciliated Fig Marygold.
Nat. of the Cape of Good Hope. Mr. *Fr. Maffon.*
Introd. 1774.
Fl. G. H. ♄.

3. M. foliis filiformi-femiteretibus diftinctis : papulis   *caducum.*
ovatis, floribus lateralibus feffilibus; terminalibus
pari foliorum cinctis.
Small-flower'd Fig Marygold.
Nat. of the Cape of Good Hope. Mr. *Fr. Maffon.*
Introd. 1774:
Fl. July and Auguft. G. H. ♂.

4. M. foliis alternis ovatis papulofis, floribus feffilibus,   *cryftalli-*
calycibus late ovatis acutis retufis. *Syft. veget.*   *num.*
468.
Diamond Fig Marygold, or Ice-plant.
Nat. of Greece : near Athens. *John Sibthorp*, M. D.
Cult. 1727. *Bradl. fucc.* 5. *p.* 15. *t.* 48.
Fl. July and Auguft. G. H. ⊙.

5. M. foliis amplexicaulibus fpathulatis carinatis : papu-   *humifu-*
lis conicis fcabris, petalis minutiffimis.   *fum.*
Narrow-leav'd Icy Fig Marygold:
Nat. of the Cape of Good Hope.
Introd. 1774, by Mr. Francis Maffon.
Fl. July and Auguft. G. H. ♄.

6. M.

apetalum. 6. M. foliis amplexicaulibus diftinctis linearibus fupra
planis internodiis longioribus papulofis : papulis
oblongis, floribus pedunculatis, calycibus quinquefidis.

Mefembryanthemum apetalum.   *Linn. fuppl.* 258.

Mefembryanthemum copticum.   *Jacqu. hort.* 3. *p.* 7.
t. 6.

Dwarf-fpreading Fig Marygold.

Nat. of the Cape of Good Hope.

Introd. 1774, by Mr. Francis Maffon.

Fl. July and Auguft.   G. H. ☉.

geniculi- 7. M. foliis femiteretibus papulofis diftinctis, floribus
florum.   feffilibus axillaribus, calycibus quadrifidis.   *Sp. pl.*
688.

Jointed Fig Marygold.

Nat. of the Cape of Good Hope.

Cult. 1727.   *Bradl. fucc.* 5. *p.* 17. *t.* 34.

Fl. June——Auguft.   G. H. ♄.

noctiflo- 8. M. foliis femicylindricis impunctatis diftinctis, flori-
rum.   bus pedunculatis, calycibus quadrifidis. *Sp. pl.* 689.

Night-flowering Fig Marygold.

Nat. of the Cape of Good Hope.

Cult. 1714, by the Dutchefs of Beaufort.   *Br. Muf.*
H. S. 142. *fol.* 72.

Fl. June——Auguft.   G. H. ♄.

fplendens. 9. M. foliis fubteretibus impunctatis recurvis diftinctis
congeftis, calycibus digitiformibus terminalibus.
Syft. veget. 468.

Shining Fig Marygold.

Nat. of the Cape of Good Hope.

Cult. 1716.   *Bradl. fucc.* 1. *p.* 7. *t.* 6.

Fl. June——Auguft.   G. H. ♄.

                                                       10. M.

10. M. foliis fubulatis fcabrido-punctatis connatis apice   *umbella-*
    patulo, caule erecto, corymbo trichotomo. *Sp. pl.*   *tum.*
    689.
    Umbel'd Fig Marygold.
    Nat. of the Cape of Good Hope.
    Cult. 1727. *Bradl. fucc.* 4. *p.* 12. *t.* 44.
    Fl. June——September.   G. H. ♄.

11. M. foliis planiufculis lanceolatis impunctatis paten-   *expanfum*
    tibus diftinctis oppofitis alternifque remotis. *Sp.*
    pl. 697.
    Houfeleek-leav'd Fig Marygold.
    Nat. of the Cape of Good Hope.
    Cult. 1705, by Dr. Uvedale. *Pluk. amalth.* 90.
    Fl. July and Auguft.   G. H. ♃.

12. M. foliis quatuor decuffatis fupra planis.   *tefticu-*
    Short white-leav'd Fig Marygold.   *lare.*
    Nat. of the Cape of Good Hope. Mr. *Fr. Maffon.*
    Introd. 1774.
    Fl.   G. H. ♄.

13. M. acaule, foliis fubteretibus adfcendentibus im-   *calami-*
    punctatis connatis, floribus octogynis. *Sp. pl.* 690.   *forme.*
    Quill-leav'd Fig Marygold.
    Nat. of the Cape of Good Hope.
    Cult. 1717. *Bradl. fucc.* 2. *p.* 10. *f.* 19.
    Fl. July——September.   G. H. ♄.

14. M. fubacaule, foliis alternis teretibus obtufis, flori-   *digita-*
    bus axillaribus feffilibus.   *tum.*
    Blunt-leav'd Fig Marygold.
    Nat. of the Cape of Good Hope. Mr. *Fr. Maffon.*
    Introd. 1775.
    Fl.   G. H. ♄.

15. M.

pallens.   15. M. foliis oppofitis amplexicaulibus diftinctis oblongo-
lanceolatis acutis obtufe carinatis : papulis mi-
nutis.
Channel-leav'd Fig Marygold.
Nat. of the Cape of Good Hope. Mr. *Fr. Maffon.*
Introd. 1774.
Fl. July and Auguft.   G. H. ♄.

Tripoli-   16. M. foliis alternis lanceolatis planis impunctatis,
um.   caulibus laxis fimplicibus, calycibus pentagonis.
Sp. pl. 690.
Plain-leav'd Fig Marygold.
Nat. of the Cape of Good Hope.
Cult. 1700, in Chelfea Garden. *Pluk. mant.* 77.
t. 329. *f.* 4.
Fl. June——September.   G. H. ♂.

** *Rubicundis corollis.*

papulo-   17. M. foliis oppofitis diftinctis ovato-fpathulatis: pa-
fum.   ‹pulis fubglobofis, calycibus angulatis quinquefi-
dis, ramis angulatis.
Mefembryanthemum papulofum. *Linn. fuppl.* 259.
Mefembryanthemum Aitonis. *Jacqu. hort.* 3. *p.* 8.
t. 7.
Angular-ftalk'd Fig Marygold.
Nat. of the Cape of Good Hope.
Introd. 1774; by Mr. Francis Maffon.
Fl. April——October.   G. H. ♂.

cordifoli- 18. M. foliis oppofitis petiolatis cordatis, calycibus qua-
um.   drifidis, caule tereti.
Mefembryanthemum cordifolium. *Linn. fuppl.* 260.
Gloxin obf. bot. p. 22. *tab.* 1. *fig. a.*
Heart-leav'd Fig Marygold.
Nat. of the Cape of Good Hope. Mr. *Fr. Maffon.*
Introd.

Introd. 1774.
Fl. May——September.   G. H. ♄.

19. M. foliis oppofitis fpathulatis obtufis fcabridis : pa-   *limpidum.*
pulis oblongis, foliolis calycinis oblongis obtufis
medio coarctatis.
Tranfparent Fig Marygold.
Nat. of the Cape of Good Hope. Mr. *Fr. Maſſon.*
Introd. 1774.
Fl. July.   G. H. ☉.

20. M. acaule, foliis triquetris linearibus impunctatis   *bellidiflo-*
apice trifariam dentatis. *Sp. pl.* 690.   *rum.*
Daify-flower'd Fig Marygold.
Nat. of the Cape of Good Hope.
Cult. 1717. *Bradl. ſucc.* 2. *p.* 9. *f.* 18.
Fl. June——Auguft.   G. H. ♄.

21. M. foliis deltoidibus triquetris dentatis impunctatis   *deltoides.*
diftinctis. *Sp. pl.* 690.
α Mefembr. deltoides, & dorfo & lateribus muricatis,
minus. *Dill. elth.* 255. *t.* 195. *f.* 246.
Delta-leav'd Fig Marygold.
β Mefembr. deltoides, & dorfo & lateribus muricatis,
majus. *Dill. elth.* 254. *t.* 196. *f.* 247.
Great Delta-leav'd Fig Marygold.
γ Mefembr. deltoides, non dorfo, fed lateribus muricatis. *Dill. elth.* 253. *t.* 195. *f.* 243 & 244.
Small Delta-leav'd Fig Marygold.
Nat. of the Cape of Good Hope.
Cult. 1714, by the Dutchefs of Beaufort. *Br. Muſ.*
H. S. 131. *fol.* 40.
Fl. May——Auguft.   G. H. ♄.

N 4   22. M.

barba-   22. M. foliis fubovatis papulofis diftinctis apice barbatis.
tum.   *Sp. pl.* 691.
α Mefembr. radiatum, ramulis prolixis recumbentibus.
    Dill. elth. 245. *t.* 190. *f.* 234.
    Shrubby bearded Fig Marygold.
β Mefembr. radiatum humile, foliis minoribus. *Dill.*
    elth. 246. *t.* 190. *f.* 235.
    Small dwarf-bearded Fig Marygold.
γ Mefembr. radiatum humile, foliis majoribus. *Dill.*
    elth. 248. *t.* 190. *f.* 236.
    Great dwarf-bearded Fig Marygold.
    Nat. of the Cape of Good Hope.
    Cult. 1714, by the Dutchefs of Beaufort. *Br. Muf.*
    H. S. 133. *fol.* 17.
    Fl. June——Auguft.   G. H. ♄.

hifpidum.   23. M. foliis cylindricis papulofis diftinctis, caule hifpi-
            do. *Sp. pl.* 691.
α Mefembr. pilofum micans, flore faturanter purpureo.
    Dill. elth. 289. *t.* 214. *f.* 277 & 278.
    Purple-flower'd briftly Fig Marygold.
β Mefembr. pilofum micans, flore purpureo pallidiore.
    Dill. elth. 290. *t.* 214. *f.* 279, 280.
    Pale-flower'd briftly Fig Marygold.
γ Mefembr. pilofum micans, flore purpureo ftriato.
    Dill. elth. 291. *t.* 215. *f.* 281.
    Striped-flower'd briftly Fig Marygold.
    Nat. of the Cape of Good Hope.
    Cult. 1705, by Charles du Bois, Efq. *Pluk. àmalth.*
    89.
    Fl. moft part of the Year.   G. H. ♄.

villofum.   24. M. foliis pubefcentibus connatis impunctatis, caule
            pilofo. *Sp. pl.* 692.

                                                Hairy-

Hairy-ftalk'd Fig Marygold.
Nat. of the Cape of Good Hope.
Cult. 1759, by Mr. Philip Miller. *Mill. dict. edit.* 7.
n. 16.
Fl.   G. H. ♄.

25. M. foliis fubacinaciformibus punctatis apice recurvo,   *bractea-*
bracteis amplexicaulibus lato-ovatis carinatis.   *tum.*
Mefembryanthemum uncinatum fcabrum, petalis
purpureis circumactis. *Dill. elth.* 258. *t.* 197.
f. 249.
Bracteated Fig Marygold.
Nat. of the Cape of Good Hope.
Cult. 1732, by James Sherard, M.D. *Dill. elth.*
loc. cit.
Fl. July——October.   G. H. ♄.

26. M. foliis fubulatis diftinctis fubtus undique punctato-   *fcabrum.*
muricatis, calycibus muticis. *Syft. veget.* 469.
Rough Fig Marygold.
Nat. of the Cape of Good Hope.
Cult. 1732, by James Sherard, M. D. *Dill. elth.*
260. *t.* 197. *f.* 251.
Fl. June——Auguft.   G. H. ♄.

27. M. foliis triquetris acutis fcabris, caule reptante.   *reptans.*
Creeping Fig Marygold.
Nat. of the Cape of Good Hope. Mr. *Fr. Maffon.*
Introd. 1774.
Fl. July and Auguft.   G. H. ♄.

28. M. foliis fubulatis congeftis fubfcabris, calycibus   *emargi-*
fpinofis, petalis emarginatis. *Sp. pl.* 692.   *natum.*
Notched-flower'd Fig Marygold.
Nat. of the Cape of Good Hope.
                                             Cult.

Cult. 1732, by James Sherard, M. D.   *Dill. elth.*
    259. *t.* 197. *f.* 250.
    Fl. June——Auguft.   G. H. ♄.

uncina-   29. M. articulis caulinis terminatis foliis connatis acu-
tum.   minatis punctatis fubtus dentatis. *Syft. veget.* 469.
    α Mefembr. perfoliatum, foliis minoribus diacanthis.
        Dill. elth. 250. *t.* 193. *f.* 239.
        Small hook'd-leav'd Fig Marygold.
    β Mefembr. perfoliatum, foliis majoribus triacanthis.
        Dill. elth. 251. *t.* 193. *f.* 240.
        Great hook'd-leav'd Fig Marygold.
        Nat. of the Cape of Good Hope.
        Cult. 1714, by the Dutchefs of Beaufort.   *Br. Muf.*
        H. S. 142. *fol.* 64.
        Fl. June——Auguft.   G. H. ♄.

fpinofum.   30. M. foliis tereti-triquetris punctatis diftinctis, fpinis
            ramofis. *Sp. pl.* 693.
        Thorny Fig Marygold.
        Nat. of the Cape of Good Hope.
        Cult. 1714, by the Dutchefs of B.aufort.   *Br. Muf.*
        H. S. 142. *fol.* 67.
        Fl. June——Auguft.   G. H. ♄.

tubero-   31. M. foliis fubulatis papulofis diftinctis apice patulis,
fum.   radice capitata. *Sp. pl.* 693.
        Tuberous-rooted Fig Marygold.
        Nat. of the Cape of Good Hope.
        Cult. 1714, by the Dutchefs of Beaufort.   *Br. Muf.*
        H. S. 142. *fol.* 67.
        Fl. June——September.   G. H. ♄.

tenuifoli- 32. M. fubfiliformibus glabris diftinctis internodio lon-
um.   gioribus, caulibus procumbentibus. *Syft. veget.*
            469.

                                              Slender-

Slender-leav'd Fig Marygold.
Nat. of the Cape of Good Hope.
Cult. 1700, in Chelfea Garden. *Pluk. mant.* 77.
Fl. June——September.   G. H. ♄.

33. M. foliis fubtriquetris compreffis incurvatis punĉta- *ftipula-*
tis diftinctis congeftis bafi marginatis. *Sp. pl.* 693. *ceum.*
Upright fhrubby Fig Marygold.
Nat. of the Cape of Good Hope.
Cult. 1732, by James Sherard, M. D.   *Dill. elth.*
279 *t.* 209. *f.* 267 & 268.
Fl. June——Auguft.   G. H. ♄.

34. M. foliis cylindraceis obtufis amplexicaulibus lævi- *læve.*
bus, calycibus quinquefidis: laciniis oblongis ob-
tufis.
Upright white-wooded Fig Marygold.
Nat. of the Cape of Good Hope. Mr. *Fr. Maffon.*
Introd. 1774.
Fl. July——September.   G. H. ♄.

35. M. foliis triquetris acutis glaucis: punĉtis obfoletis *deflexum.*
fcabriufculis, laciniis calycinis interioribus mem-
branaceis.
Bending Fig Marygold.
Nat. of the Cape of Good Hope. Mr. *Fr. Maffon.*
Introd. 1774.
Fl. July.   G. H. ♄.

36. M. foliis fubtriquetris punĉtulatis connatis obtufiuf- *auftrale.*
culis, caule tereti repente, pedunculis obtufe anci-
pitibus folitariis.
New Zealand Fig Marygold.
Nat. of New Zealand. Sir *Jofeph Banks,* Bart.
Introd.

Introd. 1773.
Fl. July and Auguſt.   G. H. ♃.

craſſifoli-
um.
 37. M. foliis femicylindricis impunctatis connatis apice triquetris, caule repente femicylindrico. *Sp. pl.* 693.
Thick-leav'd Fig Marygold.
Nat. of the Cape of Good Hope.
Cult. 1727, by Profeſſor Richard Bradley. *Bradl. fucc.* 4. *p.* 16. *t.* 38.
Fl. July and Auguſt.   G. H. ♄.

falcatum.
 38. M. foliis fubacinaciformibus incurvis punctatis diſtinctis, ramis teretibus. *Sp. pl.* 694.
Sickle-leav'd Fig Marygold.
Nat. of the Cape of Good Hope.
Cult. 1727. *Bradl. fucc.* 5. *p.* 9. *t.* 42.
Fl. June——Auguſt.   G. H. ♄.
Obs. In *Syſt. veget.* 469, nomen triviale Mefembr. falcati præfixum eſt differentiæ fpecificæ Mefembr. acinaciformis.

glomera-
tum.
 39. M. foliis teretiufculis compreſſis punctatis diſtinctis, caule paniculato multifloro. *Syſt. veget.* 469.
Cluſter'd Fig Marygold.
Nat. of the Cape of Good Hope.
Cult. 1732, by James Sherard, M.D. *Dill. elth.* 287. *t.* 213. *f.* 274.
Fl. June——Auguſt.   G. H. ♄.

brevifoli-
um.
 40. M. foliis cylindraceis obtuſiſſimis papuloſis patentibus, ramis diffuſis.
Short-leav'd Fig Marygold.
Nat. of the Cape of Good Hope. Mr. *Fr. Maſſon.*
Introd.

Introd. 1774.
Fl. July and Auguft. G. H. ♄.

41. M. foliis femicylindricis recurvis congeftis bafi in- *loreum.*
teriore gibbis connatis, caule pendulo. *Sp. pl.* 694.
Leathery-ftalk'd Fig Marygold.
Nat. of the Cape of Good Hope.
Cult. 1732, by James Sherard, M. D. *Dill. elth.*
264. *t.* 200. *f.* 255.
Fl. Auguft and September. G. H. ♃.

42. M. foliis æquilateri-triquetris acutis fubpunctatis *filaments-*
fubconnatis : angulis fcabris, ramis hexagonis. *Sp.* *fum.*
pl. 694.
Thready Fig Marygold.
Nat. of the Cape of Good Hope.
Cult. 1732, by James Sherard, M. D. *Dill. elth.*
285. *t.* 212. *f.* 273.
Fl. July and Auguft. G. H. ♃.

43. M. foliis acinaciformibus impunctatis connatis : an- *acinaci-*
gulo carinali fcabris, petalis lanceolatis. *Sp. pl.* *forme.*
695.
Cymeter-leav'd Fig Marygold.
Nat. of the Cape of Good Hope.
Cult. 1714, by the Dutchefs of Beaufort. *Br. Muf.*
H. S. 133. *fol.* 19.
Fl. Auguft and September. G. H. ♄.

44. M. foliis acinaciformibus obtufis impunctatis conna- *forfica-*
tis apice fpinofis, caule ancipiti. *Sp. pl.* 695. *Jacqu.* *tum.*
hort. 1. *p.* 9 *t.* 26.
Forked Fig Marygold.
Nat. of the Cape of Good Hope.

Cult.

Cult. 1758, by Mr. Philip Miller.

Fl.   G. H. ♃.

*** *Luteis corollis.*

edule.   45. M. foliis æquilateri-triquetris acutis ſtriƈtis impunc-
tatis connatis carina ſubſerratis, caule ancipiti.
Sp. pl. 695.
Eatable Fig Marygold.
Nat. of the Cape of Good Hope.
Introd. 1690, by Mr. Bentick. *Br. Muſ. Sloan.*
mſſ. 3370.
Fl. July and Auguſt.   G. H. ♄.

bicolo-   46. M. foliis ſubulatis lævibus punƈtatis diſtinƈtis, caule
rum.   fruteſcente, corollis bicoloribus.   *Sp. pl.* 695.
Curtis magaz. 59.
Two-colour'd Fig Marygold.
Nat. of the Cape of Good Hope.
Cult. 1696, in the Royal Garden at Hampton-court.
Catal. mſſ.
Fl. May——September.   G. H. ♄.

aureum.   47. M. fol. cylindrico-triquetris punƈtatis diſtinƈtis, piſ-
tillis atro-purpuraſcentibus.   *Syſt. nat. ed.* 10.
p. 1060.
Golden Fig Marygold.
Nat. of the Cape of Good Hope.
Cult. 1750, by Mr. Philip Miller.
Fl. June——Auguſt.   G. H. ♄.

micans.   48. M. foliis ſubcylindricis papuloſis diſtinƈtis, caule
ſcabro. *Sp. pl.* 696.
Glittering Fig Marygold.
Nat. of the Cape of Good Hope.

Cult.

Cult. 1716, by Profeffor Richard Bradley. *Bradl. fucc.* 1. *p.* 9. *t.* 8.
Fl. May——-Auguft.   G. H. ♄.

49. M. foliis fubcylindricis confertis papulofis, caudice   *groffum.*
 bafi incraffato, ramis diffufis glabris.
 Gouty Fig Marygold.
 Nat. of the Cape of Good Hope. Mr. *Fr. Maffon.*
 Introd. 1774.
 Fl. Auguft——Oftober.   G. H. ♄.

50. M. caulibus foliifque cylindricis papulofis, ramis tri-   *brachia-*
 chotomis.   *tum.*
 Three-fork'd Fig Marygold.
 Nat. of the Cape of Good Hope. Mr. *Fr. Maffon.*
 Introd. 1774.
 Fl. July and Auguft.   G. H. ♄.

51. M. acaule, foliis femicylindricis connatis externe   *roftra-*
 tuberculatis. *Sp. pl.* 696.   *tum.*
 Heron-beak'd Fig Marygold.
 Nat. of the Cape of Good Hope.
 Cult. 1732, by James Sherard, M. D. *Dill. elth.*
 240. *t.* 186. *f.* 229.
 Fl.   G. H. ♃.

52. M. acaule, foliis connatis punctatis femiteretibus   *compac-*
 apice triquetris fubreflexis acutis, floribus feffili-   *tum.*
 bus, calyce fubcylindraceo fexfido.
 Dotted thick-leav'd Fig Marygold.
 Nat. of the Cape of Good Hope. Mr. *William
 Paterfon.*
 Introd. 1780, by the Countefs of Strathmore.
 Fl. November.   G. H. ♄.

53. M.

verueula-
tum.

53. M. foliis triquetro-cylindricis acutis connatis ar-
cuatis impunctatis diſtinctis. *Sp. pl.* 696.

Spit-leav'd Fig Marygold.

Nat. of the Cape of Good Hope.

Cult. 1732, by James Sherard, M. D. *Dill. elth.*
268. *t.* 203. *f.* 259.

Fl. May and June.   G. H. ♄.

malle.

54. M. foliis triquetris connatis erectis glaucis impunctatis, ramis femiteretibus, pedunculis axillaribus
compreſſis.

Soft Fig Marygold.

Nat. of the Cape of Good Hope. Mr. *Fr. Maſſon.*
Introd. 1774.

Fl.   G. H. ♄.

glaucum.

55. M. foliis triquetris acutis punctatis diſtinctis, calycinis foliolis ovato-cordatis. *Sp. pl.* 696.

Glaucous-leav'd Fig Marygold.

Nat. of the Cape of Good Hope.

Cult. 1696, in the Royal Garden at Hampton-court.
Catal. mſſ.

Fl. June and July.   G. H. ♄.

cornicu-
latum.

56. M. foliis triquetro-femicylindricis ſcabrido-puncta-
tis fupra bafin linea elevatis connatis. *Sp. pl.* 697.

α Mefembr. foliis corniculatis longioribus. *Dill. elth.*
262. *t.* 199. *f.* 253, 254.

Long-leav'd horn'd Fig Marygold.

β Mefembr. foliis corniculatis brevioribus. *Dill. elth.*
261. *t.* 198. *f.* 252.

Short-leav'd horn'd Fig Marygold.

Nat. of the Cape of Good Hope.

Cult. 1727. *Bradl. fucc.* 4. *p.* 18. *t.* 40.

Fl. March——May.   G. H. ♃.

57. M.

57. M. foliis planis oblongis pinnatifidis.　　　　　*pinnatifi-*
　　Mefembryanthemum pinnatifidum. *Linn. fuppl.* 260.　*dum.*
　　Pinnated Fig Marygold.
　　Nat. of the Cape of Good Hope.　Mr. *Fr. Maffon.*
　　Introd. 1774.
　　Fl. July and Auguft.　　　　　　　　　G. H. ☉.

58. M. foliis planis fpathulatis caulibufque papulofis,　*feffiliflo-*
　　ramis divaricatis, floribus feffilibus.　　　　　*rum.*
　　Seffile-flower'd Fig Marygold.
　　Nat. of the Cape of Good Hope.　Mr. *Fr. Maffon.*
　　Introd. 1774.
　　Fl. July.　　　　　　　　　　　　　　G. H. ☉.

59. M. foliis planiufculis oblongo-ovatis fubpapillofis　*tortuo-*
　　confertis connatis, calycibus triphyllis bicornibus.　*fum.*
　　Sp. pl. 697.
　　Twifted-leav'd Fig Marygold.
　　Nat. of the Cape of Good Hope.
　　Cult. 1705, by Dr. Uvedale.　*Pluk. amalth.* 90.
　　Fl. June——Octoberober.　　　　　　　　G. H. ♃.

60. M. foliis amplexicaulibus diftinctis fpathulatis gla-　*glabrum,*
　　berrimis, pedunculis longitudine foliorum, calyci-
　　bus hemifphæricis.
　　Smooth-leav'd Fig Marygold.
　　Nat. of the Cape of Good Hope.　Mr. *Fr. Maffon.*
　　Introd. 1787.
　　Fl. July.　　　　　　　　　　　　　　G. H. ☉.

61. M. foliis fpathulatis planis lævibus, pedunculis lon-　*Helian-*
　　giffimis, calycibus bafi planis angulatis.　　　*thoides.*
　　Spatula-leav'd Fig Marygold.
　　Nat. of the Cape of Good Hope.　Mr. *Fr. Maffon.*

Introd. 1774.

Fl. July and Auguft.   G. H. ⊙.

pomeri-   62. M. foliis planiufculis lato-lanceolatis lævibus fub-
dianum.   ciliatis diftinƈtis, caule pedunculis germinibufque
                  hirtis. *Sp. pl.* 698.

             Great Yellow-flower'd Fig Marygold.

             Nat. of the Cape of Good Hope.

             Introd. 1774, by Mr. Francis Maffon.

             Fl. July and Auguft.   G. H. ⊙.

echina-   63. M. foliis oblongo-ovatis fubtriquetris gibbis ramen-
tum.   taceo-hifpidis, laciniis calycinis foliiformibus.

             α flore luteo.

             Yellow echinated Fig Marygold.

             β flore albo.

             White echinated Fig Marygold.

             Nat. of the Cape of Good Hope.   Mr. *Fr. Maffon.*

             Introd. 1774.

             Fl. July——Oƈtober.   G. H. ♄.

ringens.   64. M. fubacaule, foliis ciliato-dentatis punƈtatis. *Sp.*
                  pl. 698.

caninum.   α Mefembryanthemum riƈtum caninum referens. *Dill.*
                  elth. 241. *t.* 188. *f.* 231.

             Dog's Chop Fig Marygold.

felinum.   β Mefembryanthemum riƈtum felinum repræfentans.
                  Dill. elth. 240. *t.* 187. *f.* 230.

             Cat's Chop Fig Marygold.

             Nat. of the Cape of Good Hope.

             Cult. 1717.   *Bradl. fucc.* 2. *p.* 8. *t.* 17.

             Fl. May——July.   G. H. ♄.

dolabri-   65. M. acaule, foliis dolabriformibus punƈtatis. *Syf.*
forme.   *veget.* 470. *Curtis magaz.* 32.

                                                         Hatchet-

Hatchet-leav'd Fig Marygold.
Nat. of the Cape of Good Hope.
Cult. 1714, by the Dutchefs of Beaufort. *Br. MuJ.*
H. S. 142. *fol.* 72.
Fl. May——July.   G. H. ♄.

66. M. acaule, foliis difformibus punctatis connatis.   *difforme.*
Syft. veget. 470.
Deform'd Fig Marygold.
Nat. of the Cape of Good Hope.
Cult. 1732, by James Sherard, M. D. *Dill. eith.*
252. *t.* 194. *f.* 241 & 242.
Fl. Auguft.   G. H. ♄.

67. M. acaule, foliis triquetris integerrimis. *Sp. pl.* 699.   *albidum.*
White Fig Marygold.
Nat. of the Cape of Good Hope.
Cult. 1714, by the Dutchefs of Beaufort. *Br. Muf.*
H. S. 131. *fol.* 23.
Fl. July and Auguft.   G. H. ♄.

68. M. acaule, foliis linguiformibus altero margine craf-   *lingui-*
fioribus impunctatis. *Sp. pl.* 699.   *forme.*
α Mefembr. folio fcalprato. *Dill. elth.* 235. *t.* 183.
f. 224.
Broad Tongue-leav'd Fig Marygold.
β Mefembr. folio linguiformi anguftiore. *Dill. elth.*
237. *t.* 184. *f.* 226.
Narrow Tongue-leav'd Fig Marygold.
γ Mefembr. folio linguiformi longiore. *Dill. elth.*
238. *t.* 185. *f.* 227.
Long Tongue leav'd Fig Marygold.
Nat. of the Cape of Good Hope.
Cuit. 1714, by the Dutchefs of Beaufort. *Br. Mrf.*
H. S. 131. *fol.* 22.
Fl. Auguft——October.   G. H. ♃.
          O 2   69. M.

pugioni-
forme.

69. M. foliis alternis confertis fubulatis triquetris longiffimis impunctatis. *Syft. veget.* 470.

Dagger-leav'd Fig Marygold.

Nat. of the Cape of Good Hope.

Cult. 1714, by the Dutchefs of Beaufort. *Br. Muf.*
H. S. 142. *fol.* 75.

Fl. May——Auguft.   G. H. ♃.

**** *Viridibus corollis.*

viridiflo-
rum.

70. M. foliis femicylindraceis papulofo-pilofis, calycibus
quinquefidis hirfutis.

Green-flowered Fig Marygold.

Nat. of the Cape of Good Hope. *Mr. Fr. Maffon.*
Introd. 1774.

Fl. July and Auguft.   G. H. ♃.

A I Z O O N. *Gen. pl.* 629.

Cal. 5-partitus. *Petala* o. *Capf.* fupera, 5-locularis,
5-valvis.

canari-
enfe.

1. A. foliis cuneiformi-ovatis, floribus feffilibus. *Sp. pl.*
700.

Purflane-leav'd Aizoon.

Nat. of the Canary Iflands.

Cult. 1731, by Mr. Philip Miller. *Mill. dict. edit.* 1.
Ficoides 41.

Fl. July and Auguft.   G. H. ☉.

Glinoides.

2. A. foliis fubrotundo-cuneiformibus pilofis, floribus
feffilibus, calycibus hirfutis.

Aizoon Glinoides. *Linn. fuppl.* 261.

Hairy Aizoon.

Nat. of the Cape of Good Hope. *Mr. Fr. Maffon.*
Introd. 1774.

Fl. June——Auguft.   G. H. ♃.

3. A.

3. A. foliis lanceolatis, floribus feffilibus. *Sp. pl.* 700.   *hifpani-*
Spanifh Aizoon.   *cum.*
Nat. of Spain and Africa.
Cult. 1728, by James Sherard, M. D. *Dill. elth.*
143. *t.* 117. *f.* 143.
Fl. July and Auguft.   G. H. ☉.

4. A. foliis lanceolatis, floribus paniculatis. *Syft. veget.*   *lanceola-*
471.   *tum.*
Aizoon paniculatum. *Sp. pl.* 700.
Panicled Aizoon.
Nat. of the Cape of Good Hope.
Cult. 1759, by Mr. Ph. Miller. *Mill. dict. edit.* 7. *n.* 3.
Fl. Auguft.   G. H. ♂.

S P I R Æ A.   *Gen. pl.* 630.

Cal. 5-fidus. *Petala* 5. *Capf.* polyfpermæ.

* *Fruticofæ.*

1. S. foliis lanceolatis integerrimis feffilibus, racemis com-   *lævigata.*
pofitis. *Syft. veget.* 471.
Spiræa altaica. *Pallas roff.* 1. *p.* 37. *t.* 23.
Smooth-lcav'd Spiræa.
Nat. of Siberia.
Introd. 1774, by Daniel Charles Solander, LL.D.
Fl. April——June.   H. ♄.

2. S. foliis oblongis ferratis glabris, racemis decompo-   *falicifo-*
fitis.   *lia.*

α foliis lanceolatis, racemis fubfpicatis, petalis carneis,   carnea.
cortice ramorum lutefcente.
Spiræa falicifolia. *Sp. pl.* 700. *Pallas roff.* 1. *p.* 36.
t. 21, 22.
Flefh-colour'd Willow-leav'd Spiræa.

                     β foliis

paniculata.

β foliis lanceolatis, racemis paniculatis divaricatis, petalis albis, cortice ramorum rubro.

Spiræa foliis lanceolatis acute ferratis, floribus paniculatis, caule fruticofo. *Mill. dict. edit.* 7. *n.* 8. *ic.* 171. *t.* 257. *f.* 2.

Panicled Willow-leav'd Spiræa.

latifolia.

γ foliis ovato-oblongis, racemis paniculatis, petalis albis, cortice ramorum rufefcente.

Broad Willow-leav'd Spiræa.

Nat. α. of Siberia ; β. γ. of North America.

Cult. 1665. *Rea's Flora,* 22.

Fl. June——Auguft.   H. ♄.

tomentofa.

3. S. foliis lanceolatis inæqualiter ferratis fubtus tomentofis, floribus duplicato-racemofis. *Sp. pl.* 701.

Scarlet Spiræa.

Nat. of Penfylvania.

Introd. 1736, by Peter Collinfon, Efq. *Coll. mff.*

Fl. Auguft and September.   H. ♄.

hypericifolia.

4. S. foliis obovatis integerrimis, umbellis feffilibus. *Sp. pl.* 701.

Hypericum Frutex, or Hypericum-leav'd Spiræa.

Nat. of Canada.

Cult. 1640. *Park. theat.* 573. *f.* 7.

Fl. April and May.   H. ♄.

crenata.

5. S. foliis oblongiufculis apice ferratis, corymbis lateralibus. *Sp. pl.* 701. *Pallas roff.* 1. *p.* 35. *t.* 19.

Hawthorn-leav'd Spiræa.

Nat. of Spain and Siberia.

Cult. 1739, by Mr. Philip Miller. *Rand. chel. n.* 4.

Fl. April and May.   H. ♄.

opulifolia.

6. S. foliis lobatis ferratis, corymbis terminalibus. *Sp. pl.* 702.

Virginian

Virginian Gilder-rofe, or Spiræa.
Nat. of Canada and Virginia.
Cult. 1713, by Bifhop Compton. *Philofoph. tranf.* 337.
p. 220. *n.* 158.
Fl. June and July.   H. ♄.

7. S. foliis pinnatis: foliolis uniformibus ferratis, caule   *forbifolia.*
   fruticofo, floribus paniculatis. *Sp. pl.* 702. *Pallas*
   rofs. 1. *p.* 38. *t.* 24, 25.
   Service-Tree-leav'd Spiræa.
   Nat. of Siberia.
   Cult. 1759, by Mr. Ph. Miller. *Mill. dict. edit.* 7. *n.* 6.
   Fl. Auguft.   H. ♄.

** *Herbaceæ.*

8. S. foliis fupradecompofitis, fpicis paniculatis, floribus   *Aruncus.*
   dioicis. *Sp. pl.* 702. *Pallas rofs.* 1. *p.* 39. *t.* 26.
   Goat's-beard Spiræa.
   Nat. of Auftria and Siberia.
   Cult. 1633, by Mr. John Tradefcant, Sen. *Ger. emac.*
   1043. *n.* 2.
   Fl. June and July.   H. ♃.

9. S. foliis interrupte pinnatis: foliolis lineari-lanceolatis   *Filipen-*
   interrupte ferratis glaberrimis, floribus cymofis.   *dula.*
   Syft. veget. 472.
   α floribus fimplicibus.
   Single-flower'd Dropwort, or Filipendula.
   β floribus plenis.
   Double-flower'd Dropwort.
   Nat. of Britain.
   Fl. June——October.   H. ♃.

10. S. foliis interrupte pinnatis: foliolis ovatis biferratis   *Ulmaria.*
    fubtus canis, floribus cymofis. *Syft. veget.* 472.
    Curtis lond.

                          O₄   α floribus

α floribus fimplicibus.

Single-flower'd Meadow-fweet.

β floribus plenis.

Double-flower'd Meadow-fweet.

Nat. of Britain.

Fl. June——September.   H. ♃.

lobata.   11. S. foliis pinnatis : impari lateralibufque lobatis biferratis, floribus cymofis. *Syft. veget. ed.* 14. *p.* 472.

Spiræa lobata. *Jacqu. hort.* 1. *p.* 38. *t.* 88.

Spiræa palmata. *Syft. veget. ed.* 13. *p.* 393. *Pallas
roff.* 1. *p.* 40. *t.* 27.

Lobe-leav'd Spiræa.

Nat. of Siberia.

Introd. 1765, by Meffrs. Kennedy and Lee.

Fl. July and Auguft.   H. ♃.

trifoliata.   12. S. foliis ternatis ferratis fubæqualibus, floribus fub-
paniculatis. *Sp. pl.* 702.

Three-leav'd Spiræa.

Nat. of North America.

Cult. 1758. *Mill. ic.* 171. *t.* 256.

Fl. June and July.   H. ♃.

POLYGYNIA.

R O S A. *Gen. pl.* 631.

Petala 5. *Cal.* urceolatus, 5-fidus, carnofus, collo
coarctatus. *Sem.* plurima, hifpida, calycis interiori
lateri affixa.

* *Germinibus fubglobofis.*

lutea.   1. R. germinibus globofis pedunculifque glabris, calycibus petiolifque fpinulofis, aculeis ramorum rectis.

α Rofa

α Rofa eglanteria. *Sp. pl.* 703. (exclufis fynonymis.)
Leyf. hal. 121.

Rofa lutea. *Mill. dict. Du Roi hort. harbecc.* 2. *p.* 344.

Rofa lutea fimplex. *Bauh. pin.* 483.

Rofa lutea flore fimplici. *Befl. eyft. vern.* 6. *t.* 5. *f.* 1.
Weinm. phytanth. 4. *p.* 232. *tab.* 870. *fig. C.*

Rofa lutea. *Dalech. hift.* 126. *cum fig. Lob. ic.* 2.
p. 209. *Tabern. hift.* 1495. *cum fig. Bauh. hift.* 2.
p. 47. *cum fig.*

Rofa fœtida. *Allion. pedem.* 1792.

Single Yellow Rofe.

β Rofa bicolor. *Jacqu. hort.* 1. *p.* 1. *t.* 1. conf. *vol.* 3.
p. 1.

Rofa fylveftris auftriaca flore phœniceo. *Park. theatr.*
1019. *n.* 6. *cum fig. in pag.* 1018. *Hort. angl. p.* 66.
tab. 18,

Rofa graveolens fimplex flore extus luteo intus rubro
holoferico. *Weinm. phytanth.* 4. *p.* 231. *t.* 868. *fig. C.*

Rofa punicea. *Corn. canad.* 11. *Mill. dict. Du Roi
hort. harbecc.* 2. *p.* 347.

Red and Yellow Auftrian Rofe.

Nat. of Germany and Italy.

Cult. 1596, by Mr. John Gerard. *Hort. Ger.*

Fl. June. H. ♄.

2. R. germinibus globofis, petiolis cauleque aculeatis: *fulphu-*
aculeis caulinis duplicibus majoribus minoribufque *rea.*
numerofis, foliis ovalibus.

Rofa lutea multiplex. *Bauh. pin.* 483. *Hort. angl.*
66. *t.* 18.

Rofa flava pleno flore. *Cluf. cur. poft.* 6. *fig.*

Double Yellow Rofe.

Nat. of the Levant.

Introd. before 1629, by Mr. John de Franqueville.
Park. parad. 420.

Fl. July. H. ♄.

3. R.

blanda.   3. R. germinibus globosis glabris, caulibus adultis pe-
            dunculisque lævibus inermibus.
            Hudson's Bay Rose.
            Nat. of Newfoundland and Hudson's-bay.
            Cult. 1773, by Mr. James Gordon.
            Fl. May——August.   H. ♄.
            DESCR. *Caules* adulti læves, inermes; juniores seu
            primi anni aculeis rectis subreflexis tenuibus arma-
            ti. *Rami* teretes, inermes, nitidi, rubicundi. *Folia*
            pinnata: *foliola* plerumque septem, oblonga, argute
            et subæqualiter serrata, glabra. *Petioli* glabri, ple-
            rumque una alterave spinula armati.

cinnamo- 4. R. germinibus globosis pedunculisque glabris, caule
mea.   aculeis stipularibus, petiolis subinermibus. *Sp. pl.*
            703.
            α Single Cinnamon Rose.
            β Double Cinnamon Rose.
            Nat. of the South of Europe.
            Cult. 1596, by Mr. John Gerard. *Hort. Ger.*
            Fl. May.   H. ♄.

arvensis. 5. R. germinibus globosis pedunculisque glabris, caule
            petiolisque aculeatis, floribus cymosis. *Linn. mant.*
            245.
            White Dog Rose.
            Nat. of Britain.
            Fl. June and July.   H. ♄.

pimpinel- 6. R. germinibus globosis pedunculisque glabris, caule
lifolia.   aculeis sparsis rectis, petiolis scabris, foliolis obtusis.
            Sp. pl. 703.
            Small Burnet-leav'd Rose.
            Nat. of the South of Europe.
            Fl. May and June.   H. ♄.
                                                            7. R.

7. R. germinibus globofis glabris, pedunculis hifpidis, *fpinofiffi-*
caule petiolifque aculeatiffimis. *Syft. veget.* 473. *ma.*

α Common Scotch Rofe.

β Rofa pimpinellæ foliis minor noftras flore eleganter
variegato. *Sutherl. hort. edin.* 298.
Strip'd-flower'd Scotch Rofe.

γ Red Scotch Rofe.
Nat. of Britain.
Fl. June and July.   H. ♄ .

8. R. germinibus globofis pedunculifque hifpidis, petio- *carolina,*
lis aculeatis, caule glabro aculeis ftipularibus, foliis
glabris.
Rofa carolina. *Sp. pl.* 703.

α Great fingle Burnet-leav'd Rofe.

β Great double Burnet-leav'd Rofe.

γ Single Penfylvanian Rofe.

δ Double Penfylvanian Rofe.

ε Spreading Carolina Rofe.

ζ Upright Carolina Rofe.
Nat. of North America.
Cult. 1726, by James Sherard, M.D. *Dill. elth.* 325.
t. 245. *f.* 316.
Fl. June and July.   H. ♄ .

9. R. germinibus globofis pedunculifque hifpidis, caule *villofa.*
aculeis fparfis, petiolis aculeatis, foliis tomentofis.
Syft. veget. 474.

α Single Apple Rofe.

β Double Apple Rofe.
Nat. of Britain.
Fl. June and July.   H. ♄ .

10. R. germinibus fubglobofis glabris, pedunculis aculea- *finica,*
tis hifpidis, caule petiolifque aculeatis, calycinis foliolis lanceolatis fubpetiolatis. *Syft. veget.* 474.
                                             Chinefe

Chinese Rose.
Nat. of China.
Cult. 1759, by Mr. Philip Miller.
Fl.   G. H. ♄.

provin-
cialis.
11. R. germinibus subrotundis pedunculis petiolisque
hispidis, aculeis ramorum sparsis rectis subreflexis,
foliolis ovatis subtus villosis : serraturis glandulosis.

Rosa provincialis. *Mill. dict. Du Roi hort harbecc.* 2.
p. 349.
α Common Provence Rose.
β Red Provence Rose.
γ Blush Provence Rose.
δ White Provence Rose.
ε Rose de meaux, or great dwarf Rose.
ζ Rose de meaux, or small dwarf Rose.
Nat. of Spain and Italy.
Cult. 1596, by Mr. John Gerard. *Hort. Ger.*
Fl. June——August.   H. ♄.

** *Germinibus ovatis.*

centifolia.
12. R. germinibus ovatis pedunculisque hispidis, caule
hispido aculeato, petiolis inermibus. *Sp. pl.* 704.
α Dutch Hundred-leav'd Rose.
β Blush Hundred-leav'd Rose.
γ Singleton's Hundred-leav'd Rose.
δ Burgundy Rose.
ε Single Velvet Rose.
ζ Double Velvet Rose.
η Sultan Rose.
θ Stepney Rose.
ι Garnet Rose.
κ Bishop Rose.
λ Lisbon Rose.

Nat.

Nat.
Cult. 1596, by Mr. John Gerard. *Hort. Ger.*
Fl. June and July.   H. ♄.

13. R. germinibus ovatis pedunculifque hifpidis, caule *gallica.*
    petiolifque hifpido-aculeatis. *Sp. pl.* 704.
α Red officinal Rofe.
β Rofa præneftina variegata plena. *Mill. ic.* 148. verfico-
   t. 221. *f.* 2.   lor.
   Mundi Rofe.
γ Marbled Rofe.
δ Virgin Rofe.
Nat. of the South of Europe.
Cult. 1596, by Mr. John Gerard. *Hort. Ger.*
Fl. June and July.   H. ♄.

14. R. calycibus femipinnatis, germinibus ovatis turgi- *damafce-*
    dis pedunculifque hifpidis, caule petiolifque acu-   *na.*
    leatis, foliolis ovatis acuminatis fubtus villofis.
    Du Roi hort. harbecc. 2. *p.* 369.
    Rofa damafcena. *Mill. dict.*
α Red Damafk Rofe.
β Blufh Damafk Rofe.
γ York and Lancafter Rofe.
δ Red Monthly Rofe.
ε White Monthly Rofe.
ζ Blufh Belgick Rofe.
η Great Royal Rofe.
Nat. of the South of France.
Cult. 1596, by Mr. John Gerard. *Hort. Ger.*
Fl. June and July.   H. ♄,

15. R. germinibus ovatis calycibus pedunculifque hifpi- *fempervi-*
    dis, caule petiolifque aculeatis, floribus fubumbel- *rens.*
    latis, bracteis lanceolatis reflexis.
                                          Rofa

Rofa fempervirem.   *Sp. pl.* 704.
Rofa fcandens.   *Mill. dict.*
Evergreen Rofe.
Nat. of Germany.
Cult. 1629.   *Park. parad.* 420.   *n.* 24.
Fl. June.   H. ♃.

pumila.   16. R. germinibus ovatis pedunculifque hifpidis, petiolis
                cauleque aculeato, foliis fubtus glaucis: ferraturis
                glandulofis, fructibus pyriformibus.
            Rofa pumila.   *Jacqu. auftr.* 2. *p.* 59. *t.* 198.   *Linn.*
                fuppl. 262.   *Allion. pedem.* 1802.
            Rofa fylveftris pumila rubens.   *Bauh. pin.* 483.
            Rofa VI. pumila.   *Cluf. hift.* 1. *p.* 117.
            Dwarf Auftrian Rofe.
            Nat. of Auftria and Italy.
            Introd. about 1773, by Meffrs. Kennedy and Lee.
            Fl. June and July.   H. ♄.

turbina-   17. R. germinibus turbinatis pedunculifque pilofis, pe-
ta.   tiolis villofis, aculeis fparfis recurvis.
            Frankfort Rofe.
            Nat.
            Cult. 1629.   *Park. parad.* 414. *n.* 11.
            Fl. June.   H. ♄.

rubigino-   18. R. germinibus ovatis pecunculifque hifpidis, petiolis
fa.   cauleque aculeatis : aculeis recurvis, foliolis ovatis
                fubtus glandulofo-pilofis.
            Rofa rubiginofa.   *Linn. mant.* 564.   *Jacqu. auftr.* 1.
                p. 31. *t.* 50.
            Rofa eglanteria.   *Mill. dict.   Du Roi bort. harbecc.* 2.
                p. 336.   *Hudf. angl* 218.
            Rofa fuavifolia.   *Lightf. fcot.* 262.   *Walc. brit.*
        α Common Sweet-brier Rofe.
                                        β Common

ε Common Double Sweet-brier Rofe.

γ Moffy Double Sweet-brier Rofe.

δ Evergreen Double Sweet-brier Rofe.

ε Marbled Double Sweet-brier Rofe.

ζ Red Double Sweet-brier Rofe.

Nat. of Britain.

Fl. May and June. H. ♄.

19. R. germinibus ovatis calycibus pedunculis petiolis *mufcofa-* ramulifque hifpidis glandulofo-vifcofis, fpinis ra- morum fparfis rectis.

Rofa mufcofa. *Mill. dict. Du Roi hort. barbecc.* 2. *p.* 368.

Rofa rubra plena, fpinofiffima, pedunculo mufcofo. *Mill. ic.* 148. *t.* 221. *f.* 1.

Rofa provincialis fpinofiffima pedunculo mufcofo. *Hort. angl. p.* 66. *n.* 14. *tab.* 18.

Mofs Provence Rofe.

Nat.

Cult. 1724. *Furber's catal.*

Fl. June and July. H. ♄.

20. R. germinibus ovatis pedunculifque villofis, caule *mofchata.* petiolifque aculeatis, foliolis oblongis acuminatis glabris, paniculis multifloris.

Rofa mofchata. *Mill. dict. Du Roi hort. barbecc.* 2. *p.* 365.

Rofa mofchata minor flore fimplici. *Bauh. hift.* 2. *p.* 45. *fig.*

α Single Mufk Rofe.

β Double Mufk Rofe.

Nat.

Cult. 1596, by Mr. John Gerard. *Hort. Ger.*

Fl. July——October. H. ♄.

21. R.

alpina. 21. R. germinibus ovatis glabris, pedunculis petiolifque
hifpidis, caule inermi. *Syfl. véget.* 474. *Jacqu.*
auflr. 3. *p.* 43. *t.* 279.
Alpine Rofe.
Nat. of the Alps of Switzerland and Auftria.
Cult. 1683, by Mr. James Sutherland. *Sutherl. hort.*
edin. 297. *n.* 4.
Fl. June and July.   H. ♄.

canina. 22. R. germinibus ovatis pedunculifque glabris, caule
petiolifque aculeatis. *Sp. pl.* 704. *Curtis lond.*
Dog Rofe, or Hip Tree.
Nat. of Britain.
Fl. June——Auguft.   H. ♄.

penduli- 23. R. inermis, germinibus oblongis, pedunculis petio-
na.   lifque hifpidis, caule ramifque glabris, fructibus
pendulis.
Rofa pendulina. *Sp. pl.* 705.
Smooth pendulous Rofe.
Nat. of North America.
Cult. 1726, by James Sherard, M.D. *Dill. elth.* 325.
t. 245. *f.* 317.
Fl. May and June.   H. ♄.

alba. 24. R. germinibus ovatis glabris, pedunculis hifpidis,
caule petiolifque aculeatis. *Sp. pl.* 705.
α Single White Rofe.
β Double White Rofe.
γ Small Maiden's-blufh Rofe.
δ Great Maiden's-blufh Rofe.
Nat. of Europe.
Cult. 1597. *Ger. herb.* 1079. *f.* 1.
Fl. June and July.   H. ♄.

RUBUS.

R U B U S. *Gen. pl.* 632.

Cal. 5-fidus. *Petala* 5. *Bacca* compofita: acinis monofpermis.

* *Frutefcentes.*

1. R. foliis quinato-pinnatis ternatifque, caule aculeato, *idæus.*
petiolis canaliculatis. *Sp. pl.* 706.

α Rubus idæus fpinofus. *Bauh. pin.* 479. ruber.
Red Rafpberry.

β Rubus idæus fructu albo. *Bauh. pin.* 479. albus.
White Rafpberry.

γ Rubus idæus lævis. *Bauh. pin.* 479. læviga-
Smooth Rafpberry. tus.
Nat. of Britain.
Fl. May and June. H. ♄.

2. R. foliis ternatis fubtus tomentofis, caule aculeato, *occidenta-*
petiolis teretibus. *Sp. pl.* 706. *lis.*
Virginian Rafpberry.
Nat. of North America.
Cult. 1696, in Chelfea Garden. *Pluk. alm.* 325.
Fl. May and June. H. ♄.

3. R. foliis ternatis nudis, caulibus petiolifque hifpidiffi- *hifpidus.*
mis ftrigis rigidulis. *Syft. veget.* 475.
Briftly Bramble.
Nat. of Canada.
Cult. 1768, by Mr. Philip Miller.
Fl. Auguft. H. ♄.

4. R. foliis ternatis fubnudis: lateralibus bilobis, caule *cæfius.*
tereti aculeato. *Sp. pl.* 706.
Dewberry Bramble.
Nat. of Britain.
Fl. June and July. H. ♄.

fruticofus. 5. R. foliis quinato-digitatis ternatifque, caule petiolifque aculeatis. *Sp. pl.* 707.

niger.   α Rubus vulgaris, five rubus fructu nigro. *Bauh. pin.*
479.
Black-fruited common Bramble.

albus.   β Rubus vulgaris major, fructu albo. *Raj. fyn.* 467.
White-fruited common Bramble.

plenus.   γ Rubus flore albo pleno. *Magn. hort.* 175.
Double-flower'd Bramble.

inermis. δ caule petiolifque inermibus.
Smooth Bramble.
Nat. of Britain.
Fl. June——September.   H. ♄.

villofus. 6. R. foliis quinatis ellipticis acuminatis argute ferratis
utrinque villofis, caulibus petiolifque aculeatis.
Hairy Bramble.
Nat. of North America.
Fl. July.   H. ♄.

odoratus. 7. R. foliis fimplicibus palmatis, caule inermi multifolio
multifloro. *Sp. pl.* 707.
Flowering Rafpberry.
Nat. of North America.
Cult. 1739, by Mr. Ph. Miller. *Rand. chel. n.* 7.
Fl. June and July.   H. ♄.

** *Herbacei.*

faxatilis. 8. R. foliis ternatis nudis, flagellis reptantibus herba-
ceis. *Syft. veget.* 476.
Stone Bramble.
Nat. of Britain.
Fl. June.   H. ♄.

arcticus. 9. R. foliis ternatis, caule inermi unifloro. *Sp. pl.* 708.
Dwarf

Dwarf Bramble.
Nat. of the North of Europe, Afia, and America.
Cult. 1759, by Mr. Ph. Millet. *Mill. dict. edit.* 7. *n.* 9.
Fl. June and July.　　　　　　　　　H. ♃.

10. R. foliis fimplicibus lobatis, caule inermi unifloro. *Chamæ-*
Sp. pl. 708.　　　　　　　　　　　　　　*morus.*
Mountain Bramble, or Clowdberry.
Nat. of Britain.
Fl. May and June.　　　　　　　　　H. ♃.

11. R. foliis fimplicibus cordatis indivifis crenatis, fcapo *Dalibar-*
aphyllo unifloro. *Sp. pl.* 708.　　　　　*da.*
Simple-leav'd Bramble.
Nat. of Canada.
Cult. 1768, by Mr. Philip Miller. *Mill. dict. edit.* 8.
Fl.　　　　　　　　　　　　　　　　H. ♃.

FRAGARIA. *Gen. pl.* 633.

Cal. 10-fidus. *Petala* 5. *Receptaculum* feminum ovatum, baccatum, deciduum.

1. F. flagellis reptantibus. *Syft. veget.* 476.　　*vefca.*
α Fragaria vulgaris. *Bauh. pin.* 326.　　　　　fylveftris.
Le Fraifier de Bois. *Duchefne fraif. p.* 61.
Wood Strawberry.
β F. fructu parvi pruni magnitudine. *Bauh. pin.* 327.　pratenfis.
Le Capiton. *Duchefne fraif. p.* 145.
Hautboy Strawberry.
γ F. chiloenfis fructu maximo, foliis carnofis hirfutis.　chiloen-
Dill. elth. 145. *t.* 120. *f.* 146.　　　　　fis.
Le Frutiller. *Duchefne fraif. p.* 165.
Chili Strawberry.
δ F. foliis oblongo-ovatis ferratis inferne incanis, caly-　virginia-
cibus longioribus, fructu fubrotundo. *Mill. dict.*　na.
Le Fraifier écarlate. *Duchefne fraif. p.* 204.
　　　　　　　　　　　　　Scarlet

Scarlet or Virginian Strawberry.

ananas. ε F. foliis ovatis crenatis nervofis, calycibus maximis.
Mill. ic. 192. *tab.* 288.
Le Fraifier-ananas. *Duchefne fraif. p.* 190.
Pine Strawberry.
Nat. α. β. of Britain; γ. of South America; δ. of
Carolina and Virginia ; and ε. of Surinam.
Fl. April and May.   H. ♃.

monophyl-   2. F. foliis fimplicibus. *Syft. veget.* 476. *Curtis magaz.*
la.   63.
Le Fraifier de Verfailles. *Duchefne fraif. p.* 124.
Simple-leav'd Strawberry.
Nat.
Introd. about 1773.
Fl. June.   H. ♃.

fterilis.   3. F. caule decumbente, ramis floriferis laxis. *Syft.*
veget. 476. *Curtis lond.*
Barren Strawberry.
Nat. of Britain.
Fl. May.   H. ♃.

POTENTILLA. *Gen. pl.* 634.

Cal. 10-fidus. *Petala* 5. *Sem.* fubrotunda, nuda, receptaculo parvo exfucco affixa.

* *Foliis pinnatis.*

fruticofa.   1. P. foliis pinnatis, caule fruticofo. *Sp. pl.* 709.
Shrubby Cinquefoil.
Nat. of England.
Fl. June——Auguft.   H. ♄.

2. P.

2. P. foliis pinnatis ferratis, caule repente, pedunculis *Anferina.*
   unifloris. *Syft. veget.* 477. *Curtis lond.*
   Silvery Cinquefoil, or Wild Tanfy.
   Nat. of Britain.
   Fl. May——September.   H. ♃.

3. P. foliis bipinnatis utrinque tomentofis : fegmentis *fericea.*
   parallelis approximatis, caulibus decumbentibus.
   Sp. pl. 710.
   Silky Cinquefoil.
   Nat. of Siberia.
   Introd. 1780, by Peter Simon Pallas, M. D.
   Fl. May and June.   H. ♃.

4. P. foliis bipinnatis : fegmentis integerrimis diftanti- *multifida.*
   bus fubtus tomentofis, caule decumbente. *Sp. pl.*
   710.
   Multifid Cinquefoil.
   Nat. of Siberia.
   Cult. 1759, by Mr. Philip Miller.
   Fl. May and June.   H. ♃.

5. P. foliis pinnatis ternatifque : extimis majoribus, fla- *fragar-*
   gellis reptantibus. *Sp. pl.* 710.   *ioides.*
   Strawberry-leav'd Cinquefoil.
   Nat. of Siberia.
   Introd. 1773, by Chevalier Murray.
   Fl. May and June.   H. ♃.

6. P. foliis pinnatis alternis : foliolis quinis ovatis cre- *rupeftris,*
   natis, caule erecto. *Sp. pl.* 711. *Jacqu. auftr.* 2.
   p. 9. *t.* 114.
   Rock Cinquefoil.
   Nat. of England.
   Fl. May——September.   H. ♃.

7. P.

bifurca.   7. P. foliis pinnatis subæqualibus : foliolis oblongis sub-
bifidis : extimis confluentibus. *Sp. pl.* 711.
Bifid-leav'd Cinquefoil.
Nat. of Siberia.
Introd. 1773, by John Earl of Bute.
Fl.   H. ♃.

pimpinel-   8. P. foliis pinnatis : foliolis subrotundis dentatis æqua-
loides.   libus, caule erecto. *Sp. pl.* 711.
Burnet-leav'd Cinquefoil.
Nat. of the Levant.
Cult. 1758, by Mr. Philip Miller.
Fl. June——Auguft.   H. ♃.

penfylva-   9. P. foliis inferioribus pinnatis ; fuperioribus ternatis :
nica.   foliolis incifo-ferratis, caule erecto pubefcente. *Linn.*
mant. 76. *Jacqu. hort.* 2. *p.* 89. *t.* 189.
Agrimony-leav'd Cinquefoil.
Nat. of North America.
Introd. 1773, by John Earl of Bute.
Fl. June——Auguft.   H. ♃.

fupina.   10. P. foliis pinnatis, caule dichotomo decumbente. *Sp.*
pl. 711. *Jacqu. auftr.* 5. *p.* 3. *t.* 406.
Trailing Cinquefoil.
Nat. of Siberia and Germany.
Introd. 1773, by Chevalier Murray.
Fl. July.   H. ♃.

*** Foliis digitatis.*

recta.   11. P. foliis feptenatis lanceolatis ferratis utrinque fubpilofis, caule erecto. *Sp. pl.* 711. *Jacqu. auftr.* 4.
p. 43. *t.* 383.
Upright Cinquefoil.
Nat. of the South of Europe.

                                                   Cult.

Cult. 1648, in Oxford Garden. *Hort. oxon. edit.* 1.
p. 39.
Fl. June and July. H. ♃.

12. P. foliis quinatis cuneiformibus incifis fubtus tomen- *argentea.*
tofis, caule erecto. *Sp. pl.* 712.
Sattin Cinquefoil.
Nat. of Britain.
Fl. June——October. H. ♃.

13. P. foliis feptenatis quinatifque cuneiformibus incifis *hirta.*
pilofis, caule erecto hirto. *Syft. veget.* 477.
Hairy Cinquefoil.
Nat. of the South of France and the Pyrenees.
Cult. 1725, in Chelfea Garden. *R. S. n.* 184.
Fl. May——September. H. ♃.

14. P. foliis radicalibus quinatis acute ferratis retufis; *verna.*
caulinis ternatis, caule declinato. *Sp. pl.* 712.
Spring Cinquefoil.
Nat. of Britain.
Fl. March——May. H. ♃.

15. P. foliis radicalibus quinatis ferratis acuminatis; *aurea.*
caulinis ternatis, caule declinato. *Sp. pl.* 712.
Golden Cinquefoil.
Nat. of the Alps of Switzerland.
Cult. 1739, by Mr. Philip Miller. *Rand. chel.*
Quinquefolium 2.
Fl. May——July. H. ♃.

16. P. foliis quinatis apice conniventi-ferratis, caulibus *alba.*
filiformibus procumbentibus, receptaculis hirfutis.
Sp. pl. 713. *Jacqu. auftr.* 2. *p.* 10. *t.* 115.
White Cinquefoil.

Nat.

Nat. of Wales.
Fl. moſt part of the Summer.   H. ♃.

caulef- 17. P. foliis quinatis apice conniventi-ſerratis, caulibus
cens.   multifloris decumbentibus, receptaculis hirſutis.
               Syſt. veget. 478. *Jacqu. auſtr.* 3. *p.* 11. *t.* 220.
               Alpine Cinquefoil.
               Nat. of Switzerland and Auſtria.
               Cult. 1759, by Mr. Ph. Miller. *Mill. dict. edit.* 7. *n.* 6.
               Fl. May and June.   H. ♃.

reptans. 18. P. foliis quinatis, caule repente, pedunculis unifloris.
               Syſt. veget. 4⁻9. *Curtis lond.*
               Common Cinquefoil.
               Nat. of Britain.
               Fl. Auguſt and September.   H. ♃.

                    *** *Foliis ternatis.*

monfpe- 19. P. foliis ternatis, caule ramoſo erecto, pedunculis
lienfis.   ſupra genicula enatis. *Sp. pl.* 714.
               Montpelier Cinquefoil.
               Nat. of the South of France.
               Cult. 1759, by Mr. Ph. Miller. *Mill. dict. edit.* 7.
               n. 7.
               Fl. July and Auguſt.   H. ♃.

tridenta- 20. P. foliis ternatis cuneiformibus apice trifidis. TAB. 9.
ta.   Trifid-leav'd Cinquefoil.
               Nat. of Newfoundland.
               Introd. 1776, by Benjamin Bewick, Eſq.
               Fl. June.   H. ♃.

grandi- 21. P. foliis ternatis dentatis utrinque ſubpiloſis, caule
flora.   decumbente foliis longiore. *Sp. pl.* 715.
               Great-flower'd Cinquefoil.

                                                            Nat.

Tab. 9. Vol. 2. Page 216.

Chret. del.

Potentilla tridentata.

M. Kenzie. sc.

Nat. of Siberia and Switzerland.

Cult. 1768, by Mr. Philip Miller. *Mill. dict. edit.* 8.

Fl. June and July. H. ♃.

TORMENTILLA. *Gen. pl.* 635.

Cal. 8-fidus. *Petala* 4. *Sem.* fubrotunda, nuda, receptaculo parvo exfucco affixa.

1. T. caule erectiufculo, foliis feffilibus. *Sp. pl.* 716.   *erecta.*
Tormentilla officinalis. *Curtis lond.*
Common Tormentil, or Upright Septfoil.
Nat. of Britain.
Fl. May——October. H. ♃.

G E U M. *Gen. pl.* 636.

Cal. 10-fidus : laciniæ alternæ minores. *Petala* 5.
Sem. ariftata.

1. G. floribus erectis, ariftis uncinatis nudis, foliis cauli-   *virginia-*
nis ternatis : fuperioribus lanceolatis, petalis calyce   *num.*
brevioribus.
Geum virginianum. *Sp. pl.* 716.
Geum canadenfe. *Jacqu. hort.* 2. *p.* 82. *t.* 175.
American Avens.
Nat. of North America.
Cult. 1739, by Mr. Philip Miller. *Rand chel.* Cary-
ophyllata 5.
Fl. July and Auguft. H. ♃.

2. G. floribus erectis, ariftis uncinatis nudis, foliis cau-   *strictum.*
linis pinnatis : foliolis ftipulifque incifo-fiffis, petalis calyce longioribus.
Geum canadenfe. *Murray nov. comm. gotting.* 5.
p. 33. *tab.* 4. *B.*
Geum aleppicum. *Jacqu. ic. collect.* 1. *p.* 88.

Upright

Upright Avens.
Nat. of North America.
Introd. 1778, by Monf. Thouin.
Fl. May and June.   H. ♃.

urbanum.   3. G. floribus erectis, aristis uncinatis nudis, foliis cau-
linis ternatis; radicalibus lyrato-pinnatis.
Geum urbanum. *Sp. pl.* 716. *Curtis lond.*
α Caryophyllata vulgaris. *Bauh. pin.* 321.
Common Avens, or Herb-bennet.
β Caryophyllata vulgaris majore flore. *Bauh. pin.* 321.
Great-flower'd Common Avens.
Nat. of Britain.
Fl. June——Auguft.   H. ♃.

rivale.   4. G. floribus nutantibus, aristis uncinatis villofis, peta-
lis retufis fubrotundo-cuneiformibus, foliis pinnatis,
Geum rivale. *Sp. pl.* 717.
Water Avens.
Nat. of Britain.
Fl. July.   H. ♃.

monta-   5. G. caule unifloro, aristis rectis villofis, foliis pinnatis
num.   pilofis : foliolo extimo maximo fubrotundo ; inferioribus fenfim minoribus.
Geum montanum. *Sp. pl.* 717. *Jacqu. auftr.* 4.
p. 38. *t.* 373.
α Caryophyllata alpina lutea. *Bauh. pin.* 322.
Great Mountain Avens.
β Caryophyllata alpina minor. *Bauh. pin.* 322. *prodr.*
139.
Small Mountain Avens.
Nat. of the Alps of Auftria and Switzerland.
Cult. 1597. *Ger. herb.* 842. *f.* 2.
Fl. May——September.   H. ♃.
6. G.

6. G. caule fubbifloro, ariftis rectis nudis, calycibus *potentil-*
fructus erectis, foliis pinnatis dentatis. *loides.*
Geum Laxmanni. *Gærtn. fem.* 1. *p.* 352. *tab.* 74.
Dryas geoides. *Pallas it.* 3. *p.* 732. *tab.* Y. *f.* 1.
Jacqu. hort. 3. *p.* 38. *tab.* 68. *Falck ruff. tab.* 11.
Caryophyllata potentilloides. *De Lamarck encycl.* 1.
p. 400.
Siberian Avens.
Nat. of Siberia.
Introd. 1780, by Peter Simon Pallas, M. D.
Fl. June. H. ♃.

7. G. caulibus unifloris, ariftis rectis villofis, foliis pin- *reptans.*
natis incifis pilofis, flagellis reptantibus.
Geum reptans. *Sp. pl.* 717. *Jacqu. auftr.* 5. *p.* 38.
tab. app. 22.
Creeping Avens.
Nat. of Switzerland.
Introd. 1775, by the Doctors Pitcairn and Fothergill.
Fl. H. ♃.

D R Y A S. *Gen. pl.* 637.

Cal. 8-partitus, æqualis. *Petala* 8. *Sem.* ariftata.

1. D. floribus octopetalis, foliis fimplicibus. *Sp. pl.* *octopeta-*
717. *la.*
Mountain Dryas.
Nat. of Scotland.
Fl. June and July. H. ♃.

C O M A R U M. *Gen. pl.* 638.

Cal. 10-fidus. *Petala* 5, calyce minora. *Recept.* feminum ovatum, fpongiofum, perfiftens.

1. COMARUM. *Sp. pl.* 718. *paluftre.*
VOL. II. P 6 Marfh

Marſh Comarum, or Cinquefoil,
Nat. of Britain.
Fl. June.　　　　　　　　　　　　　　　　　H. ♃.

C A L Y C A N T H U S.　*Gen. pl.* 639.

Cal. 1-phyllus, urceolatus, ſquarroſus : foliolis colo‑
ratis. *Cor.* calycina. *Styli* plurimi, ſtigmate glan‑
duloſo. *Sem.* plurima, caudata, intra calycem ſuc‑
culentum.

floridus.　　1. C. petalis interioribus longioribus,　*Sp. pl.* 718.
oblonga‑　　α foliis oblongis.
tus.　　　　Long-leav'd Carolina All-ſpice,
ovatus.　　β foliis ſubrotundo‑ovatis.
　　　　　　Round-leav'd Carolina All-ſpice.
　　　　　　Nat. of Carolina.
　　　　　　Introd. 1726, by Mr. Mark Cateſby.　*Ehret. pict.*
　　　　　　　tab. 13.
　　　　　　Fl. May——Auguſt,　　　　　　　　　H. ♄.

præcox.　　2. C. petalis interioribus minutis. *Sp. pl.* 718. TAB. 10,
　　　　　　Japan All-ſpice.
　　　　　　Nat. of Japan.
　　　　　　Introd. 1771, by Benjamin Torin, Eſq.
　　　　　　Fl. December and January.　　　　　G. H. ♄.

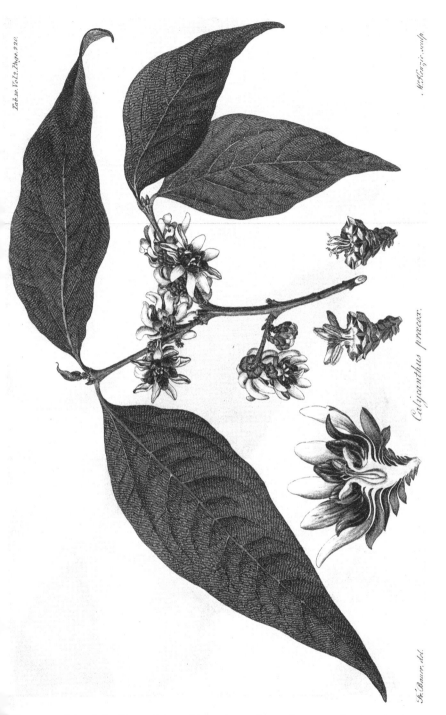

Claſſis XIII.

POLYANDRIA

MONOGYNIA.

CAPPARIS. *Gen. pl.* 643.

Cal. 4-phyllus, coriaceus. *Petala* 4. *Stam.* longa.
Bacca corticoſa, unilocularis, pedunculata.

1. C. pedunculis ſolitariis unifloris, ſtipulis ſpinoſis, foliis   *ſpinoſa.*
annuis, capſulis ovalibus. *Sp. pl.* 720.
Prickly Caper-buſh.
Nat. of Italy and the Levant.
Cult. 1596, by Mr. John Gerard. *Hort. Ger.*
Fl. May and June.   G. H. ♄.

ACTÆA, *Gen. pl.* 644.

Cor. 4-petala. *Cal.* 4-phyllus. *Bacca* 1-locularis.
Sem. ſemiorbiculata.

1. A. racemo ovato, fructibus baccatis. *Sp. pl.* 722.   *ſpicata.*
α baccis nigris.   nigra.
Common black-berried Herb-chriſtopher.
β baccis niveis.   alba.
White-berried Herb-chriſtopher.
γ baccis rubris.   rubra.
Red-berried Herb-chriſtopher.
Nat. α. of Britain; β. and γ. of North America.
Fl. April and May.   H. ♃.

2. A. racemis longiſſimis, fructibus ſiccis. *Syſt. veget.* *racemoſa.*
488.

                         American

American Herb-chriſtopher, or Black Snake-root.
Nat. of North America.
Cult. 1732, by James Sherard, M. D. *Dill. elth.* 79.
t. 67. *f.* 78.
Fl. July and Auguſt. H. ♃.

SANGUINARIA. *Gen. pl.* 645.

Cor. 8-petala. *Cal.* 2-phyllus. *Siliqua* ovata, 1-locularis.

canaden-
ſis.

1. SANGUINARIA. *Sp. pl.* 723.
Canadian Sanguinaria, or Puccoon.
Nat. of North America.
Cult. 1680, by Mr. William Walker. *Moriſ. hiſt.* 2.
p. 257. *n.* 1. *ſ.* 3. *t.* 11. *f.* 1.
Fl. March and April. H. ♃.

PODOPHYLLUM. *Gen. pl.* 646.

Cor. 9-petala. *Cal.* 3-phyllus. *Bacca* unilocularis,
coronata ſtigmate.

peltatum.

1. P. foliis peltatis lobatis. *Syſt. veget.* 489.
Duck's Foot, or May-apple.
Nat. of North America.
Cult. 1664. *Evel. kalend. hort.* 67.
Fl. May. H. ♃.

CHELIDONIUM. *Gen. pl.* 647.

Cor. 4-petala. *Cal.* 2-phyllus. *Siliqua* 1-locularis,
linearis.

majus.

1. C. pedunculis umbellatis. *Sp. pl.* 723.
α Chelidonium majus vulgare. *Bauh. pin.* 144.
Common Celandine.
β Chelidonium majus, foliis quernis. *Bauh. pin.* 144.
Cut-leav'd Celandine.

Nat.

Nat. of Britain.

Fl. April——October. H. ♃.

2. C. pedunculis unifloris, foliis amplexicaulibus finuatis, *Glauci-*
caule glabro. *Sp. pl.* 724. *um.*
Sea Celandine, or Yellow Horn'd-poppy.
Nat. of Britain.
Fl. June——October. H. ♃.

3. C. pedunculis unifloris, foliis feffilibus pinnatifidis, *cornicu-*
caule hifpido. *Sp. pl.* 724. *latum.*
Red Celandine, or Horn'd-poppy.
Nat. of England.
Fl. June and July. H. ☉.

4. C. pedunculis unifloris, foliis pinnatifidis linearibus, *hybridum.*
caule lævi, filiquis trivalvibus. *Sp. pl.* 724.
Violet-colour'd Celandine, or Horn'd-poppy.
Nat. of Wales.
Fl. Auguft. H. ☉.

P A P A V E R. *Gen. pl.* 648.

Cor. 4-petala. *Cal.* 2-phyllus. *Capfula* 1-locularis,
fub ftigmate perfiftente poris dehifcens.

* *Capfulis hifpidis.*

1. P. capfulis fubglobofis torofis hifpidis, caule foliofo *hybridum.*
multifloro. *Sp. pl.* 725.
Baftard Poppy.
Nat. of England.
Fl. June and July. H. ☉.

2. P. capfulis clavatis hifpidis, caule foliofo multifloro. *Arge-*
Sp. pl. 725. *Curtis lond.* *mone.*
Rough Poppy.
Nat. of Britain.
Fl. June and July. H. ☉.

3. P.

nudicaule.   3. P. capſulis hiſpidis, ſcapo unifloro, nudo hiſpido, foliis ſimplicibus pinnato-ſinuatis. *Sp. pl.* 725.

    α floribus albis.

    White-flower'd, naked-ſtalk'd Poppy.

    β floribus flavis.

    Yellow-flower'd, naked-ſtalk'd Poppy.

    Nat. of Siberia.

    Cult. 1730, by James Sherard, M. D. *Dill. elth.* 302. *t.* 224. *f.* 291.

    Fl. June——Auguſt.   H. ♂ .

* * *Capſulis glabris.*

Rhoeas.   4. P. capſulis glabris globoſis, caule piloſo multifloro, foliis pinnatifidis inciſis. *Sp. pl.* 726. *Curtis lond.*

    Corn, or Red Poppy.

    Nat. of Britain.

    Fl. June and July.   H. ⊙.

dubium.   5. P. capſulis oblongis glabris, caule multifloro ſetis ad-preſſis, foliis pinnatifidis inciſis. *Sp. pl.* 726. *Jacqu. auſtr.* 1. *p.* 17. *t.* 25. *Curtis lond.*

    Smooth Poppy.

    Nat. of Britain.

    Fl. June and July.   H. ⊙.

ſomnife-   6. P. calycibus capſuliſque glabris, foliis amplexicaulibus
rum.   inciſis. *Sp. pl.* 726.

    α Papaver hortenſe ſemine albo. *Bauh. pin.* 170.

    Common White Poppy.

    β Papaver hortenſe nigro ſemine. *Bauh. pin.* 170.

    Black-ſeeded White Poppy.

    γ Papaver pleno flore album. *Bauh. pin.* 171.

    Double-flower'd White Poppy.

    Nat. of England.

    Fl. July and Auguſt.   H. ⊙.

                                             7. P.

7. P. capfulis glabris oblongis, caule multifloro lævi, fo- *cambri-*
liis pinnatis incifis. *Sp. pl.* 727. *cum.*
Welch Poppy.
Nat. of Wales.
Fl. May——Auguſt. H. ♃.

8. P. capfulis glabris, caulibus unifloris fcabris foliofis, *orientale.*
foliis pinnatis ferratis. *Sp. pl.* 727. *Curtis magaz.*
57.
Oriental Poppy.
Nat. of the Levant.
Cult. before 1714, by Mr. George London. *Philofoph.*
tranſ. n. 346. *p.* 361. *n.* 117.
Fl. May and June. H. ♃.

A R G E M O N E. *Gen. pl.* 649.

Cor. 6-petala. *Cal.* 3-phyllus. *Capf.* femivalvis.

1. A. capfulis fexvalvibus, foliis fpinofis. *Syſt. veget.* *mexicana.*
490.
Prickly Argemone, or Poppy.
Nat. of Mexico and the Weſt Indies.
Cult. 1597, by Mr. John Gerard. *Ger. herb.* 993. *f.* 2.
Fl. July and Auguſt. H. ☉.

S A R R A C E N I A. *Gen. pl.* 652.

Cor. 5-petala. *Cal.* duplex, 3-phyllus & 5-phyllus.
Capf. 5-locularis, ſtylo ſtigmate clypeato.

1. S. foliis ſtrictis. *Sp. pl.* 729. *flava.*
Yellow Sidefaddle-flower.
Nat. of North America.
Cult. 1752, by Mr. Ph. Miller. *Mill. dict. edit.* 6. *n.* 2.
Fl. June and July. H. ♃.

purpurea. 2. S. foliis gibbis. *Sp. pl.* 728.

Purple Sidefaddle-flower.

Nat. of North America.

Introd. before 1640, by Mr. John Tradefcant, Jun.

Park. theat. 1235. *f.* 7.

Fl. June and July.   H. ♃.

NYMPHÆA.   *Gen. pl.* 653.

Cor. polypetala. *Cal.* 4 - 6-phyllus. *Bacca* multilocularis, truncata.

advena. 1. N. foliis cordatis integerrimis : lobis divaricatis, calyce hexaphyllo petalis longiore.

Nymphæa floribus flavis. *Clayton in Gron. virg. ed.* 1.
p. 164.

Three-colour'd Water Lily.

Nat. of North America.

Introd. 1772, by Mr. William Young.

Fl. July.   H. ♃.

DESCR. *Petioli* femiteretes, plerumque fupra aquam
eriguntur. *Folia* cordato-oblonga, verfus apicem
magis anguftata, quam in Nymphæa lutea. *Pe-
dunculus* teres. *Calyx* hexaphyllus : *foliola* fubrotunda, obtufa, concava : *tria exteriora* patentia,
extus viridia, intus fordide purpurafcentia, diametro
unciali ; *tria interiora* duplo majora, fuberecta, extus flava, intus fordide purpurafcentia. *Petala* circiter 13, lateribus receptaculi fimplici ferie inferta,
fubcuneiformia, obtufe rotundata, patenti-reflexa,
luteo-flavicantia, tres lineas lata. *Filamenta* nu-
merofiffima, multiplici ferie receptaculo inferta, linearia, lineam lata, vix femunciam longa, primum erecta, dein reflexa, flava, medio rubra. *Antheræ* lineares, binæ in fingulo filamento, margini-
bus

bus adnatæ. *Germen* columnare, profunde fulca-
tum, angulis (13) extantibus, flavum. *Stigma* peltatum, leviter umbilicatum, flavum, radiis 13 no-
tatum, margine profunde 13-dentatum, dentibus
obtufis fuperne viridibus.

2. N. foliis cordatis integris : lobis approximatis, calyce   *lutea.*
   pentaphyllo petalis longiore.
Nymphæa lutea. *Sp. pl.* 729.
Yellow Water Lily.
Nat. of Britain.
Fl. June——Auguft.   H. ♃.

3. N. foliis cordatis integerrimis : lobis imbricatis ro-   *alba,*
   tundis, calyce tetraphyllo.
Nymphæa alba. *Sp. pl.* 729.
White Water Lily.
Nat. of Britain.
Fl. June and July.   H. ♃.

4. N. foliis cordatis integris emarginatis : lobis divarica-   *odorata,*
   tis acumine obtufo, calyce tetraphyllo.
Nymphæa alba, flore pleno odorato. *Clayton in Gron.*
  virg. ed. 1. *p.* 57. *ed.* 2. *p.* 81.
Nymphæa alba minor. *Gmel. fib.* 4. *p.* 184. *tab.* 71.
  (exclufo fynonymo Morifoni.)
Sweet-fmelling Water Lily.
Nat. of North America and the Eaft of Siberia.
Introd. 1786, by William Hamilton, Éfq.
Fl. July.   H. ♃.
Obs. Omnino diverfa a Nyphæa alba minore Morifo-
  ni et aliorum, quæ a Nymphæa alba nil nifi mag-
  nitudine differt.

5. N. foliis peltatis undique integris. *Sp. pl.* 730.   *Nelumbo.*
              Q 2   Peltated

Peltated Water Lily.
Nat. of both Indies.
Introd. 1787, by Sir Joseph Banks, Bart.
Fl.   G. H. ♃.

B I X A.   *Gen. pl.* 654.

Cor. 10-petala. *Cal.* 5-dentatus. *Capf.* hifpida, 2-valvis.

Orellana.   1. BIXA.   *Sp. pl.* 730.
Heart-leav'd Bixa, or Anotta.
Nat. of the Weft Indies.
Introd. 1690, by Mr. Bentick.   *Br. Muf. Sloan. mff.*
3370. *p.* 1. *n.* 160.
Fl.   S. ♄.

A P E I B A.

Cor. 5-petala. *Cal.* 5-phyllus, deciduus. *Antheræ*
filamentis infra apicem adnatæ.   *Capf.* echinata,
depreffa, multilocularis.

Tibour-   1. A. foliis acute ferratis hirfutis.   *Swartz prodr.* 82.
bou.   Apeiba Tibourbou.   *Aublet guian.* 538. *tab.* 213.
Sloanea.   *Loefl. it.* 311.
Apeiba.   *Marcgr. braf.* 123. *cum. fig. in pag.* 124.
Hairy Apeiba.
Nat. of South America and the Weft Indies.
Cult. 1756, by Mr. Philip Miller.   *Mill. dict. edit.* 7.
Caftanea 3.
Fl.   S. ♄.

M A M M E A.   *Gen. pl.* 656.

Cor. 4-petala. *Cal.* 2-phyllus. *Bacca* maxima, 4-fperma.

america-   1. M. ftaminibus flore brevioribus.   *Sp. pl.* 731.
na.   American Mammee.

Nat.

Nat. of Jamaica and Hifpaniola.
Cult. 1739, by Mr. Ph. Miller. *Rand. chel.* Mamei.
Fl.   S. ♄ .

T I L I A.   *Gen. pl.* 660.

Cor. 5-petala. *Cal.* 5-partitus. *Bacca* ficca globofa,
5-locularis, 5-valvis, bafi dehifcens.

1. T. floribus nectario deftitutis. *Sp. pl.* 733.   *europæa.*
α Tilia fœmina folio majore. *Bauh. pin.* 426.'   commu-
   Common Lime Tree.   nis.
β Tilia foliis molliter hirfutis, viminibus rubris, fructu *corallina.*
   tetragono. *Raj. fyn.* 473.
   Red-twig'd Lime Tree.
γ Tilia fœmina folio minore. *Bauh. pin.* 426.   parviflo-
   Small-leav'd Lime Tree.   ra.
   Nat. of Britain.
   Fl. June——Auguft.   H. ♄ .

2. T. floribus nectario inftructis, foliis profunde corda- *america-*
   tis argute ferratis glabris.   *na.*
   Tilia americana. *Sp. pl.* 733.
   Broad-leav'd Lime Tree.
   Nat. of Virginia and Canada.
   Cult. 1752, by Mr. Ph. Miller. *Mill. dict. edit.* 6. *n.* 6.
   Fl. June and July.   H. ♄ .

3. T. floribus nectario inftructis, foliis bafi truncatis *pubefcens.*
   obliquis denticulato-ferratis fubtus pubefcentibus.
   Tilia caroliniana. *Mill. dict.*
   Pubefcent Lime Tree.
   Nat. of Carolina. Mr. *Mark Catefby.*
   Introd. about 1726. *Mill. dict. edit.* 8.
   Fl. July and Auguft.   H. ♄ .

alba.   4. T. foliis profunde cordatis fubfinuatis dentatis fubtus
            tomentofis.
            White Lime Tree.
            Nat. of North America.
            Cult. 1767, by Mr. James Gordon.
            Fl.   H. ♄.

DECUMARIA. *Gen. pl.* 597.

Cal. 8 - 12-phyllus, fuperus. *Petala* 8 - 12. *Capf.* 8-
        locularis, polyfperma.

barbara.   1. DECUMARIA. *Sp. pl.* 1663.
            Forfythia fcandens. *Walter carolin.* 154.
            Climbing Decumaria.
            Nat. of Carolina.
            Introd. 1785, by Baron Hake.
            Fl.   S. ♄.

LAGERSTRŒMIA. *Gen. pl.* 667.

Pet. 6, calyci inferta, unguiculata. *Cal.* 6-fidus. *Stam.*
        6 majora. *Capf.* 6-locularis, polyfperma.

indica.   1. LAGERSTROEMIA. *Sp. pl.* 734.
            Indian Lagerftrœmia.
            Nat. of the Eaft Indies.
            Introd. 1759, by Hugh Duke of Northumberland.
            Fl. Auguft——Octtober.   G. H. ♄.

T H E A. *Gen. pl.* 668.

Cor. 6- f. 9-petala. *Cal.* 5 - f. 6-phyllus. *Capf.* 3-cocca.

Bohea.   1. T. floribus hexapetalis. *Sp. pl.* 734.
laxa,   α foliis elliptico-oblongis rugofis.
                Broad-leav'd Tea.

                                                        β foliis

c. foliis lanceolatis planis.   ſtri&ta.
Narrow-leav'd Tea.
Nat. of China and Japan.
Introd. about 1768, by John Ellis, Eſq.
Fl. Auguſt and September.   H. ♄.

GORDONIA. *Linn. mant.* 556.

Cal. 5-phyllus. *Pet.* 5, mediante ne&tario baſi coalita.
Filam. ne&tario inſerta. *Capſ.* ſupera, 5-locularis.
Sem. alata.

1. G. foliis coriaceis utrinque glabris.   *Laſian-*
Gordonia Laſianthus. *Linn. mant.* 570.   *thuⅰ.*
Hypericum Laſianthus. *Sp. pl.* 1101.
Smooth Loblolly Bay.
Nat. of North America.
Introd. about 1768, by Benjamin Bewick, Eſq.
Fl. Auguſt and September.   G. H. ♄.

2. G. foliis ſubtus pubeſcentibus.   *pubeſcenⅰ.*
Gordonia pubeſcens. *de Lamarck encycl.* 2. *p.* 770.
Cavan. diſſ. 6. *p.* 308. *tab.* 162.
Pubeſcent Loblolly Bay.
Nat. of South Carolina.
Introd. 1774, by Mr. William Malcolm.
Fl. September.   H. ♄.

C I S T U S. *Gen. pl.* 673.

Cor. 5-petala. *Cal.* 5-phyllus : foliolis duobus mino-
        ribus. *Capſula.*

* *Exſtipulati, fruticoſi.*

1. C. arboreſcens exſtipulatus, foliis cordatis lævibus *populifo-*
acuminatis petiolatis. *Sp. pl.* 736.   *lius.*
α Ciſtus Ledon, foliis Populi nigræ, major. *Bauh. pin.* major.
467.

Great

Great Poplar-leav'd Ciſtus.

minor.   β Ciſtus Ledon, foliis Populi nigræ, minor.   *Bauh. pin.*
467.

Small Poplar-leav'd Ciſtus.

Nat. of Portugal.

Cult. 1656, by Mr. John Tradeſeant, Jun. *Muſ.*
Trad. 100.

Fl. May and June.   H. ♄.

laurifoli-   2. C. arboreſcens exſtipulatus, foliis oblongo-ovatis pe-
us.   tiolatis trinerviis ſupra glabris, petiolis baſi connatis. *Sp. pl.* 736.

Laurel-leav'd Ciſtus.

Nat. of Spain.

Cult. 1752, by Mr. Ph. Miller. *Mill. dict. edit.* 6. *n.* 8.

Fl. June and July.   H. ♄.

vagina-   3. C. arboreſcens exſtipulatus, foliis oblongis piloſis ſubtus.   tus reticulato-rugoſis, petiolis baſi coalitis vaginantibus ſulcatis.

Oblong-leav'd Ciſtus.

Nat. of the Iſland of Teneriffe. Mr. *Fr. Maſſon.*
Introd. 1779.

Fl. April——June.   G. H. ♄.

ladanife-   4. C. arboreſcens exſtipulatus, foliis lanceolatis ſupra
rus.   lævibus : petiolis baſi coalitis vaginantibus. *Sp.*
pl. 737.

undula-   α foliis undulatis.
tus.   Common Gum Ciſtus.

planifo-   β foliis planis.
lius,   Flat-leav'd Gum Ciſtus.

Nat. of Spain and Portugal.

Cult. 1656, by Mr. John Tradeſcant, Jun. *Muſ.*
Trad. 100.

Fl. June and July.   H. ♄.

5. C.

5. C. arboreſcens exſtipulatus, foliis lineari-lanceolatis *monſpe-*
    ſeſſilibus utrinque villoſis trinerviis. *Sp. pl.* 737. *lienſis.*
    Montpelier Ciſtus.
    Nat. of the South of France and Spain.
    Cult. 1656, by Mr. John Tradeſcant, Jun. *Muſ.*
    Trad. 101.
    Fl. July.   H. ♃.

6. C. a.boreſcens exſtipulatus, foliis ovato-lanceolatis *laxus.*
    undulatis denticulatis glabris: ſummis hirtis, fo-
    liolis calycinis ſubrotundo-cordatis.
    Ledon IV. *Cluſ. hiſt.* 1. *p.* 78. *fig.*
    Broad waved-leav'd Ciſtus.
    Nat. of Spain and Portugal.
    Cult. 1656, by Mr. John Tradeſcant, Jun. *Muſ.*
    Trad. 100.
    Fl. June and July.   H. ♃.

7. C. arboreſcens exſtipulatus, foliis ovatis petiolatis *ſalvifoli-*
    utrinque hirſutis. *Sp. pl.* 738. *us.*
    Sage-leav'd Ciſtus, or Rock-roſe.
    Nat. of the South of Europe.
    Cult. before 1551, in Sion Garden. *Turn. herb. part* 1.
    ſign. K v. conf. *edit.* 2. *p.* 145.
    Fl. June and July.   H. ♃.

8. C. arboreſcens exſtipulatus, foliis ſpathulatis tomento- *incanus.*
    ſis rugoſis: inferioribus baſi vaginantibus connatis.
    Sp. pl. 737. *Curtis magaz.* 43.
    Hoary-leav'd Ciſtus, or Rock-roſe.
    Nat. of the South of Europe.
    Cult. 1596, by Mr. John Gerard. *Hort. Ger.*
    Fl. June——Auguſt.   H. ♃.

9. C.

albidus.   9. C. arboreſcens exſtipulatus, foliis ovato-lanceolatis
tomentoſis incanis feſſilibus ſubtrinerviis. *Sp. pl.*
737.
White-leav'd Ciſtus, or Rock-roſe.
Nat. of Spain and France.
Cult. 1656, by Mr. John Tradeſcant, Jun. *Muſ.
Trad.* 100.
Fl. June and July.   H. ♄.

creticus.   10. C. arboreſcens exſtipulatus, foliis ſpathulato-ovatis
petiolatis enerviis ſcabris, calycinis lanceolatis.
Sp. pl. 738. *Jacqu. ic. collect.* 1. *p.* 80.
Cretan Ciſtus.
Nat. of the Levant.
Cult. 1731, by Mr. Philip Miller. *Mill. dict. edit.* 1.
n. 11.
Fl. June and July.   H. ♄.

criſpus.   11. C. arboreſcens exſtipulatus, foliis lanceolatis pubeſ-
centibus trinerviis undulatis. *Sp. pl.* 738.
Curl'd-leav'd Ciſtus, or Rock-roſe.
Nat. of Portugal.
Cult. 1731, by Mr. Ph. Miller. *Mill. dict. edit.* 1. *n.* 4.
Fl. June and July.   H. ♄.

*halimifo-
lius.*   12. C. arboreſcens exſtipulatus, foliolis duobus calycinis
linearibus. *Sp. pl.* 738.
α Ciſtus fœmina, portulacæ marinæ folio latiore obtu-
ſo. *Bauh. pin.* 465.
Broad Sea Purſlane-leav'd Ciſtus.
β Ciſtus fœmina, portulacæ marinæ folio anguſtiore
mucronato. *Bauh. pin.* 465.
Narrow Sea Purſlane-leav'd Ciſtus.
Nat. of Portugal.

Cult.

Cult. 1656, by Mr. John Tradeſcant, Jun. *Muſ. Trad.* 101.

Fl. June and July. H. ♄.

13. C. arboreſcens exſtipulatus, foliis revoluto-lineari- *libanotis.* bus, floribus umbellatis. *Sp. pl.* 739.
Roſemary-leav'd Ciſtus.
Nat. of Spain.
Introd. 1783, by P. M. A. Brouſſonet, M. D.
Fl. June. G. H. ♄.

** *Exſtipulati, ſuffruticoſi.*

14. C. ſuffruticoſus procumbens exſtipulatus, foliis oppo- *umbella-* ſitis linearibus, floribus umbellatis. *Sp. pl.* 739. *tus.*
Umbel'd Ciſtus.
Nat. of France and Spain.
Cult. 1731, by Mr. Philip Miller. *Mill. dict. edit.* 1.
Helianthemum 10.
Fl. June——Auguſt. H. ♄.

15. C. ſuffruticoſus adſcendens exſtipulatus, foliis alternis *lævipes.* faſciculatis filiformibus glabris, pedunculis racemo-
ſis. *Syſt. veget.* 497. *Jacqu. hort.* 2. *p.* 74. *t.* 158.
Cluſter-leav'd Ciſtus.
Nat. of the South of France.
Cult. 1690, in the Royal Garden at Hampton-court.
Catal. mſſ.
Fl. June——Auguſt. G. H. ♄.

16. C. ſuffruticoſus erectus exſtipulatus, foliis lanceola- *ſyriacus.* tis ad oras revolutis, floribus racemoſis. *Jacqu. ic. collect.* 1. *p.* 98.
Syrian Ciſtus.
Nat. of the Levant.

Introd.

Introd. 1788, by Monſ. Thouin.
Fl. G. H. ♄.

Fumana. 17. C. ſuffruticoſus procumbens exſtipulatus, foliis alternis linearibus margine ſcabris, pedunculis unifloris. *Sp. pl.* 740. *Jacqu. auſtr.* 3. *p.* 29. *t.* 252.
Heath-leav'd Ciſtus.
Nat. of France and Switzerland.
Cult. 1739, by Mr. Philip Miller. *Mill. dict. vol.* 2.
Helianthemum 6.
Fl. June and July. G. H. ♄.

ſcabroſus. 18. C. ſuffruticoſus exſtipulatus, foliis oppoſitis ovatis piloſo-ſcabris trinerviis, calycibus triphyllis.
Rough Ciſtus.
Nat. of Italy and Portugal.
Introd. 1775, by Meſſrs. Kennedy and Lee.
Fl. June and July. H. ♄.
DESCR. *Caules* decumbentes, teretes, pilis brevis ſtellatis denſe adſperſi, ſcabri. *Rami* breves. *Folia* ſubpetiolata, trinervia, uncialia. *Flores* terminales, ſubpaniculati. *Calyx* triphyllus : foliola ovato-lanceolata, acuminata, extus pubeſcentia pilis ſtellatis et villo longo, æqualia, quadrilinearia. *Petala* obovata, ſubretuſa, calyce fere duplo longiora, intenſe flava, baſi lutea.

canus. 19. C. ſuffruticoſus exſtipulatus procumbens, foliis oppoſitis obovatis villoſis ſubtus tomentoſis, floribus ſubumbellatis. *Sp. pl.* 740.
Myrtle-leav'd dwarf Ciſtus.
Nat. of the South of Europe.
Introd. 1772, by Monſ. Richard.
Fl. June and July. H. ♄.
20. C.

20. C. ſuffruticoſus exſtipultatus, foliis oppoſitis oblongis *marifo-*
petiolatis planis ſubtus incanis. *Sp. pl.* 741. *lius.*
Marum-leav'd Ciſtus.
Nat. of the South of Europe.
Cult. 1731, by Mr. Philip Miller. *Mill. dict. edit.* 1.
Helianthemum 3.
Fl. May and June. H. ♃.

21. C. ſuffruticoſus exſtipulatus procumbens, foliis op- *anglicus.*
poſitis oblongis revolutis piloſis, floribus racemo-
ſis. *Linn. mant.* 245.
Ciſtus hirſutus. *Hudſ. angl.* 232.
Hairy Ciſtus.
Nat. of England.
Fl. June and July. H. ♃.

*** *Exſtipulati, herbacei.*

22. C. exſtipulatus perennis, foliis radicalibus ovatis *Tubera-*
trinerviis tomentoſis; caulinis glabris lanceola- *ria.*
tis: ſummis alternis. *Sp. pl.* 741.
Plantain-leav'd Ciſtus.
Nat. of the South of Europe.
Cult. 1748, by Mr. Philip Miller. *Mill. dict. edit.* 5.
Helianthemum 43.
Fl. June and July. H. ♃.

23. C. herbaceus exſtipulatus, foliis oppoſitis lanceolatis *guttatus.*
trinerviis, racemis ebracteatis. *Sp. pl.* 741.
Annual ſpotted-flower'd Ciſtus.
Nat. of England.
Fl. June and July. H. ☉.

**** *Stipulati, herbacei.*

24. C. herbaceus erectus glaber ſtipulatus, floribus ſo- *ledifolius.*
litariis ſubſeſſilibus, folio ternato oppoſitis. *Sp.*
pl. 742.

Ledum-

Ledum-leav'd Ciftus.
Nat. of the South of France.
Cult. 1731, by Mr. Philip Miller. *Mill. dict. edit.* 1.
Helianthemum 14.
Fl. June and July.   H. ☉.

falicifo-
lius.   25. C. herbaceus patulus villofus ftipulatus, floribus racemofis erectis : pedicellis horizontalibus. *Sp. pl.*
742.
Willow-leav'd Annual Ciftus.
Nat. of England.
Fl. June——Auguft.   H. ☉.

ægyptia-
cus.   26. C. herbaceus erectus ftipulatus, foliis lineari-lanceo-
latis petiolatis, calycibus inflatis corolla majoribus.
Sp. pl. 742.
Egyptian Ciftus.
Nat. of Egypt.
Cult. 1768, by Mr. Philip Miller. *Mill. dict. edit.* 8.
Helianthemum 23.
Fl. June and July.   H. ☉.

***** *Stipulati, fuffruticofi.*

furreja-
nus.   27. C. fuffruticofus procumbens ftipulatus, foliis ovato-
oblongis fubpilofis, petalis lanceolatis. *Sp. pl.* 743.
Small-flower'd Ciftus.
Nat. of England.
Fl. July——October.   H. ♄.

ferpyllifo-
lius.   28. C. fuffruticofus ftipulatus, foliis oblongis, calycibus
lævibus. *Sp. pl.* 743.
Serpyllum-leav'd Ciftus.
Nat. of the Alps of Auftria.
Cult. 1759, by Mr. Philip Miller. *Mill. dict. edit.* 7.
Helianthemum 8.
Fl. May——September.   H. ♄.

                               29. C.

29. C. fuffruticofus ftipulatus procumbens, foliis ovali-   *thymifo-*
    linearibus oppofitis breviffimis congeftis. *Syft.* *lius.*
    veget. 500.
    Thyme-leav'd Ciftus.
    Nat. of France and Spain.
    Cult. 1714, by the Dutchefs of Beaufort. *Br. Muf.*
    H. S. 133. *fol.* 61.
    Fl. June and July.   H. ♄.

30. C. fuffruticofus procumbens, ftipulis lanceolatis, fo-   *Helian-*
    liis oblongis revolutis fubpilofis. *Sp. pl.* 744. *themum.*
    Curtis lond.
    α flore luteo.
    Yellow-flower'd Dwarf Ciftus.
    β flore albo.
    White-flower'd Dwarf Ciftus.
    γ flore rofeo.
    Rofe-flower'd Dwarf Ciftus.
    Nat. of Britain.
    Fl. May——September.   H. ♄.

31. C. fuffruticofus ftipulatus patulus, foliis lanceolatis   *apenni-*
    hirtis. *Syft. veget.* 500.   *nus.*
    Apennine Ciftus.
    Nat. of the Alps of Italy.
    Cult. 1731, by Mr. Philip Miller. *Mill. dict. edit.* 1.
    Helianthemum 8.
    Fl. June——Auguft.   H. ♄.

32. C. fuffruticofus ftipulatus procumbens, foliis oblongo-   *polifolius.*
    ovatis incanis, calycibus lævibus, petalis ferratis.
    Sp. pl. 745.
    Mountain Ciftus.
    Nat. of England.
    Fl. May and June.   H. ♄.

CORCHO-

CORCHORUS. *Gen. pl.* 675.

Cor. 5-petala. *Cal.* 5-phyllus, deciduus. *Capf.* pluri-
　　valvis, loculamentofa.

olitorius.　1. C. capfulis oblongis ventricofis, foliorum ferraturis
　　　　infimis fetaceis. *Sp. pl.* 746.
　　　　Briftly-leav'd Corchorus.
　　　　Nat. of Afia, Africa, and America.
　　　　Cult. 1731, by Mr. Ph. Miller. *Mill. dict. edit.* 1. *n.* 1.
　　　　Fl. June——Auguft.　　　　　　　　　　S. ☉.

æftuans.　2. C. capfulis oblongis trilocularibus trivalvibus fexful-
　　　　catis fexcufpidatis, foliis cordatis : ferraturis infimis
　　　　fetaceis. *Syft. veget.* 501. *Jacqu. hort.* 1. *p.* 37.
　　　　t. 85.
　　　　Hornbeam-leav'd Corchorus.
　　　　Nat. of South America.
　　　　Cult. 1739, by Mr. Philip Miller. *Rand. chel. n.* 2.
　　　　Fl. June and July.　　　　　　　　　　S. ☉.

capfula-　3. C. capfulis fubrotundis depreffis rugofis, foliis ferraturis
ris.　　　　infimis fetaceis. *Syft. veget.* 501.
　　　　Heart-leav'd Corchorus.
　　　　Nat. of the Eaft Indies.
　　　　Cult. 1731, by Mr. Ph. Miller. *Mill. dict. edit.* 1. *n.* 3.
　　　　Fl. June and July.　　　　　　　　　　S. ☉.

filiquofus.　4. C. capfulis linearibus compreffis bivalvibus, foliis
　　　　lanceolatis æqualiter ferratis. *Syft. veget.* 501.
　　　　Jacqu. hort. 3. *p.* 34. *t.* 59.
　　　　Germander-leav'd Corchorus.
　　　　Nat. of the Weft Indies.
　　　　Cult. 1732, by Mr. Philip Miller. *R. S. n.* 519.
　　　　Fl. June——Auguft.　　　　　　　　　　S. ♄.

DIGYNIA.

D I G Y N I A.

P Æ O N I A. *Gen. pl.* 678.

Cal. 5-phyllus. *Petala* 5. *Styli* 0. *Capſ.* polyſpermæ.

1. P. foliolis oblongis. *Sp. pl.* 747.
   Common Pæony.
   Nat. of Switzerland.
   Cult. 1562. *Turn. herb. part* 2, *fol.* 84.
   Fl. May and June.   H. ♃.

officina-lis.

2. P. foliolis linearibus multipartitis. *Sp. pl.* 748.
   Slender-leav'd Pæony.
   Nat. of Ucraine.
   Introd. 1765, by Mr. William Malcolm.
   Fl. May.   H. ♃.

tenuifolia.

FOTHERGILLA. *Linn. ſuppl.* 42.

Cal. Amentum ovatum: *Squamis* 1-floris. *Cor.* caly-
   ciformis, 1-petala, 5-fida.

1. FOTHERGILLA. *Linn. ſuppl.* 267.
α foliis obtuſis.
  Broad-leav'd Fothergilla.
β foliis acutis.
  Fothergilla Gardeni. *Jacqu. ic. collect.* 1. *p.* 97.
  Narow-leav'd Fothergilla:
  Nat. of North America.
  Introd. 1765, by Mr. John Buſh.
  Fl. April——June.   H. ♄.

alnifolia.
obtuſa.

acuta.

VOL. II.   R   CALLI-

CALLIGONUM. *Gen. pl.* 680.

Cal. 5-phyllus. *Pet.* 0. *Styli* 0. *Fructus* alatus, mo-
nospermus.

Pallasia.   1. C. fructibus alatis, alis membranaceis crispis. *L'Herit.*
stirp. nov. tom. 2.
Calligonum polygonoides. *Pallas it.* 3. *p.* 536.
Pallasia caspica. *Linn. suppl.* 252.
Pterococcus aphyllus. *Pallas it.* 2. *p.* 738. *tab. S.*
Nat. of Siberia.
Introd. 1780, by Peter Simon Pallas, M. D.
Fl. August.   H. ♄.

T R I G Y N I A.

D E L P H I N I U M. *Gen. pl.* 681.

Cal. 0.   *Petala* 5.   *Nectarium* bifidum, postice cor-
nutum.   *Siliquæ* 3 f. 1.

* *Unicapsularia.*

Consolida.   1. D. nectariis monophyllis, caule subdiviso. *Sp. pl.*
748.
Branching Larkspur.
Nat. of England.
Fl. June.   H. ☉.

Ajacis.   2. D. nectariis monophyllis, caule simplici. *Sp. pl.* 748.
α Consolida regalis hortensis, flore majore & simplici.
Bauh. pin. 142.
Single Upright Larkspur.
β Consolida regalis flore majore & multiplici. *Bauh.*
pin. 142.
Double

Double Upright Larkfpur.
Nat.
Cult. 1596, by Mr. John Gerard. *Hort. Ger.*
Fl. June and July. H. ☉.

** *Tricapfularia.*

3. D. nectariis diphyllis, corollis enneapetalis, capfulis *peregri-*
ternis, foliis multipartitis obtufis. *Sp. pl.* 749. *num.*
Broad-leav'd Annual Larkfpur.
Nat. of the South of Europe and the Levant.
Cult. 1731, by Mr. Ph. Miller. *Mill. dict. edit.* 1. *n.* 2.
Fl. June and July. H. ☉.

4. D. nectariis diphyllis: labellis integris, floribus fub- *grandi-*
folitariis, foliis compofitis lineari-multipartitis. *Sp.* *florum.*
pl. 749.
Great-flower'd Larkfpur.
Nat. of Siberia.
Cult. 1758, by Mr. Philip Miller. *Mill. ic.* 167. *t.* 250.
f. 1.
Fl. June——September. H. ♃.

5. D. nectariis diphyllis: labellis ovatis bifidis: laci- *interme-*
niis ovatis, foliis tripartitis: laciniis trifidis incifis. *dium.*
Delphinium elatum. *Mattufchka enum. ftirp. filef.*
p. 132.
Delphinium nectariis diphyllis, labellis integris, flori-
bus fpicatis, foliis palmatis multifidis glabris. *Mill.*
ic. p. 79. *t.* 119.
Palmated Bee Larkfpur.
Nat. of Silefia. *Jacquin.*
Cult. 1756, by Mr. Philip Miller. *Mill. ic. loc. cit.*
Fl. July. H. ♃.

6. D. nectariis diphyllis: labellis ovatis emarginatis: *elatum.*

laciniis

laciniis breviffimis inæqualibus, foliis fubpeltatis tripartitis: laciniis multifidis.

Delphinium elatum. *Sp. pl.* 749.

Common Bee Larkfpur.

Nat. of Siberia and Switzerland.

Cult. 1656, by Mr. John Tradefcant, Jun. *Muf. Trad.* 75.

Fl. June —— September.   H. ♃.

exalta-
tum.
7. D. nectariis diphyllis: labellis oblongis bifidis: laciniis lanceolatis æqualibus, foliis tripartitis: laciniis trifidis.

Delphinium nectariis diphyllis: labellis bifidis apice barbatis, foliis trilobis incifis, caule erecto. *Mill. ic. p.* 167. *t.* 250. *f.* 2.

American Larkfpur.

Nat. of North America. Mr. *John Bartram.*

Cult. 1758, by Mr. Philip Miller. *Mill. ic. loc. cit.*

Fl. July and Auguft.   H. ♃.

OBS. Maxime affine Delphinio urceolato *Jacqu. ic. collect.* 1. *p.* 153; differt vero foliis glabris, planis; caule glabro, purpureo.

Staphi-
fagria.
8. D. nectariis tetraphyllis petalo brevioribus, foliis palmatis: lobis obtufis. *Syft. veget.* 503.

Palmated Larkfpur, or Stavefacre.

Nat. of the South of Europe.

Cult. 1596, by Mr. John Gerard. *Ger. hort.*

Fl. April——Auguft.   G. H. ♂.

ACONITUM. *Gen. pl.* 682.

Cal. o. *Petala* 5: fupremo fornicato. *Nectaria* 2, pedunculata, recurva. *Siliquæ* 3 f. 5.

lycocto-
num.
1. A. folis palmatis multifidis villofis. *Sp. pl.* 750. *Jacqu. auftr.* 4. *p.* 41. *t.* 380.

Great

Great Yellow Wolf's-bane.
Nat. of the Alps of Europe.
Cult. 1596, by Mr. John Gerard. *Hort. Ger.*
Fl. July and Auguft.   H. ♃.

2. A. foliorum laciniis linearibus fuperne latioribus linea   *Napellus.*
    exaratis. *Sp. pl.* 751. *Jacqu. auftr.* 4. *p.* 42. *t.* 381.
α Aconitum cæruleum f. Napellus 1.   *Bauh. pin.* 183.
    Common Wolf's-bane, or Monk s-hood.
β Aconitum *pyramidale* foliis multipartitis, fpicis florum
    longiffimis feffilibus. *Mill. dict.*
Long-fpiked Common Wolf's-bane.
Nat. of Germany, France, and Switzerland.
Cult. 1596, by Mr. John Gerard. *Hort. Ger.*
Fl. May——July.   H. ♃.

3. A. foliis multipartitis: laciniis linearibus incumben-   *pyrenai-*
    tibus fquarrofis. *Sp. pl.* 751.   *cum.*
Pyrenean Wolf's-bane.
Nat. of the Pyrenees and Siberia.
Cult. 1739, by Mr. Philip Miller. *Rand. chel. n.* 5.
Fl. June and July.   H. ♃.

4. A. floribus pentagynis, foliorum laciniis linearibus.   *Anthora.*
    Sp. pl. 751. *Jacqu. auftr.* 4. *p.* 43. *t.* 382.
Wholefome Wolf's-bane.
Nat. of the Pyrenees and the Alps of Switzerland.
Cult. 1596, by Mr. John Gerard. *Hort. Ger.*
Fl. June and July.   H. ♃.

5. A. floribus pentagynis, foliorum laciniis femipartitis   *varicga-*
    fuperne latioribus. *Syft. veget.* 504.   *tum.*
Variegated Wolf's-bane.
Nat. of the South of Europe.

R 3   *Cult.*

Cult. 1752, by Mr. Ph. Miller. *Mill. dict. edit.* 6. *n.* 8.
Fl. June——Auguft.   H. ♃.

album.   6. A. floribus pentagynis, foliis glabris tripartitis : laci-
niis acute incifis, petali fupremi ungue lateralibus
longiore.
Aconitum orientale. *Mill. dict.*
Aconitum lycoctonum orientale flore magno albo.
Tourn. cor. 30.
White Wolf's-bane, or Monk's-hood.
Nat. of the Levant.
Cult. 1739, by Mr. Ph. Miller. *Mill. dict. vol.* 2. *n.* 19.
Fl. July and Auguft.   H. ♃.

Camma-   7. A. floribus fubpentagynis, foliorum laciniis cuneifor-
rum.   mibus incifis acutis. *Sp. pl.* 751. *Jacqu. auftr.* 5.
p. 11. *tab.* 424.
Purple Wolf's-bane.
Nat. of Germany.
Cult. 1748, by Mr. Ph. Miller. *Mill. dict. edit.* 5. *n.* 15.
Fl. July.   H. ♃.

uncina-   8. A. floribus fubpentagynis, foliis multilobis, corolla-
tum.   rum galea longius extenfa. *Syft. veget.* 504.
American Wolf's-bane.
Nat. of Penfylvania.
Cult. 1770, by Mr. James Gordon.
Fl. July and Auguft.   H. ♃.

TETRAGY-

TETRAGYNIA.

CIMICIFUGA. *Syſt. veget.* 505.

Cal. 5-phyllus. *Cor.* nectariis 4, urceolatis. *Capſ.* 4.
Sem. ſquamoſa.

1. CIMICIFUGA. *Syſt. veget.* 505.
   Actæa Cimicifuga. *Sp. pl.* 722.
   Fœtid Cimicifuga.
   Nat. of Siberia.
   Introd. 1777, by Meſſrs. Gordon and Græfer.
   Fl. June.   H. ♃.

fœtida.

PENTAGYNIA.

AQUILEGIA. *Gen. pl.* 684.

Cal. 0.   *Petala* 5.   *Nectaria* 5, corniculata, inter petala. *Capſ.* 5, diſtinctæ.

1. A. nectariis incurvis. *Sp. pl.* 752.   *vulgaris.*
   α Aquilegia hortenſis ſimplex. *Bauh. pin.* 144.
   Common Columbine.
   β Aquilegia hortenſis multiplex, flore magno. *Bauh.*
   pin. 144.
   Double-flower'd Columbine.
   γ Aquilegia flore roſeo multiplici. *Bauh. pin.* 145.
   Roſe Columbine.
   δ nectariis cæruleis apice luteis.   *ſpecioſa.*
   Great-flower'd Columbine.
   Nat. of Britain; δ. of Siberia.
   Fl. May and June.   H. ♃.

R 4   2. A.

alpina.   2. A. nectariis rectis petalo lanceolato brevioribus. *Sp.*
       pl. 752.
       Alpine Columbine.
       Nat. of Switzerland.
       Cult. 1731, by Mr. Ph. Miller. *Mill. dict. edit.* 1. *n.* 4.
       Fl. May and June.   H. ♂ .

canaden-   3. A. nectariis rectis, ftaminibus corolla longioribus.
fis.   *Sp. pl.* 753.
       Canadian Columbine.
       Nat. of Virginia and Canada.
       Introd. before 1640, by Mr. John Tradefcant, Sen.
          Park. theat. 1367. *n.* 1.
       Fl. April and May.   H. ♃ .

viridiflo- 4. A. nectariis rectis apice incraffato fubinflexo, ftami-
ra.   nibus corollam fubæquantibus.
       Aquilegia viridiflora. *Murray in commentat. gotting.*
          1780. *p.* 8. *tab.* 2. *Retz. obf. bot.* 3. *p.* 34. *n.* 64.
          Jacqu. ic. collect. 1. *p.* 35.
       Green-flower'd Columbine.
       Nat. of Siberia.
       Introd. 1780, by Peter Simon Pallas, M. D.
       Fl. May and June.   H. ♃ .

            N I G E L L A. *Gen. pl.* 685.

       Cal. ◦. *Petala* 5. *Nectaria* 5, trifida, intra corollam.
              Capfulæ 5, connexæ.

                    * *Pentagynæ.*

damafce-   1. N. floribus involucro foliofo cinctis. *Sp. pl.* 753.
na.   *Curtis magaz.* 22.
       α Nigella anguftifolia, flore majore fimplici cæruleo.
          Bauh. pin. 145.
       Common Fennel-flower.
                                              β Nigella

β Nigella flore majore pleno cæruleo. *Bauh. pin.* 145.
Great double Fennel-flower.
Nat. of the South of Europe.
Cult. 1570. *Lobel. adv.* 329.
Fl. June——September.　　　　　　　　H. ⊙.

2. N. piftillis quinis, capfulis muricatis fubrotundis, fo-　*fativa.*
liis fubpilofis. *Sp. pl.* 753.
Small Fennel-flower.
Nat. of Candia and Egypt.
Cult. 1658, in Oxford Garden. *Hort. oxon, edit.* 2.
p. 117.
Fl. June——September.　　　　　　　　H. ⊙.

3. N. piftillis quinis, petalis integris, capfulis turbinatis.　*arvenfis.*
Sp. pl. 753.
Field Fennel-flower.
Nat. of Germany and France.
Cult. 1713. *Philofoph. tranf. n.* 337. *p.* 206. *n.* 106.
Fl. June——September.　　　　　　　　H. ⊙.

✿✿ *Decagynæ.*

4. N. piftillis denis corollam æquantibus. *Sp. pl.* 753.　*hifpanica.*
Spanifh Fennel-flower.
Nat. of Spain and the South of France.
Cult. 1629. *Park. parad.* 285. *f.* 9.
Fl. June——September.　　　　　　　　H. ⊙.

5. N. piftillis denis corolla longioribus. *Sp. pl.* 753.　*orienta-*
Yellow Fennel-flower.　　　　　　　　　　　　*lis.*
Nat of Syria.
Cult. 1731, by Mr. Ph. Miller. *Mill. dict. edit.* 1. *n.* 8.
Fl. July——September.　　　　　　　　H. ⊙.

HEXAGY-

HEXAGYNIA.

STRATIOTES. *Gen. pl.* 687.

Spatha diphylla. *Perianth.* 3-fidum. *Petala* 3. *Bacca* 6-locularis, infera.

Aloides.   1. S. foliis enfiformi-triangulis ciliato-aculeatis. *Sp. pl.* 754.
Water Aloe, or Frefh Water Soldier.
Nat. of England.
Fl. June.   H. ♃.

POLYGYNIA.

ILLICIUM. *Linn. mant.* 167.

Cal. 6-phyllus. *Petala* 27. *Capf.* plures, in orbem digeftæ, bivalves, monofpermæ.

florida-   1. I. floribus rubris. *Syft. veget.* 507.
num.   Red-flower'd Anifeed Tree.
Nat. of Florida.
Introd. 1766, by John Ellis, Efq.
Fl. April——July.   G. H. ♄.

LIRIODENDRON. *Gen. pl.* 689.

Cal. 3-phyllus. *Petala* 9. *Sem.* imbricata in ftrobilum.

Tulipife-   1. L. foliis lobatis. *Sp. pl.* 755.
ra.   Common Tulip Tree.
Nat. of North America.
                                                          Cult.

Cult. 1688, by Bifhop Compton. *Raj. hift.* 2. *p.* 1798.
Fl. June and July.   H. ♄ .

MAGNOLIA. *Gen. pl.* 690.

Cal. 3-phyllus. *Pet.* 9. *Capf.* 2-valves, imbricatæ.
Sem. baccata, pendula.

1. M. foliis perennantibus oblongis, petalis obovatis.   *grandiflo-*
Magnolia grandiflora. *Sp. pl.* 755.   *ra.*
α foliis oblongo-ellipticis coriaceis, floribus fubcon-   elliptica.
tractis.
Common Laurel-leav'd Magnolia.
β foliis obovato-oblongis, floribus expanfis.   obovata.
Broad Laurel-leav'd Magnolia.
γ foliis oblongo-lanceolatis apice flexis, floribus fubcon-   lanceola-
tractis.   ta.
Long Laurel-leav'd Magnolia.
Nat. of Florida and Carolina.
Cult. before 1737, by Sir John Colliton, *Catefb. carol.*
2. *p.* 61.
Fl. June——September.   H. ♄ .

2. M. foliis ovato-oblongis fubtus glaucis. *Sp. pl.* 755.   *glauca.*
α foliis deciduis.   latifolia.
Deciduous Swamp Magnolia.
β foliis perennantibus.   longifo-
Evergreen Swamp Magnolia.   lia.
Nat. of North America.
Cult. 1688, by Bifhop Compton. *Raj. hift.* 2. *p.* 1798.
Fl. June——September.   H. ♄ .

3. M. foliis ovato-oblongis acuminatis. *Sp. pl.* 756.   *acumina-*
Blue Magnolia.   *ta.*
Nat. of North America.
                                                   Introd.

Introd. 1736, by Peter Collinfon, Efq.   *Coll. mff.*
Fl. May and June.   H. ♄.

tripetala.   4. M. foliis lanceolatis, petalis exterioribus dependenti-
bus. *Sp. pl.* 756.
Umbrella Magnolia.
Nat. of Carolina.
Cult. 1752. *Mill. dict. ed.* 6. *n.* 4.
Fl. May and June.   H. ♄.

A N N O N A.   *Gen. pl.* 693.

Cal. 3-phyllus.   *Petala* 6.   *Bacca* polyfperma, fubrotunda, cortice fquamato.

muricata.   1. A. foliis ovali-lanceolatis glabris acutis, fructibus
muricatis, petalis ovatis: interioribus obtufis bre-
vioribus.
Annona muricata. *Sp. pl.* 756.
Rough-fruited Cuftard-apple, or Sour-fop.
Nat. of the Weft Indies.
Cult. 1656, by Mr. John Tradefcant, Jun. *Muf.*
Trad. 155.
Fl.   S. ♄.

tripetala.   2. A. foliis ovatis acutis fubtus pubefcentibus, floribus
tripetalis: petalis lanceolatis coriaceis tomentofis.
Annona Cherimola. *Mill. dict.*
Anona foliis ovatis acutis, flore albido ungue purpu-
reo, fructu coniformi tuberofo nigricante. *Trew.*
Ehret. p. 16. *t.* 49.
Guanabanus Perfeæ folio, flore intus albo, exterius
virefcente, fructu nigricante fquamato, vulgo Che-
rimolia. *Feuillée it.* 3. *p.* 24. *t.* 17.
Broad-leav'd Cuftard-apple.

Nat.

Nat. of South America.
Cult. 1739, by Mr. Philip Miller. *Rand. chel. Addenda.* Guanabanus.
Fl. July and Auguft.   S. ♄.

3. A. foliis oblongis acutis glabris, fructibus obtufe fqua- *fquamofa.*
mofis, petalis exterioribus lanceolatis ; interiori-
bus minutis.
Annona fquamofa. *Sp. pl.* 757.
Undulated Cuftard-apple.
Nat. of South America.
Cult. 1739, by Mr. Philip Miller. *Rand. chel. n.* 2.
Fl.   S. ♄.

4. A. foliis oblongo-lanceolatis acutis glabris, fructibus *reticula-*
ovatis reticulato-areolatis, petalis exterioribus lan- *ta.*
ceolatis ; interioribus minutis.
Annona reticulata. *Sp. pl.* 757.
Netted Cuftard-apple.
Nat. of South America.
Cult. 1690, in the Royal Garden at Hampton-court.
Catal. mff.
Fl.   S. ♄.

5. A. foliis elliptico-oblongis acutis glabris, petalis fpa- *hexapeta-*
thulatis æqualibus acutis. *la.*
Annona hexapetala. *Linn. fuppl.* 270.
Long-leav'd Cuftard-apple.
Nat. of China and the Eaft Indies.
Introd. 1758, by Hugh Duke of Northumberland.
Fl. June and July.   S. ♄.

6. A. foliis oblongis obtufiufculis glabris, fructibus areo- *paluſris.*
latis. *Sp. pl.* 757.
Shining-leav'd Cuftard-apple.
Nat. of the Weft Indies.
                                              Introd.

Introd. 1778, by Thomas Clarke, M.D.
Fl.   S. ♄ .

triloba.   7. A. foliis ellipticis acutis glabris, floribus pendulis
campanulatis, calycibus ovatis, petalis pluribus ova-
libus.
Annona triloba. *Sp. pl.* 758.
Trifid-fruited Cuſtard-apple.
Nat. of North America.
Introd. 1736, by Peter Collinſon, Eſq.   *Coll. mſſ.*
Fl. Auguſt.   H. ♄ .

A N E M O N E.   *Gen. pl.* 694.
Cal. o.   *Petala* 6—9.   *Sem.* plura.

* Hepaticæ, *flore ſubcalyculato.*

Hepatica. 1. A. foliis trilobis integerrimis.   *Sp. pl.* 758.
α flore cæruleo ſimplici.
Blue Hepatica.
β flore cæruleo pleno.
Double Blue Hepatica.
γ flore rubro ſimplici. *Curtis magaz.* 10.
Red Hepatica.
δ flore rubro pleno.
Double Red Hepatica.
ε flore albo ſimplici.
White Hepatica.
Nat. of Europe.
Cult. 1596, by Mr. John Gerard.   *Hort. Ger.*
Fl. February——April.   H. ♃ .

** Pulſatillæ, *pedunculo involucrato, ſeminibus caudatis.*
patens. 2. A. pedunculo involucrato, foliis digitatis multifidis.
Sp. pl. 759.

Woolly-

Woolly-leav'd Anemone.

Nat. of Siberia.

Cult. 1759, by Mr. Philip Miller. *Mill. dict. edit.* 7.
  Pulſatilla 4.

Fl. June and July.   H. ♃.

3. A. pedunculo involucrato, petalis rectis, foliis bipin-   *Pulſatil-*
    natis. *Sp. pl.* 759.   *la.*
  Paſque-flower Anemone.
  Nat. of England.
  Fl. April and May.   H. ♃.

4. A. pedunculo involucrato, petalis apice reflexis, fo-   *pratenſis.*
    liis bipinnatis. *Sp. pl.* 760.
  Meadow Anemone.
  Nat. of Germany.
  Cult. 1731, by Mr. Philip Miller. *Mill. diſt. edit.* 1.
    Pulſatilla 3.
  Fl. May.   H. ♃.

   *** Anemones, *caule folioſo, ſeminibus caudatis.*

5. A. foliis caulinis ternis connatis ſupradecompoſitis   *alpina,*
    multifidis, ſeminibus hirſutis caudatis. *Sp. pl.* 760.
    Jacqu. auſtr. 1. *p.* 53. *t.* 85.
  Alpine Anemone.
  Nat. of Switzerland and Auſtria.
  Cult. 1731, by Mr. Philip Miller. *Mill. dict. edit.* 1.
    Pulſatilla 5.
  Fl. July.   H. ♃.

6. A. foliis radicalibus ternato-decompoſitis, involucro   *corona-*
    folioſo. *Sp. pl.* 760.   *ria.*
  Narrow-leav'd Garden Anemone.
  Nat. of the Levant.

                                                  Cult.

Cult. 1596, by Mr. John Gerard. *Hort. Ger.*
Fl. April and May.   H. ♃.

hortenſis. 7. A. foliis digitatis, ſeminibus lanatis. *Syſt. veget.* 510.
Broad-leav'd Garden Anemone.
Nat. of Italy.
Cult. 1597, by Mr. John Gerard. *Ger. herb.* 303. *f.* 5.
Fl. April.   H. ♃.

**** Anemonoideæ, *flore nudo, ſeminibus ecaudatis.*

ſylveſtris. 8. A. pedunculo nudo, ſeminibus ſubrotundis hirſutis
muticis. *Syſt. veget.* 510. *Curtis magaz.* 54.
Large white-flower'd Wood Anemone.
Nat. of Germany.
Cult. 1596, by Mr. John Gerard. *Hort. Ger.*
Fl. April and May.   H. ♃.

virginia- 9. A. pedunculis alternis longiſſimis, fructibus cylin-
na.   dricis, ſeminibus hirſutis muticis. *Sp. pl.* 761.
Virginian Anemone.
Nat. of North America.
Cult. 1722, in Chelſea Garden. *R. S. n.* 1.
Fl. May and June.   H. ♃.

penſylva- 10. A. caule dichotomo, foliis ſeſſilibus amplexicaulibus :
nica.   infimis ternis trifidis inciſis. *Syſt. veget.* 510.
Penſylvanian Anemone.
Nat. of Canada and Penſylvania.
Cult. 1766, by Mr. James Gordon.
Fl. May and June.   H. ♃.

nemoroſa. 11. A. ſeminibus acutis, foliolis inciſis, caule unifloro.
Sp. pl. 762. *Curtis lond.*
Wood Anemone.

Nat.

Nat. of Britain.

Fl. March——May. H. ♃.

12. A. feminibus acutis, foliolis incifis, petalis lanceo-   *apennina.*
latis numerofis. *Sp. pl.* 762.
Mountain Wood Anemone.
Nat. of England.
Fl. March. H. ♃.

13. A. feminibus acutis, foliolis incifis, petalis fubrotun-   *ranuncu-*
dis, caule fub-bifloro. *Sp. pl.* 762.   *loides.*
Yellow Wood Anemone.
Nat. of England.
Fl. March. H. ♃.

14. A. floribus umbellatis, feminibus depreffo-ovalibus   *narciffi-*
nudis. *Sp. pl.* 763. *Jacqu. auftr.* 2. *p.* 38. *t.* 159.   *flora.*
Narciffus-flower'd Anemone.
Nat. of Siberia, Auftria, and Switzerland.
Introd. 1773, by John Earl of Bute.
Fl. May. H. ♃.

15. A. floribus umbellatis, foliis caulinis fimplicibus ver-   *Thalic-*
ticillatis; radicalibus biternatis. *Syft. veget.* 511.   *troides.*
Meadow Rue-leav'd Anemone.
Nat. of North America.
Cult. 1768, by Mr. Philip Miller. *Mill. dict. edit.* 8.
Fl. April and May. H. ♃.

A T R A G E N E. *Gen. pl.* 695.
Cal. 4-phyllus. *Petala* 12. *Sem.* caudata.

1. A. foliis duplicato-ternatis ferratis, petalis exteriori-   *alpina.*
bus quaternis. *Sp. pl.* 764. *Jacqu. auftr.* 3. *p.* 24.
t. 241.

Vol. II.   S   Alpine

Alpine Atragene.

Nat. of the Alps of Siberia and Auſtria.

Cult. 1759, by Mr. Philip Miller. *Mill. dict. edit.* 7. Clematis 9.

Fl. June and July.   H. ♄.

C L E M A T I S.   *Gen. pl.* 696.

Cal. o.   *Petala* 4—6.   *Sem.* caudata.

* *Scandentes.*

cirrhoſa.   1. C. foliis ſimplicibus, caule cirrhis oppoſitis ſcandente, pedunculis unifloris lateralibus. *Syſt. veget.* 512.

Evergreen Virgin's Bower.

Nat. of Spain.

Cult. 1596, by Mr. John Gerard.   *Hort. Ger.*

Fl. December——February.   H. ♄.

florida.   2. C. foliis decompoſitis: foliolis binatis ternatiſque, petalis ovatis. *Thunb. japon.* 240. *Syſt. veget.* 512.

Japan Virgin's Bower.

Nat. of Japan.

Introd. about 1776, by John Fothergill, M.D.

Fl. moſt part of the Year.   S. ♄.

Viticella.   3. C. foliis compoſitis decompoſitiſque: foliolis ovatis ſublobatis integerrimis. *Syſt. veget.* 512.

α floribus ſimplicibus.

Purple Virgin's Bower.

β floribus plenis.

Double purple Virgin's Bower.

Nat. of Spain and Italy.

Cult. 1569, by Mr. Hugh Morgan.   *Lobel. adv.* 276.

Fl. June——September.   H. ♄.

4. C.

4. C. foliis compofitis decompofitifque : foliolis quibuf- *Viorna.*
    dam trifidis. *Sp. pl.* 765.
Leathery-flower'd Virgin's Bower.
Nat. of Carolina and Virginia.
Cult. 1732, by James Sherard, M. D. *Dill. elth.*
    144. *t.* 118. *f.* 144.
Fl. June——September.   H. ♄.

5. C. involucro calycino approximato, foliis ternatis : *calycina.*
    intermedio tripartito. *L'Herit. ſtirp. nov. tom.* 2.
    tab. 26.
Clematis balearica. *De Lamarck encycl.* 2. *p.* 43.
Minorca Virgin's Bower.
Nat. of Minorca.
Introd. 1783, by Monf. Thouin.
Fl. February.   G. H. ♄.

6. C. foliis fimplicibus ternatifque : foliolis integris tri- *criſpa.*
    lobifve. *Sp. pl.* 765.
Curl'd Virgin's Bower.
Nat. of Carolina and Florida.
Introd. 1726, by Mr. Philip Miller. *Mill. dict. edit.* 8.
Fl. July and Auguft.   H. ♄.

7. C. foliis compofitis : foliolis incifis angulatis lobatis *orientalis.*
    cuneiformibus, petalis interne villofis. *Sp. pl.* 765.
Oriental Virgin's Bower.
Nat. of the Levant.
Cult. 1732, by James Sherard, M. D. *Dill. elth.* 144.
    t. 119. *f.* 145.
Fl. July——October.   H. ♄.

8. C. foliis ternatis : foliolis cordatis fublobato-angula- *virginia-*
    tis fcandentibus, floribus dioicis. *Sp. pl.* 766.   *na.*
Virginian Virgin's Bower.

          Nat.

Nat. of North America.
Cult. 1767, by James Gordon.
Fl. June——Auguſt.   H. ♄.

Vitalba.   9. **C.** foliis pinnatis: foliolis cordatis ſcandentibus. *Sp.*
pl. 766. *Jacqu. auſtr.* 4. *p.* 4. *t.* 308. *Curtis lond.*
Common Virgin's Bower, or Traveller's Joy.
Nat. of England.
Fl. July——September.   H. ♄.

Flammu-   10. **C.** foliis inferioribus pinnatis lacinatis ; ſummis ſim-
la.   plicibus integerrimis lanceolatis. *Syſt. veget.* 512.
Sweet-ſcented Virgin's Bower.
Nat. of the South of France.
Cult. 1597, by Mr. John Gerard. *Ger. herb.* 741.
f. 1.
Fl. July——Oĉtober.   H. ♄.

** *Erecĉtæ.*

recĉta.   11. **C.** foliis pinnatis: foliolis ovato-lanceolatis integerrimis, caule erecĉto, floribus pentapetalis tetrape-
taliſque. *Sp. pl.* 767. *Jacqu. auſtr.* 3. *p.* 49.
t. 291.
Upright Virgin's Bower.
Nat. of Hungary, Auſtria, and France.
Cult. 1597, by Mr. John Gerard. *Ger. herb.* 741.
f. 2.
Fl. June——Auguſt.   H. ♃.

ochroleu-   12. **C.** foliis ſimplicibus ovatis pubeſcentibus integerri-
ca.   mis, floribus erecĉtis.
Clematis erecĉta humilis non ramoſa, foliis ſubrotundis, flore unico ochroleuco. *Pluk. mant.* 51. *t.* 379.
f. 5.
Yellow-flower'd Virgin's Bower.

Nat.

Nat. of North America.
Cult. 1767, by Mr. James Gordon.
Fl. June and July.   H. ♃.

13. C. foliis fimplicibus feffilibus ovato-lanceolatis, flori-   *integrif-*
bus cernuis. *Sp. pl.* 767. *Jacqu. auftr.* 4. *p.* 33.   *lia.*
t. 363. *Curtis magaz.* 65.
Intire-leav'd Virgin's Bower.
Nat. of Hungary and Auftria.
Cult. 1596, by Mr. John Gerard.   *Hort. Ger.*
Fl. June——Auguft.   H. ♃.

THALICTRUM. *Gen. pl.* 697.

Cal. 0.   *Petala* 4 f. 5.   *Sem.* ecaudata.

1. T. caule fimpliciffimo fubnudo, racemo fimplici ter-   *alpinum.*
minali. *Sp. pl.* 767.
Alpine Meadow Rue.
Nat. of Britain.
Fl. May——July.   H. ♃.

2. T. caule paniculato filiformi ramofiffimo foliofo. *Sp.*   *fœtidum.*
pl. 768.
Fœtid Meadow Rue.
Nat. of France and Switzerland.
Cult. 1739, by Mr. Philip Miller.   *Rand. chel. n.* 6.
Fl. May——July.   H. ♃.

3. T. floribus pentapetalis, radice tuberofa. *Sp. pl.* 768.   *tubero-*
Tuberous-rooted Meadow Rue.   *fum.*
Nat. of Spain.
Cult. 1713. *Philofoph. tranf. n.* 337. *p.* 198. *n.* 73.
Fl. June.   H. ♃.

S 3   4. T.

Cornuti.   4. T. floribus dioicis, foliolis ovatis trifidis, paniculis
        terminalibus.
        Thaliċtrum Cornuti.   *Sp. pl.* 768.
        Canadian Meadow Rue.
        Nat. of North America.
        Cult. 1683, by Mr. James Sutherland.   *Sutherl. hort.*
        edin. 332.
        Fl. May——July.   H. ♃.

dioicum.   5. T. floribus dioicis, foliolis fubrotundis cordatis loba-
        tis : lobis obtufis, pedunculis axillaribus folio bre-
        vioribus.
        Thaliċtrum dioicum.   *Sp. pl.* 768.
        Dioecious Meadow Rue.
        Nat. of North America.
        Cult. 1759, by Mr. Ph. Miller. *Mill. dict. edit.* 7. *n.* 9.
        Fl. June and July.   H. ♃.

minus.   6. T. foliis fexpartitis, floribus cernuis.   *Sp. pl.* 769.
        Jacqu. auſtr. 5. *p.* 9. *t.* 419.
        Small Meadow Rue.
        Nat. of Britain.
        Fl. May——July.   H. ♃.

rugoſum.   7. T. caule ſtriato, foliolis rugofis venofis : lobulis
        obtufis.
        Rough Meadow Rue.
        Nat. of North America.
        Introd. 1774, by John Fothergill, M. D.
        Fl. July.   H. ♃.

ſibiricum.   8. T. foliis tripartitis : foliolis fubreflexis argute incifis,
        floribus cernuis.   *Sp. pl.* 769.
        Siberian Meadow Rue.
        Nat. of Siberia.

                                           Introd.

Introd. 1775, by Monf. Thouin.
Fl. June and July.   H. ♃.

9. **T.** foliolis lanceolato-linearibus integerrimis. *Sp.*   *angufti-*
    pl. 769. *Jacqu. hort.* 3. *p.* 25. *t.* 43.   *folium.*
    Narrow-leav'd Meadow Rue.
    Nat. of Germany.
    Cult. 1739, by Mr. Ph. Miller. *Rand. chel. n.* 4.
    Fl. June and July.   H. ♃.

10. **T.** caule foliofo fulcato, panicula multiplici erecta. *flavum.*
    Sp. pl. 770.
    Common Meadow Rue.
    Nat. of Britain.
    Fl. May——July.   H. ♃.

11. **T.** caule foliofo fimpliciffimo angulato. *Linn. mant.* *fimplex.*
    78.
    Simple-ftalk'd Meadow Rue.
    Nat. of Sweden.
    Introd. 1778, by Monf. Thouin.
    Fl. May and June.   H. ♃.

12. **T.** caule foliofo fulcato, foliis linearibus carnofis. *Sp.* *lucidum.*
    pl. 770.
    Shining-leav'd Meadow Rue.
    Nat. of France and Spain.
    Cult. 1739, by Mr. Ph. Miller. *Rand. chel. n.* 5.
    Fl. May——July.   H. ♃.

13. **T.** fructibus pendulis triangularibus rectis, caule te- *aquilegi-*
    reti. *Sp. pl.* 770. *Jacqu. auftr.* 4. *p.* 10. *t.* 318.   *folium.*
    Columbine-leav'd Meadow Rue, or Feather'd Co-
    lumbine.
    Nat. of Switzerland and Auftria.

                    S 4   *Cult.*

Cult. 1731, by Mr. Ph. Miller. *Mill. dict. edit.* 1. *n.* 1.
Fl. May——July.   H. ♃.

A D O N I S. *Gen. pl.* 698.

Cal. 5-phyllus. *Petala* quinis plura, abfque nectario.
Sem. nuda.

æftivalis. 1. A. floribus pentapetalis, fructibus ovatis. *Sp. pl.* 771.
         Tall Adonis.
         Nat. of the South of Europe.
         Cult. 1768, by Mr. Ph. Miller. *Mill. dict. edit.* 8.
         Fl. June and July.   H. ☉.

autum- 2. A. floribus octopetalis, fructibus fubcylindricis. *Sp.*
nalis.   *pl.* 771. *Curtis lond.*
         Common Flos Adonis.
         Nat. of England.
         Fl. June and July.   H. ☉.

vernalis. 3. A. flore dodecapetalo, fructu ovato. *Sp. pl.* 771.
         Perennial, or Spring Adonis.
         Nat. of Bohemia and Oeland.
         Cult. 1731, by Mr. Ph. Miller. *Mill. dict. edit.* 1. *n.* 3.
         Fl. March and April.   H. ♃.

veficato- 4. A. floribus decapetalis, follis biternatis: foliolis fer-
ria.   ratis glabris. *Linn. fuppl.* 272.
         Imperatoria Ranunculoides Africana enneaphyllos,
           Laferpitii lobatis foliis rigidis margine fpinofis.
           Pluk. alm. 198. *phyt. tab.* 95. *fig.* 2.
         Nat. of the Cape of Good Hope.
         Cult. 1691, in the Royal Garden at Hampton-court.
           Pluk. phyt. loc. cit.
         Fl. February——April.   G. H. ♃.

RANUN-

RANUNCULUS. *Gen. pl.* 699.

Cal. 5-phyllus. *Petala* 5, intra ungues poro mellifero. *Sem.* nuda.

* *Foliis fimplicibus.*

1. R. foliis ovato-lanceolatis petiolatis, caule declinato. *Sp. pl.* 772.
Small Spear-wort Crowfoot.
Nat. of Britain.
Fl. June——Auguft. H. ♃.

Flammula.

2. R. foliis lanceolatis, caule erecto. *Sp. pl.* 773.
Great Spear-wort Crowfoot.
Nat. of Britain.
Fl. June——Auguft. H. ♃.

Lingua.

3. R. foliis lanceolato-linearibus indivifis, caule erecto læviffimo paucifloro. *Syft. veget.* 515.
Grafs-leav'd Crowfoot.
Nat. of France and Spain.
Cult. 1596, by Mr. John Gerard. *Hort. Ger.*
Fl. April and May. H. ♃.

gramineus.

4. R. foliis fubovatis nervofis lineatis integerrimis petiolatis, floribus umbellatis. *Syft. veget.* 515.
Parnaffia-leav'd Crowfoot.
Nat. of the South of Europe.
Introd. 1769, by Meffrs. Kennedy and Lee.
Fl. H. ♃.

parnaffifolius.

5. R. foliis ovatis acuminatis amplexicaulibus, caule multifloro, radice fafciculata. *Sp. pl.* 774.
Plantain-leav'd Crowfoot.
Nat. of the Apennines and Pyrenees.

amplexicaulis.

Cult.

Cult. 1633. *Ger. emac.* 963. *f.* 2.
Fl. April and May.   H. ♃.

Ficaria.   6. R. foliis cordatis angulatis petiolatis, caule unifloro.
         Syſt. veget. 515. *Curtis lond.*
        α floribus ſimplicibus.
        Pilewort Crowfoot, or Leſſer Celandine.
        β floribus plenis.
        Double Pilewort.
        Nat. of Britain.
        Fl. April.   H. ♃.

Thora.   7. R. foliis reniformibus ſubtrilobis crenatis : caulino
         ſeſſili ; floralibus lanceolatis, caule ſubbifloro. *Sp.*
         pl. 775. *Jacqu. auſtr.* 5. *p.* 21. *t.* 442.
        Kidney-leav'd Crowfoot.
        Nat. of the Alps of Switzerland and Auſtria.
        Cult. 1710. *Salmon's herb.* 616.
        Fl. May and June.   H. ♃.

              * * *Foliis diſſectis & diviſis.*

auricо-   8. R. foliis radicalibus reniformibus crenatis inciſis ;
mus.   caulinis digitatis linearibus, caule multifloro. *Sp.*
         pl. 775. *Curtis lond.*
        Wood Crowfoot, or Goldilocks.
        Nat. of Britain.
        Fl. April.   H. ♃.

ſcelera-   9. R. foliis inferioribus palmatis ; ſummis digitatis,
tus.   fruct̄ibus oblongis. *Sp. pl.* 776. *Curtis lond.*
        Marſh Crowfoot.
        Nat. of Britain.
        Fl. May and June.   H. ☉.

                          10. R.

10. R. foliis omnibus quinatis lanceolatis incifo-ferratis. *aconitifo-*
Sp. pl. 776 *lius.*
α floribus fimplicibus.
Aconite-leav'd Crowfoot, or Fair maids of France.
β floribus plenis.
Double Aconite-leav'd Crowfoot.
Nat. of the Alps of Europe.
Cult. 1597. *Ger. herb.* 812. *f.* 1.
Fl. May and June. H. ♃.

11. R. foliis palmatis lævibus incifis, caule erecto, brac- *platanifo-*
teis linearibus. *Linn. mant.* 79. *lius.*
Platanus-leav'd Crowfoot.
Nat. of the Alps of Germany and Italy.
Introd. 1769, by Meffrs. Kennedy and Lee.
Fl. June and July. H. ♃.

12. R. foliis ternatis integerrimis lanceolatis. *Sp. pl.* *illyricus.*
776. *Jacqu. auftr.* 3. *p.* 13. *t.* 222.
Illyrian Crowfoot.
Nat. of the South of Europe.
Cult. 1596, by Mr. John Gerard. *Hort. Ger.*
Fl. May and June. H. ♃.

13. R. foliis ternatis biternatifque: foliolis trifidis inci- *afiaticus.*
fis, caule inferne ramofo. *Sp. pl.* 777.
Perfian Crowfoot, or Common Garden Ranunculus.
Nat. of the Levant.
Cult. 1596, by Mr. John Gerard. *Hort. Ger.*
Fl. May and June. H. ♃.

14. R. foliis fupradecompofitis, caule fimpliciffimo uni- *rutæfoli-*
folio unifloro, radice tuberofa. *Sp. pl.* 777. *us.*
Rue-leav'd Crowfoot.
Nat. of Auftria.

Cult.

Cult. 1759, by Mr. Philip Miller. *Mill. dict. edit.* 7. *n.* 6.
Fl. May.   H. ♃.

glacialis. 15. R. calycibus hirfutis, caule bifloro, foliis multifidis. *Sp. pl.* 777.
Two-flower'd Crowfoot.
Nat. of Lapland and Switzerland.
Introd. 1775, by the Doctors Pitcairn and Fothergill.
Fl. June.   H. ♃.

nivalis. 16. R. calyce hirfuto, caule unifloro, foliis radicalibus palmatis ; caulinis multipartitis feffilibus. *Sp. pl.* 778. *Jacqu. auftr.* 4. *p.* 13. *t.* 325, 326.
Alpine Yellow Crowfoot.
Nat. of Lapland, Auftria, and Switzerland.
Introd. 1775, by the Doctors Pitcairn and Fothergill.
Fl. June.   H. ♃.

penfylva- 17. R. calycibus reflexis, caule erecto, foliis ternatis tri-
nicus.   fidis incifis fubtus pilofis. *Linn. fuppl.* 272.
Ranunculus canadenfis. *Jacqu. ic. mifc.* 2. *p.* 343.
Penfylvanian Crowfoot.
Nat. of North America.
Introd. 1785, by Chevalier Thunberg.
Fl. June.   H. ♂.

bulbofus. 18. R. calycibus retroflexis, pedunculis fulcatis, caule erecto, foliis compofitis. *Sp. pl.* 778. *Curtis lond.*
Bulbous Crowfoot.
Nat. of Britain.
Fl. May and June.   H. ♃.

birfutus. 19. R. radice fibrofa annua, caule hirfuto, calycibus pa-pillofo-hifpidis acuminatis, demum reflexis. *Curtis lond.*

Pale-

Pale-leav'd Crowfoot.

Nat. of England.

Fl. June——October. H. ⊙.

20. R. calycibus patulis, pedunculis fulcatis, ftolonibus   *repens.*
repentibus, foliis compofitis. *Syft. veg.* 517. *Curtis
lond.*

Creeping Crowfoot.

Nat. of Britain.

Fl. May. H. ♃.

21. R. calycibus patulis, pedunculis teretibus, foliis tri-   *acris.*
partito-multifidis : fummis linearibus. *Sp. pl.*
779. *Curtis lond.*

α floribus fimplicibus.

Upright Crowfoot.

β floribus plenis.

Double Upright Crowfoot.

Nat. of Britain.

Fl. June and July. H. ♃.

22. R. calycibus patulis, pedunculis teretibus, caule pe-   *lanugino-*
tiolifque hirtis, foliis trifidis lobatis crenatis holo-   *fus.*
fericeis. *Sp. pl.* 779.

Broad-leav'd Crowfoot.

Nat. of Switzerland and the South of France.

Cult. 1683, by Mr. James Sutherland. *Sutherl. hort.*
edin. 284. *n.* 4.

Fl. June and July. H. ♃.

23. R. feminibus aculeatis, foliis fuperioribus decompo-   *arvenfis.*
fitis linearibus. *Sp. pl.* 780.

Corn Crowfoot.

Nat. of Britain.

Fl. June. H. ⊙.

24. R.

murica-   24. R. feminibus aculeatis, foliis fimplicibus lobatis ob-
tus.   tufis glabris, caule diffufo. *Sp. pl.* 780.
              Spreading prickly-capful'd Crowfoot.
              Nat. of the South of Europe.
              Cult. 1683, by Mr. James Sutherland. *Sutherl. hort.*
                  edin. 285. *n.* 4.
              Fl. July and Auguft.   H. ☉.

parviflo-   25. R. feminibus muricatis, foliis fimplicibus laciniatis
rus.   acutis hirfutis, caule diffufo. *Sp. pl.* 780.
              Small-flower'd Crowfoot.
              Nat. of England.
              Fl. May.   H. ☉.

falcatus.   26. R. foliis filiformi-ramofis, fcapo nudo unifloro, femi-
                      nibus falcatis. *Sp. pl.* 781. *Jacqu. auftr.* 1. *p.* 30.
                  t. 48.
              Crooked-podded Crowfoot.
              Nat. of the South of Europe and the Levant.
              Introd. 1782, by P. M. A. Brouffonet, M.D.
              Fl. May and June.   H. ☉.

hederace-   27. R. foliis fubrotundis trilobis integerrimis, caule re-
us.   pente. *Sp. pl.* 781. *Curtis lond.*
              Ivy Crowfoot.
              Nat. of Britain.
              Fl. May.   H. ♃.

aquatilis. 28. R. foliis fubmerfis capillaceis : emerfis peltatis. *Syft.*
                  veget. 518.
              α Ranunculus foliis inferioribus capillaceis, fuperioribus
                  peltatis. *Roy. lugdb.* 492.
              Various-leav'd Water Crowfoot.
              β Ranunculus foliis omnibus capillaceis circumfcrip-
                  tione rotundis. *Roy. lugdb.* 492.
                                                        Fennel-

Fennel-leav'd Water Crowfoot.

γ Ranunculus foliis omnibus capillaceis circumfcrip-
tione oblongis. *Roy. lugdb.* 492.

Long-leav'd Water Crowfoot.

Nat. of Britain.

Fl. April——Auguft.　　　　　　　　　　H. ♃.

T R O L L I U S.　*Gen. pl.* 700.

Cal. o.　*Petala* circiter 14.　*Capf.* plurimæ, ovatæ,
polyfpermæ.

1. T. corollis conniventibus, nectariis longitudine fta-　*europæus.*
minum. *Sp. pl.* 782.

European Globe-flower.

Nat. of Britain.

Fl. May and June.　　　　　　　　　　H. ♃.

2. T. corollis patulis, nectariis ftamine longioribus. *Syft.*　*afiaticus.*
veget. 518.

Afiatic Globe-flower.

Nat. of Siberia.

Cult. 1759, by Mr. Ph. Miller. *Mill. dict. edit.* 7. *n.* 2.

Fl. June.　　　　　　　　　　　　　H. ♃.

I S O P Y R U M.　*Gen. pl.* 701.

Cal. o.　*Petala* 5.　*Nectaria* 3-fida, tubulata.　*Capf.*
recurvæ, polyfpermæ.

1. I. ftipulis fubulatis, petalis acutis. *Sp. pl.* 783.　*fumarioi-*
Fumatory-leav'd Ifopyrum.　　　　　　　　*des.*

Nat. of Siberia.

Cult. 1759, by Mr. Ph. Miiler. *Mill. dict. edit.* 7. *n.* 1.

Fl. June.　　　　　　　　　　　　　H. ☉.

H E L L E B O-

HELLEBORUS. *Gen. pl.* 702.

Cal. o. *Petala* 5 f. plura. *Nectaria* bilabiata, tubulata. *Capf.* polyfpermæ, erectiufculæ.

hyemalis.   1. H. flore folio infidente. *Sp. pl.* 783. *Jacqu. auftr.* 3.
p. 1. *t.* 202. *Curtis magaz.* 3.
Winter Hellebore, or Aconite.
Nat. of Italy and Auftria.
Cult. 1596, by Mr. John Gerard.   *Hort. Ger.*
Fl. January——March.   H. ♃.

niger.   2. H. fcapo fubbifloro fubnudo, foliis pedatis. *Syft. veg.*
519. *Jacqu. auftr.* 3. *p.* 1. *t.* 201. *Curtis magaz.* 8.
Black Hellebore, or Chriftmas Rofe.
Nat. of Auftria and Italy.
Cult. 1596, by Mr. John Gerard.   *Hort. Ger.*
Fl. January——March.   H. ♃.

viridis.   3. H. caule bifido, ramis foliofis bifloris, foliis digitatis.
Syft. veget. 519. *Jacqu. auftr.* 2. *p.* 4. *t.* 106.
Green Hellebore.
Nat. of Britain.
Fl. March.   H. ♃.

fœtidus.   4. H. caule multifloro foliofo, foliis pedatis. *Sp. pl.* 784.
Fœtid Hellebore, or Bear's-foot.
Nat. of England.
Fl. November and December.   H. ♃.

lividus.   5. H. caule multifloro foliofo, foliis ternatis.
Helleborus fœtidus β.   *Sp. pl.* 784.
Helleborus niger trifoliatus. *Aldini hort. farnef.* 93.
t. 92. *Morif. hift.* 3. *p.* 460. *f.* 12. *t.* 4. *f.* 7.
Great three-leav'd black Hellebore.

Nat.

Nat.
Cult. 1731, by Mr. Ph. Miller. *Mill. dict. edit.* 1. *n.* 4.
Fl. January——May. H. ♃.

6. H. ſcapo unifloro, foliis ternatis. *Sp. pl.* 784. *trifolius.*
Small three-leav'd Hellebore.
Nat. of Canada, Hudſon's-bay, and Siberia.
Introd. 1782, by the Hudſon's-bay Company.
Fl. June and July. H. ♃.

C A L T H A. *Gen. pl.* 703.
Cal. o. *Petala* 5. *Nectaria* o. *Capſ.* plures, polyſpermæ.

1. CALTHA. *Sp. pl.* 784. *Curtis lond.* *paluſtris.*
α Caltha paluſtris, flore ſimplici. *Bauh. pin.* 276.
Marſh Marygold.
β Caltha paluſtris, flore pleno. *Bauh. pin.* 276.
Double Marſh Marygold.
Nat. of Britain.
Fl. April. H. ♃.

H Y D R A S T I S. *Gen. pl.* 704.
Cal. o. *Petala* 3. *Nectaria* o. *Bacca* compoſita
acinis monoſpermis.

1. HYDRASTIS. *Sp. pl.* 784. *canaden-*
Canadian Hydraſtis, or Yellow Root. *ſis.*
Nat. of Canada.
Cult. 1759, by Mr. Ph. Miller. *Mill. ic.* 190. *t.* 285.
Fl. May and June. H. ♃.

Claſſis XIV.

DIDYNAMIA

GYMNOSPERMIA.

A J U G A. *Gen. pl.* 705.

Corollæ labium ſuperius minimum. *Stamina* labio ſuperiore longiora.

orientalis. 1. A. floribus reſupinatis. *Sp. pl.* 785.
Oriental Bugle.
Nat. of the Levant.
Cult. 1732, by James Sherard, M.D. *Dill. elth.* 60.
t. 53. *f.* 61.
Fl. May and June. H. ♃.

pyrami- 2. A. tetragono-pyramidalis villoſa, foliis radicalibus
dalis. maximis. *Syſt. veget.* 525.
Mountain Bugle.
Nat. of Britain.
Fl. June. H. ♃.

alpina. 3. A. caule ſimplici, foliis caulinis radicalia æquantibus.
Lin. mant. 80.
Alpine Bugle.
Nat. of Switzerland.
Introd. 1775, by the Doctors Pitcairn and Fothergill.
Fl. May and June. H. ♃.

geneven- 4. A. foliis tomentoſis lineatis, calycibus hirſutis. *Syſt.*
ſis. *veget.* 525.

Fleſh-

Flesh-colour'd Bugle.

Nat. of Switzerland.

Cult. 1759, by Mr. Philip Miller. *Mill. dict. edit.* 7.
  Bugula 4.

Fl. May and June.   H. ♃.

5. A. glabra, stolonibus reptantibus. *Syst. veget.* 525.   *reptans.*
  Curtis lond.
α floribus cæruleis.
  Common Bugle.
β floribus albis.
  White Bugle.
  Nat. of Britain.
  Fl. May and June.   H. ♃.

TEUCRIUM. *Gen. pl.* 706.

Corollæ labium superius (nullum) ultra basin 2-parti-
  tum, divaricatum ubi stamina.

1. T. foliis multifidis, floribus lateralibus solitariis. *Syst.*   *campa-*
  veget. 525.   *nulatum.*
  Small-flower'd Germander.
  Nat. of the Levant.
  Cult. 1728, by Mr. Philip Miller. *R. S. n.* 310.
  Fl. July and August.   H. ♃.

2. T. foliis multifidis, floribus lateralibus ternis pedun-   *Botrys.*
  culatis. *Syst. veget.* 525.
  Cut-leav'd annual Germander.
  Nat. of the South of Europe.
  Cult. 1633. *Ger. emac.* 525. *f.* 2.
  Fl. July——September.   H. ☉.

3. T. foliis trifidis linearibus integerrimis, floribus ses-   *Chamæ-*
                         T 2   silibus   *pitys.*

filibus lateralibus folitariis, caule diffufo. *Sp. pl.*
787.
Ground-pine Germander.
Nat. of England.
Fl. June and July.   H. ⊙.

niffolia-   4. T. foliis trifidis quinquefidifque filiformibus, floribus
num.   pedunculatis folitariis oppofitis, caule decumbente.
                  Sp. pl. 786.
              Trifid-leav'd Germander.
              Nat. of Spain and Portugal.
              Cult. 1768, by Mr. Philip Miller.   *Mill. dict. edit* 8.
              Fl. June and July.   G. H. ♃.

fruticans.   5. T. foliis integerrimis ellipticis fubtus tomentofis, flo-
                  ribus lateralibus folitariis pedunculatis. *Syft. veget.*
                  526.
              Narrow-leav'd Tree Germander.
              Nat. of Spain and Sicily.
              Cult. 1640.   *Park. theat.* 110. *f.* 3.
              Fl. June——September,   G. H. ♄.

latifoli-   6. T. foliis integerrimis rhombeis acutis villofis : fubtus
um.   tomentofis. *Sp. pl.* 788.
              Broad-leav'd Tree Germander.
              Nat. of Spain.
              Cult. 1714, by the Dutchefs of Beaufort. *Br. Muf.*
              H. S. 134. *fol.* 15.
              Fl. June——September.   G. H. ♄.

Marum.   7. T. foliis integerrimis ovatis acutis petiolatis fubtus
                  tomentofis, floribus racemofis fecundis. *Syft. veget.*
                  526.
              Marum Germander, or Cat-thyme.
              Nat. of Spain.
                                                            Cult.

Cult. 1640. *Park. theat.* 17. *f.* 2.
Fl. July——September.   G. H. ♄.

8. T. foliis ovatis fuperne glabris ferrato-dentatis, flori- *multifla-*
    bus racemofis : verticillis fexfloris. *Syft. veget.* 526.   *rum.*
    Many-flower'd Germander.
    Nat. of Spain.
    Cult. 1731, by Mr. Philip Miller. *Mill. dict. edit.* 1.
      Chamædrys 4.
    Fl. July——September.   H. ♃.

9. T. foliis lanceolatis repando-ferratis bafi rectangulis, *afiaticum.*
    floribus folitariis. *Linn. mant.* 80. *Jacqu. hort.* 3.
    p. 24. *t.* 41.
    Afiatic Germander.
    Nat.
    Introd. 1777, by Monf. Thouin.
    Fl. June——October.   G. H. ♄.

10. T. foliis ovato-lanceolatis ferratis, caule erecto, race- *cana-*
    mo tereti terminali, verticillis hexaphyllis. *Syft.*   *denfe.*
    veget. 527.
    Nettle-leav'd Germander.
    Nat. of North America.
    Cult. 1768, by Mr. Philip Miller. *Mill. dict. edit.* 8.
    Fl. Auguft and September.   H. ♃.

11. T. foliis oblongis acuminatis inæqualiter ferratis pu- *infla-*
    befcentibus, fpicis feffilibus terminalibus, calycibus   *tum.*
    inflatis villofis. *Swartz prodr.* 88.
    Teucrium fubhirfutum, foliis ovatis dentato-ferratis,
      fpicis ftrictioribus craffis terminalibus. *Browne*
      jam. 257.
    Thick-fpiked Germander.
    Nat. of Jamaica.

              T 3   *Introd.*

Introd. 1778, by William Wright, M.D.
Fl. Auguft——October.   S. ♃.

hircani-   12. T. foliis cordato-oblongis obtufis; caule brachiato
cum.   dichotomo, fpicis longiffimis terminalibus feffilibus
                fpiralibus. *Sp. pl.* 789.
            Betony-leav'd Germander.
            Nat. of Perfia.
            Introd. 1763, by John Earl of Bute.
            Fl. Auguft——October.   H. ♃.

Abutiloi- 13. T. foliis cordatis dentatis acuminatis, racemis latera-
des.   libus nutantibus. *L'Herit. ftirp. nov. p.* 84.
            Mulberry-leav'd Germander.
            Nat. of Madeira.   Mr. *Francis Maſſon.*
            Introd. 1777.
            Fl. April and May.   G. H. ♄.

Scorodo-   14. T. foliis cordatis ferratis petiolatis, racemis latera-
nia.   libus fecundis, caule erecto. *Sp. pl.* 789. *Curtis
                lond.*
            Wood Germander, or Sage.
            Nat. of Britain.
            Fl. July.   H. ♃.

maſſili-   15. T. foliis ovatis rugofis incifo-crenatis incanis, cauli-
enſe.   bus erectis, racemis rectis fecundis. *Syſt. veget.*
                527. *Jacqu. hort.* 1. *p.* 41. *t.* 94.
            Sweet-fcented Germander.
            Nat. of the South of France.
            Cult. 1731, by Mr. Philip Miller.   *Mill. dict. edit.* 1.
                Chamædrys 6.
            Fl. June and July.   G. H. ♄.

16. T.

16. T. foliis lanceolatis crenatis tomentofis fubtus inca-   *betoni-*
nis, racemis terminalibus, caule florifero brachiato.   *cum.*
Teucrium betonicum. *L'Herit. ftirp. nov..p.* 83.
tab. 40.
Salvia major folio glauco ferrato. *Sloan. hift.* 1. *p.* 17.
t. 3. *f.* 3.
Hoary Germander.
Nat. of Madeira.
Introd. 1775, by Sir Jofeph Banks, Bart.
Fl. May——Auguft.   G. H. ♄ .

17. T. foliis oblongis feffilibus dentato-ferratis, floribus   *Scordi-*
geminis axillaribus pedunculatis, caule diffufo.   *um.*
Syft. veget. 527.
Water Germander.
Nat. of England.
Fl. Auguft.   H. ♃ .

18. T. foliis cuneiformi-ovatis incifis crenatis petiola-   *Chamæ-*
tis, floribus ternis, caulibus procumbentibus fub-   *drys.*
pilofis. *Syft. veget.* 527.
Creeping, or Common Germander.
Nat. of England.
Fl. May and June.   H. ♃ .

19. T. foliis ellipticis crenatis, floribus lateralibus foli-   *hetero-*
tariis, labio corollæ extus lanato, ramis hetero-   *phyllum.*
phyllis. *L'Herit. ftirp. nov. p.* 84.
White-leav'd Tree Germander.
Nat. of Madeira.
Cult. 1759, by Mr. Philip Miller.
Fl. June.   G. H. ♄ .

20. T. foliis ovatis acute incifo-ferratis glabris, floribus   *lucidum.*
axillaribus ternis, caule erecto lævi. *Syft. veget.*
527.

             Shining

Shining Germander.
Nat. of the South of Europe.
Cult. 1730, by Mr. Philip Miller.   *R. S. n.* 416.
Fl. June——September.   H. ♃.

flavum.   21. T. foliis cordatis obtuſe ſerratis, bracteis integerrimis
concavis, caule fruticoſo, floribus racemoſis ternis.
Syſt. veget. 527.
Yellow-flower'd ſhrubby Germander.
Nat. of the South of Europe.
Cult. 1640.   *Park. theat.* 109. *f.* 1.
Fl. July——September.   H. ♄.

monta-   22. T. corymbo terminali, foliis lanceolatis integerrimis
num.   ſubtus tomentoſis.   *Sp. pl:* 791.
Dwarf Germander.
Nat. of France, Switzerland, and Germany.
Cult. 1711.   *Salmon's herb. p.* 880. *n.* 3.
Fl. July——October.   H. ♄.

ſupinum.   23. T. corymbo terminali, foliis linearibus margine re-
volutis. *Sp. pl.* 791. *Jacqu. auſtr.* 5. *p.* 8. *t.* 417.
Procumbent Germander.
Nat. of Auſtria.
Cult. 1739, by Mr. Philip Miller. *Mill. dict. vol.* 2.
Polium 2.
Fl. June——October.   H. ♂.

pyrenai-   24. T. corymbo terminali, foliis cuneiformi-orbiculatis
cum.   crenatis. *Syſt. veget.* 528.
Pyrenean Germander.
Nat. of the Pyrenees.
Cult. 1731. *Mill. dict. edit.* 1.   Polium 5.
Fl. June——Auguſt.   H. ♃.

25. T,

25. T. capitulis fubrotundis, foliis oblongis obtufis cre- *Polium.*
    natis tomentofis feffilibus, caule proftrato. *Sp. pl.*
    792.

α Polium montanum album. *Bauh. pin.* 221.
    Poley, or White Mountain Germander.

β Polium maritimum fupinum venetum. *Bauh. pin.*
    221.
    Poley, or Sea Germander.
    Nat. of the South of Europe and the Levant.
    Cult. 1562. *Turn. herb. part* 2. *fol.* 96.
    Fl. July——September.   H. ♄.

26. T. capitulis pedunculatis, foliis lanceolatis crenatis *capita-*
    tomentofis, caule erecto. *Sp. pl.* 792.   *tum.*
    Round-headed Germander.
    Nat. of Spain.
    Cult. 1731. *Mill. dict. edit.* 1.   Polium 6.
    Fl. July and Auguft.   H. ♄.

SATUREJA. *Gen. pl.* 707.

Corollæ laciniæ fubæquales. *Stamina* diftantia.

1. S. verticillis faftigiatis, foliis lineari-lanceolatis. *Sp.* *juliana.*
    pl. 793.
    Linear-leav'd Savory.
    Nat. of Italy.
    Cult. 1739, by Mr. Ph. Miller. *Rand. chel.* Thym-
    bra 2.
    Fl. May——September.   G. H. ♃.

2. S. floribus verticillatis hifpidis, foliis oblongis acu- *Thymbra.*
    tis. *Syft. veget.* 528.
    Verticil'd Savory.
    Nat. of the Ifland of Candia.

                                     Cult.

Cult. 1739, by Mr. Ph. Miller. *Rand. chel.* Thymbra 1.

Fl. May——July.   G. H. ♄.

montana. 3. S. pedunculis lateralibus folitariis, floribus fafcicula-
tis faftigiatis, foliis mucronatis lineari-lanceolatis.
Syft. veget. 528.
Winter Savory.
Nat. of the South of France and Italy.
Cult. 1562. *Turn. herb. part* 2. *fol.* 127 *verfo.*
Fl. June and July.   H. ♄.

hortenfis. 4. S. pedunculis bifloris. *Sp. pl.* 795.
Summer Savory.
Nat. of the South of France and Italy.
Cult. 1562. *Turn. herb. part* 2. *fol.* 127 *verfo.*
Fl. June——Auguft.   H. ☉.

capitata. 5. S. floribus fpicatis, foliis carinatis punctatis ciliatis.
Sp. pl. 795.
Ciliated Savory.
Nat. of the Levant.
Cult. 1640. *Park. theat.* 7. *f.* 1.
Fl. June——October.   H. ♄.

viminea. 6. S. pedunculis axillaribus trifloris, involucris linearibus,
foliis lanceolato-ovatis integerrimis. *Sp. pl.* 795.
Twiggy Savory.
Nat. of Jamaica.
Introd. 1783, by Mathew Wallén, Efq.
Fl.   S. ♄.

THYM-

THYMBRA. *Gen. pl.* 708.

Calyx fubcylindricus, bilabiatus, utrinque linea villofa exaratus. *Stylus* femibifidus.

1. T. floribus fpicatis. *Sp. pl.* 795.   *ffcata.*
Spiked Thymbra.
Nat. of the Levant.
Cult. 1699, by Mr. Jacob Bobart. *Morif. hift.* 3. *p.*
361. *n.* 11.
Fl. June and July.   G. H. ♄.

2. T. floribus verticillatis. **Sp. pl. 796.**   *verticil-*
Verticil'd Thymbra.   *lata.*
Nat. of Spain and Italy.
Cult. 1702, by the Dutchefs of Beaufort. *Br. Muf.*
H. S. 137. *fol.* 58.
Fl. June and July.   G. H. ♄.

HYSSOPUS. *Gen. pl.* 709.

Corollæ labium inferius lacinula intermedia crenata. *Stamina* recta, diftantia.

1. H. fpicis fecundis, foliis lanceolatis. *Syft. veget.* 529. *officinalis.*
Jacqu. auftr. 3. *p.* 30. *t.* 254.
α foliis glabris, floribus cæruleis.
Common Hyffop.
β foliis glabris, floribus rubris.
Red-flower'd Hyffop.
γ foliis glabris, floribus albis.
White-flower'd Hyffop.
♂ foliis pilofis.
Hairy Hyffop.
Nat. of the South of Europe.

Cult.

Cult. 1596, by Mr. John Gerard.   *Hort. Ger.*

Fl. June——September.   H. ♄.

Lophan-   2. H. corollis reſupinatis, ſtaminibus inferioribus co-
thus.   rolla brevioribus, foliis cordatis.   *Syſt. veget.* 529.
Jacqu. hort. 2.· *p.* 85. *t.* 182.

Mint-leav'd Hyſſop.

Nat. of Siberia.

Cult. 1759, by Mr. Philip Miller.   *Mill. dict. edit.* 7.
n. 5.

Fl. Auguſt and September.   H. ♃.

Nepetoi-   3. H. caule acuto quadrangulo.   *Sp. pl.* 796.   *Jacqu.*
des.   *hort.* 1. *p.* 28. *t.* 69.

Square-ſtalked Hyſſop.

Nat. of Virginia and Canada.

Cult. 1692, in Oxford Garden.   *Pluk. phyt. t.* 150. *f.* 3.

Fl. Auguſt——October.   H. ♃.

N E P E T A.   *Gen. pl.* 710.

Corollæ labii inferioris *Lobus intermedius* ſubrotundús
crenatus : *laterales* abbreviati e margine faucis reflexo.   *Stamina* approximata.

Cataria.   1. N. floribus ſpicatis, verticillis ſubpedicellatis, foliis petiolatis cordatis dentato-ſerratis.   *Sp. pl.* 796.

Common Catmint.

Nat. of Britain.

Fl. July——September.   H. ♃.

pannoni-   2. N. cymis pedunculatis multifloris, foliis lanceolato-
ca.   oblongis cordatis nudis, corollæ lobis lateralibus
reflexis.

Nepeta pannonica.   *Sp. pl.* 797.   *Jacqu. auſtr.* 2.
p. 18. *t.* 129.

Hungarian

Hungarian Catmint.
Nat. of Hungary and Auſtria.
Cult. 1683, by Mr. James Sutherland. *Sutherl. hort.*
edin. 228. *n.* 2.
Fl. Auguſt —— October. H. ♃.

3. 'N. cymis pedunculatis multifloris hirtis, foliis oblon- *cærulea.*
gis cordatis villoſis ſubſeſſilibus, corollæ lobis la-
teralibus reflexis.
Blue-flower'd Catmint.
Nat.
Introd. 1777, by Caſimir Gomez Ortega, M. D.
Fl. May and June. H. ♃.

4. N. cymis pedunculatis multifloris piloſis, foliis corda- *violacea.*
tis ſubpetiolatis nudiuſculis, corollæ lobis lateralibus
patentibus.
Nepeta violacea. *Sp. pl.* 797.
Violet-colour'd Catmint.
Nat. of Spain.
Cult. 1723, in Chelſea Garden. *R. S. n.* 64.
Fl. July —— September. H. ♃.

5. N. cymis pedunculatis multifloris, foliis petiolatis ob- *incana.*
longis ſubcordatis crenatis tomentoſis.
Hoary Catmint.
Nat.
Introd. 1778, by Mr. Thomas Blackie.
Fl. Auguſt. H. ♃.

6. N. cymis pedunculatis, foliis cordatis oblongo-lanceo- *Nepetel-*
latis profunde ſerratis tomentoſis. *la.*
Nepeta Nepetella. *Sp. pl.* 797.
Small Catmint.
Nat. of the South of Europe.

Cult.

Cult. 1758, by Mr. Philip Miller.
Fl. July——September.   H. ♃.

nuda.   7. N. racemis verticillatis nudis, foliis cordato-oblongis
          feffilibus ferratis. *Syft. veget.* 529. *Jacqu. auftr.* 1.
          p. 17. *t.* 24.
          Spanifh Catmint.
          Nat. of the South of Europe.
          Cult. 1758, by Mr. Philip Miller.
          Fl. June——Auguft.   H. ♃.

italica. 8. N. floribus feffilibus verticillato-fpicatis, bracteis lan-
           ceolatis longitudine calycis, foliis petiolatis. *Sp. pl.*
           798. *Jacqu. hort.* 2. *p.* 51. *t.* 112.
           Italian Catmint.
           Nat. of Italy.
           Cult. 1739, by Mr. Philip Miller. *Rand. chel.* Ca-
           taria 8.
           Fl. June——Auguft.   H. ♃.

tuberofa. 9. N. fpicis terminalibus, bracteis oblongis acuminatis
            nervofo-lineatis coloratis, foliis cordatis pubefcenti-
            bus, corollæ lobis lateralibus reflexis.
            Nepeta tuberofa. *Sp. pl.* 798.
            Tuberous-rooted Catmint.
            Nat. of Spain and Portugal.
            Cult. 1683, by Mr. James Sutherland. *Sutherl. hort.*
            edin. 228. *n.* 1.
            Fl. June——Auguft.   H. ♃.

lanata.   10. N. fpicis terminalibus, bracteis ovatis nervofo-rugofis
           fubfcariofis, foliis oblongis cordatis villofis, corollæ
           lobis lateralibus patentibus.
           Nepeta lanata. *Jacqu. obf.* 3. *p.* 21. *t.* 75.
           Woolly Catmint.
                                                         Nat.

Nat.
Introd. 1774, by Jofeph Nicholas de Jacquin, M.D.
Fl. May and June.   H. ♃.

11. N. capitulis terminalibus, ftaminibus flore longiori-   *virgini-*
bus, foliis lanceolatis. *Sp. pl.* 799.   *ca.*
American Catmint.
Nat. of North America.
Introd. 1788, by Thomas Walter, Efq.
Fl. Auguft.   G. H. ♃.

12. N. floribus fpicatis, corollæ lobis lateralibus fubpatu-   *botryoi-*
lis, foliis pinnatifidis : laciniis linearibus fubæqua-   *des.*
libus.
Nepeta multifida. *Linn. fuppl.* 273. (non *fp. pl.*)
Cut-leav'd Catmint.
Nat. of Siberia.
Introd. 1779, by Meffrs. Kennedy and Lee.
Fl. June and July.   H. ☉.

LAVANDULA. *Gen. pl.* 711.

Calyx ovatus, fubdentatus, bractea fuffultus.   *Corolla*
refupinata. *Stamina* intra tubum.

1. L. foliis feffilibus lanceolato-linearibus margine revo-   *Spica.*
lutis, fpica interrupta nuda. *Syft. veget.* 530.
α Lavandula anguftifolia flore cæruleo. *Bauh. pin.* 216.
Narrow-leav'd blue-flower'd common Lavender.
β Lavandula anguftifolia flore albo. *Bauh. pin.* 216.
Narrow-leav'd white-flower'd common Lavender.
γ Lavandula latifolia. *Bauh. pin.* 216.
Broad-leav'd common Lavender.
Nat. of the South of Europe.
Cult. 1568. *Turn. herb. part* 3. *p.* 38.
Fl. July——September.   H. ♄.

Stœchas.   2. L. foliis feffilibus linearibus tomentofis margine re-
volutis, fpica coarctata comofa, bracteis fubtrilobis.
Lavandula Stœchas. *Sp. pl.* 800.
French Lavender.
Nat. of the South of Europe.
Cult. 1568. *Turn. herb. part* 2. *fol.* 148.
Fl. May——July.   G. H. ♃.

viridis.   3. L. foliis feffilibus linearibus rugofis villofis margine
revolutis, fpica comofa, bracteis indivifis. *L'Herit.*
fert. angl. tab. 21.
Madeira Lavender.
Nat. of Madeira. Mr. *Francis Maſſon.*
Introd. 1777.
Fl. May——July.   G. H. ♃.

dentata.   4. L. foliis feffilibus linearibus pectinato-pinnatis, fpica
coarctata comofa. *Syſt. veget.* 530.
Tooth'd-leav'd Lavender.
Nat. of Spain and the Levant.
Cult. 1597. *Ger. herb.* 470. *f.* 3.
Fl. June——September.   G. H. ♃.

pinnata.   5. L. foliis petiolatis pinnatis: foliolis cuneiformibus,
fpica imbricata. *Linn. fil. diſſ. de Lavandula, n.* 4.
tab. 1. *Jacqu. ic. mifcell.* 2. *p:* 318.
Lavandula maritima Canarienfis, fpica multiplici cæ-
rulea. *Pluk. alm.* 209. *t.* 303. *f.* 5.
Pinnated Lavender.
Nat. of Madeira. Mr. *Francis Maſſon.*
Introd. 1777.
Fl. April——October.   G. H. ♃.

multifida.   6. L. foliis petiolatis pinnatis: foliolis decurfive pinnatifidis, fpica quadrangulari: angulis fpiralibus. *Syſt.*
veget. 531.

Canary

Canary Lavender.
Nat. of Spain and the Canary Iflands.
Cult. 1597.   *Ger. herb.* 469. *f.* 2.
Fl. July——November.   G. H. ♂ .

7. L. foliis petiolatis ovato-cordatis ferratis carnofis,   *carnofa.*
    fpica tetragona, calycibus recurvatis.   *Linn. fil. diff.*
    de Lavandula, n. 6. *tab.* 2. *Linn. fuppl.* 273.
    Nepeta indica rotundiore folio. *Morif. hift.* 3. *p.* 415.
    f. 11. *t.* 6. *f.* 7.
    Katu-kurka. *Rheed. mal.* 10. *p.* 179. *t.* 90.
    Thick-leav'd Lavender.
    Nat. of the Eaft Indies.
    Introd. 1778, by Sir Jofeph Banks, Bart.
    Fl. June.   S. ♂ .

                S I D E R I T I S. *Gen. pl.* 712.

Stamina intra tubum corollæ.   *Stigma* brevius invol-
                        vens alterum.

                   * *Ebracteatæ.*

1. S. fruticofa villofa, foliis cordato-oblongis acutis pe-   *canarien-*
    tiolatis, fpicis verticillatis ante florefcentiam nutan-   *fis.*
    tibus, ramis divaricatis. *Syft. veget.* 531. *Jacqu.*
    hort. 3. *p.* 18. *t.* 30.
    Canary Iron-wort.
    Nat. of the Canary Iflands and Madeira.
    Cult. 1697, by the Dutchefs of Beaufort. *Br. Muf.*
    Sloan. mff. 3357. *fol.* 66.
    Fl. May——Auguft.   G. H. ♄ .

2. S. fruticofa tomentofa, foliis ovato-lanceolatis corda-   *candi-*
    tis apice attenuatis fubtus niveis, verticillis fubocto-   *cans.*
    floris remotis.

VOL. II.   U   Stachys

Stachys canarienfis frutefcens verbafci foliis. *Commel.*
   hort. 2. *p.* 197. *t.* 99.
Mullein-leav'd Iron-wort.
Nat. of Madeira.   Mr. *Francis Maſſon.*
Introd. 1777.
Fl. April——July.   G. H. ♄.

ſyriaca.   3. S. ſuffruticoſa tomentoſo-lanata, foliis lanceolatis in-
               tegerrimis, floribus verticillatis.   *Sp. pl.* 801.
            Sage-leav'd Iron-wort.
            Nat. of the Levant.
            Cult. 1739, by Mr. Philip Miller.   *Rand. chel.* Sta-
               chys 10.
            Fl. June——September.   G. H. ♄.

perfolia- 4. S. herbacea hiſpido-piloſa, foliis ſuperioribus am-
ta.   plexicaulibus.   *Sp. pl.* 802.
            Perfoliate Iron-wort.
            Nat. of the Levant.
            Cult. 1731, by Mr. Ph. Miller. *Mill. dict. edit.* 1. *n.* 3.
            Fl. Auguſt——November.   H. ♃.

montana. 5. S. herbacea ebracteata, calycibus corolla majoribus
               ſpinoſis : labio ſuperiore trifido.   *Syſt. veget.* 531.
               Jacqu. auſtr. 5. *p.* 16. *t.* 434.
            Mountain Iron-wort.
            Nat. of Italy and Auſtria.
            Cult. 1758, by Mr. Philip Miller.
            Fl. July and Auguſt.   H. ☉.

elegans. 6. S. herbacea ebracteata villoſa, caule diffuſo, calycum
               laciniis ſubæqualibus ſpinuloſis. *Murray in commen-*
               tat. gotting. 1. (1778) *p.* 92. *tab.* 4.
            Dark-flower'd Iron-wort.
            Nat.

                                              Introd.

Introd. 1787, by Mr. Zier.
Fl. July.   H. ☉.

7. S. herbacea decumbens ebracteata, calycibus fpinofis :   *romana.*
   labio fuperiori ovato. *Syft. veget.* 531.
   Roman Iron-wort.
   Nat. of Italy.
   Cult. 1759, by Mr. Ph. Miller. *Mill. dict. edit.* 7. *n.* 2.
   Fl. June——Auguft.   G. H. ♂.

** *Bracteatæ: bracteis dentatis.*

8. S. fuffruticofa tomentofa, foliis lanceolato-linearibus   *incana.*
   integerrimis, floribus bracteifque dentatis. *Sp. pl.*
   802.
   Lavender-leav'd Iron-wort.
   Nat. of Spain.
   Cult. 1748. *Mill. dict. edit.* 5. *n.* 7.
   Fl. July and Auguft.   H. ♄.

9. S. foliis lanceolatis glabris integerrimis, bracteis cor-   *hyffopifo-*
   datis dentato-fpinofis, calycibus æqualibus. *Sp. pl.*   *lia.*
   803.
   Hyffop-leav'd Iron-wort.
   Nat. of Italy and the Pyrenees.
   Cult. 1731, by Mr. Ph. Miller. *Mill. dict. edit.* 1. *n.* 2.
   Fl. June——November.   H. ♃.

10. S. foliis lanceolatis fubdentatis fupra glabris, bracteis   *fcordioi-*
    ovatis dentato-fpinofis, calycibus æqualibus. *Syft.*   *des.*
    veget. 532.
    Crenated Iron-wort.
    Nat. of the South of France.
    Cult. 1739, by Mr. Ph. Miller. *Mill. dict. vol.* 2. *n.* 3.
    Fl. Auguft——November.   H. ♃.

hirfuta.   11. S. foliis lanceolatis obtufis dentatis pilofis, bracteis
dentato-fpinofis, caulibus hirtis decumbentibus.
Syft. veget. 532.
Hairy Iron-wort.
Nat. of the South of Europe.
Cult. 1731. *Mill. dict. edit.* 1. *n.* 1.
Fl. June and July.   H. ♃.

BYSTROPOGON. *L'Herit. fert. angl.*

Cal. 5-fubulatus, fauce barbatus. *Corollæ* lab. fup. bifidum ; lab. inf. trifidum. *Stamina* diftantia.

pectina-   1. B. paniculis compactis, floribus fecundis, foliis ovatis.
tum.   *L'Herit. fert. angl. n.* 1.
Nepeta pectinata. *Sp. pl.* 799.
Balm-leav'd Byftropogon.
Nat. of Jamaica.
Introd. 1776, by Mr. Gilbert Alexander.
Fl. December and January.   S. ♄.

plumofum.   2. B. paniculis dichotomis, calycibus plumofis, foliis
ovatis fubferratis fubtus tomentofis. *L'Herit. fert.*
angl. n. 4. *tab.* 22.
Mentha plumofa. *Linn. fuppl.* 273.
Woolly-flower'd Byftropogon.
Nat. of the Canary Iflands. Mr. *Francis Maffon.*
Introd. 1779.
Fl. June and July.   G. H. ♄.

canari-   3. B. pedunculis dichotomis, floribus capitatis, foliis
enfe.   ovatis crenatis fubtus villofioribus. *L'Herit. fert.*
angl. n. 6.
Mentha canarienfis. *Sp. pl.* 807.
Canary Byftropogon.

                                               Nat.

Nat. of Madeira and the Canary Iflands.

Cult. 1714, by the Dutchefs of Beaufort. *Br. Muf.*
H. S. 142. *fol.* 26.

Fl. June——Auguft.   G. H. ♄.

4. B. pedunculis dichotomis, floribus capitatis, foliis   *punƈla-*
ovatis dentatis glabris punƈtulatis. *L'Herit. fert.*   *tum.*
angl. n. 7. *tab.* 23.

Clufter-flower'd Byftropogon.

Nat. of Madeira.

Introd. 1775, by Sir Jofeph Banks, Bart.

Fl. July——September.   G. H. ♄.

M E N T H A.   *Gen. pl.* 713.

Cor. fubæqualis, 4-fida : lacinia latiore emarginata.
Stamin. ereƈta, diftantia.

1. M. fpicis oblongis, foliis oblongis tomentofis ferratis   *fylveſtris.*
feffilibus, ftaminibus corolla longioribus. *Sp. pl.*
804.

Horfe Mint.

Nat. of Britain.

Fl. July——September.   H. ♃.

2. M. fpicis oblongis, foliis lanceolatis nudis ferratis   *viridis.*
feffilibus, ftaminibus corolla longioribus. *Sp. pl.*
804.

Spear-mint.

Nat. of England.

Fl. Auguft.   H. ♃.

3. M. fpicis oblongis, foliis fubrotundis rugofis crenatis   *rotundi-*
feffilibus. *Sp. pl.* 805.   *foli*

Round-leav'd Mint.

Nat.

Nat. of Britain.
Fl. Auguſt.   H. ♃.

criſpa.   4. M. floribus capitatis, foliis cordatis dentatis undulatis
ſeſſilibus, ſtaminibus corollam æquantibus. *Syſt.*
veget. 532.
Curled Mint.
Nat. of Siberia.
Cult. 1658, in Oxford Garden. *Hort. oxon. edit.* 2.
p. 108. *n.* 3.
Fl. July and Auguſt.   H. ♃.

hirſuta.   5. M. floribus capitatis, foliis ovatis ſerratis ſubſeſſilibus
pubeſcentibus, ſtaminibus corolla longioribus. *Syſt.*
veget. 532.
Round-headed Mint.
Nat. of England.
Fl. Auguſt.   H. ♃.

aquatica.   6. M. floribus capitatis, foliis ovatis ſerratis petiolatis,
ſtaminibus corolla longioribus. *Syſt. veget.* 532.
Water Mint.
Nat. of Britain.
Fl. July.   H. ♃.

piperita.   7. M. floribus capitatis, foliis ovatis petiolatis, ſtamini-
bus corolla brevioribus. *Syſt. veget.* 532..
Pepper-mint.
Nat. of England.
Fl. Auguſt.   H. ♃.

ſativa.   8. M. floribus verticillatis, foliis ovatis acutiuſculis ſer-
ratis, ſtaminibus corolla longioribus. *Sp. pl.* 805.
Marſh Mint.
Nat. of Britain.
Fl. July and Auguſt.   H. ♃.
                                                 9. M.

9. M. floribus verticillatis, foliis ovatis acutis ferratis, *gentilis.*
ftaminibus corolla brevioribus. *Sp. pl.* 805.
Red Mint.
Nat. of England.
Fl. June and July. H. ♃.

10. M. floribus verticillatis, foliis ovatis acutis ferratis, *arvenfis.*
ftaminibus corollam æquantibus. *Sp. pl.* 508.
Corn Mint.
Nat. of Britain.
Fl. Auguft and September. H. ♃.

11. M. floribus verticillatis, foliis ovatis obtufis fub- *Pulegi-*
crenatis, caulibus fubteretibus repentibus, ftami- *um.*
nibus corolla longioribus. *Sp. pl.* 807.
Pennyroyal Mint.
Nat. of Britain.
Fl. Auguft and September. H. ♃.

12. M. floribus verticillatis, bracteis palmatis, foliis li- *cervina.*
nearibus, ftaminibus corolla longioribus. *Syft.*
veget. 533.
Hyffop-leav'd Mint.
Nat. of France.
Cult. 1648, in Oxford Garden. *Hort. oxon. edit.* 1.
p. 43. *n.* 9.
Fl. June——Auguft. H. ♃.

PERILLA. *Gen. pl. pag.* 578.

Calycis lacinia fuprema breviffima. *Stamina* diftantia.
Styli duo connexi.

1. PERILLA. *Syft. veget.* 533. *ocymoides.*
Balm-leav'd Perilla.
Nat. of India.

U 4 *Introd.*

Introd. 1770, by Monf. Richard.
Fl. July and Auguft.   S. ☉.

G L E C O M A.   *Gen. pl.* 714.

Antherarum fingulum par in formam crucis connivens.
Calyx 5-fidus.

hedera-   1. G. foliis reniformibus crenatis. *Sp. pl.* 807. *Curtis*
cea.   *lond.*
            Ground Ivy.
            Nat. of Britain.
            Fl. March——May.   H. ♃.

L A M I U M.   *Gen. pl.* 716.

Corollæ lab. fup. integrum, fornicatum; lab. inf. 2-
lobum; faux utrinque margine dentata.

Orvala.   1. L. foliis cordatis inæqualiter arguteque ferratis, co-
              rollis fauce inflata, calyce colorato. *Sp. pl.* 808.
            Balm-leav'd Archangel.
            Nat. of Italy and Hungary.
            Cult. 1629. *Park. parad.* 387. *f.* 7.
            Fl. May——July.   H. ♃.

læviga-   2. L. foliis cordatis rugofis, caule lævi, calycibus gla-
tum.   bris longitudine tubi corollæ. *Syft. veget.* 534.
            Smooth Archangel.
            Nat. of Italy and Siberia.
            Introd. 1775, by the Doctors Pitcairn and Fothergill.
            Fl. March——October.   H. ♃.

rugofum.   3. L. foliis cordatis acutis rugofis caulibufque pilofis,
              verticillis multifloris, faucis dente unico fetaceo.
                                                            Lamium

Lamium fubrotundo rugofo folio, flore rubro. *Bocc.*
  muf. p. 35. *t.* 23.
Wrinkled Archangel,
Nat. of Italy.
Cult. 1766, in Oxford Garden.
Fl. July.   H. ♃.

4. L. foliis cordatis pubefcentibus, corollis fauce inflata,   *gargani-*
  tubo recto, dente utrinque gemino. *Sp. pl.* 808.   *cum.*
Woolly Archangel.
Nat. of Italy.
Cult. 1729, by Mr. Philip Miller. *R. S. n.* 375.
Fl. July.   H. ♃.

5. L. foliis cordatis acuminatis ferratis petiolatis, verti-   *album.*
  cillis vigintifloris. *Syft. veget.* 534. *Curtis lond.*
White Archangel, or Dead-nettle.
Nat. of Britain.
Fl. April——September.   H. ♃.

6. L. foliis omnibus cordatis petiolatis crenatis obtufis.   *purpure-*
  Lamium purpureum. *Sp. pl.* 809. *Curtis lond.*   *um.*
Purple Archangel.
Nat. of Britain.
Fl. May.   H. ☉.

7. L. foliis petiolatis fubdentatis : inferioribus cordatis ;   *molle.*
  fuperioribus ovatis.
Lamium parietariæ facie. *Mor. blæf.* 278.
Pellitory-leav'd Archangel.
Nat.
Cult. 1683, by Mr. James Sutherland. *Sutherl. hort.*
  edin. 181. *n.* 1.
Fl. April and May.   H. ♃.
Obs. Facile dignofcitur foliis fubintegerrimis, nec
  ferratis, nec crenatis. *Flores* albi.
                                          8. L.

amplexi-
caule.

8. L. foliis floralibus feffilibus amplexicaulibus obtufis.
   Sp. pl. 809. *Curtis lond.*
   Perfoliate Archangel, or great Henbit.
   Nat. of Britain.
   Fl. June.   H. ☉.

G A L E O P S I S. *Gen. pl.* 717.

Corollæ lab. fuper. fubcrenatum, fornicatum; lab. inf.
   fupra 2-dentatum.

Lada-
num.

1. G. internodiis caulinis æqualibus, verticillis omni-
   bus remotis, calycibus inertibus. *Syft. veget.* 534.
   Red Dead-nettle.
   Nat. of England.
   Fl. July and Auguft.   H. ☉.

Tetrahit.

2. G. internodiis caulinis fuperne incraffatis, verticillis
   fummis fubcontiguis, calycibus fubpungentibus.
   Syft. veget. 535.
   α Lamium cannabino folio vulgare. *Raj. fyn.* 240.
   Common Dead-nettle, or Day-nettle.
   β Lamium cannabino folio, flore amplo luteo, labio
   purpureo. *Raj. fyn.* 241.
   Strip'd-flower'd Dead-nettle.
   Nat. of Britain.
   Fl. Auguft.   H. ☉.

Galeobdo-
lon.

3. G. verticillis fexfloris: involucro tetraphyllo. *Sp. pl.*
   810.
   Galeobdolon Galeopfis. *Curtis lond.*
   Yellow Dead-nettle.
   Nat. of Britain.
   Fl. May and June.   H. ♃.

BETO-

BETONICA. *Gen. pl.* 718.

Cal. ariftatus. *Corollæ* lab. fuper. adfcendens, planiuf-
culum; tubus cylindricus.

1. B. fpica interrupta, corollarum galea integra; lacinia   *officina-*
   intermedia labii inferioris emarginata, calycibus   *lis.*
   glabriufculis.
   Betonica officinalis. *Sp. pl.* 810. *Curtis lond.*
   Wood Betony.
   Nat. of Britain.
   Fl. July and Auguft.   H. ♃.

2. B. fpica oblonga, corollarum galea integra; lacinia in-   *ftriƈta.*
   termedia labii inferioris crenato-undulata, calycibus
   pilofis.
   Betonica danica. *Mill. diƈt.*
   Danifh Betony.
   Nat.
   Cult. 1759, by Mr. Ph. Miller. *Mill. diƈt. ed.* 7. *n.* 2.
   Fl. June and July.   H. ♃.

3. B. fpica interrupta, corollarum galea bifida; lacinia   *incana.*
   intermedia labii inferioris crenata, tubo tomentofo
   incurvo.
   Betonica incana. *Mill. diƈt.*
   Hoary Betony.
   Nat.
   Cult. 1759, by Mr. Ph. Miller. *Mill. diƈt. edit.* 7. *n.* 5.
   Fl. June and July.   H. ♃.

4. B. fpica integra, corollarum lacinia labii intermedia   *orientalis.*
   integerrima. *Sp. pl.* 811.
   Oriental Betony.
   Nat. of the Levant.
                                                       Cult.

Cult. 1739, by Mr. Philip Miller. *Rand. chel. n. 7.*
Fl. June and July.   H. ♃.

alopecu- 5. B. fpica bafi foliofa, corollis galea bifida. *Sp. pl.* 811.
ros.   *Jacqu. auftr.* 1. *p.* 50. *t.* 78.
           Fox-tail Betony.
           Nat. of the South of Europe.
           Cult. 1759, by Mr. Ph. Miller. *Mill. dict. edit.* 7. *n.* 7.
           Fl. July.   H. ♃.

hirfuta. 6. B. fpica bafi foliofa, corollis galea integra. *Linn. mant.*
           248.
           Betonica alpina. *Mill. dict.*
           Hairy Betony.
           Nat. of the Alps of Italy.
           Cult. 1739, by Mr. Philip Miller. *Rand. chel. n.* 3.
           Fl. June and July.   H. ♃.

                    S T A C H Y S. *Gen. pl.* 719.

           Corollæ lab. fuper. fornicatum; lab. inferius lateribus
                reflexum: lac. intermedia majore emarginata. *Sta-
                mina* deflorata verfus latera reflexa.

fylvatica. 1. S. verticillis fexfloris, foliis cordatis petiolatis. *Sp.
                pl.* 811. *Curtis lond.*
           Wood Stachys, or Hedge-nettle.
           Nat. of Britain.
           Fl. July and Auguft.   H. ☉.

circinata. 2. S. verticillis fexfloris, foliis cordato-rotundatis crena-
                tis. *L'Herit. ftirp. nov. p.* 51. *tab.* 26.
           Blunt-leav'd Stachys.
           Nat. of Barbary.
           Introd. about 1777, by Meffrs. Gordon and Græfer.
           Fl. May——July.   H. ♃.
                                                         3. S.

3. S. verticillis fubfexfloris, foliis lineari-lanceolatis femi-   *paluftris,*
    amplexicaulibus feffilibus   *Syft. veget.* 535. *Curtis*
    lond.
Marfh Stachys, or Clown's Allheal.
Nat. of Britain.
Fl. Auguft.   H. ♃.

4. S. verticillis multifloris, foliorum ferraturis apice car-   *alpina.*
    tilagineis, corollis labio plano.   *Sp. pl.* 812.
Alpine Stachys.
Nat. of Germany and Switzerland.
Cult. 1759, by Mr. Ph. Miller. *Mill. dict. edit.* 7. *n.* 5.
Fl. June——Auguft.   H. ♃.

5. S. verticillis multifloris, foliorum ferraturis imbrica-   *germani-*
    tis, caule lanato. *Sp. pl.* 812. *Jacqu. auftr.* 4.   *ca.*
    p. 10. *t.* 319.
German Stachys, or Bafe Horehound.
Nat. of England.
Fl. July.   H. ♃.

6. S. verticillis multifloris, calycibus fubpungentibus, fo-   *interme-*
    liis oblongis fubcordatis crenatis, caule fublanato.   *dia.*
Oblong-leav'd Stachys.
Nat. of Carolina.
Introd. about 1762.
Fl. June and July.   H. ♃.

7. S. verticillis multifloris, foliis lanatis, caulibus bafi   *lanata.*
    procumbentibus et radicantibus. *Jacqu. ic. mifcell.* 2.
    p. 342.
Woolly Stachys.
Nat. of Siberia.
Introd. 1782, by William Pitcairn, M. D.
Fl. moft part of the Summer.   H. ♃.
                                       8. S.

cretica.   8. S. verticillis triginta-floris, calycibus pungentibus, caule hirto. *Syſt. veget.* 536.

Cretan Stachys.

Nat. of Candia.

Cult. 1723, in Chelſea Garden.   *R. S. n.* 89.

Fl. June——Auguſt.   H. ♃.

maritima.   9. S. foliis cordatis obtuſis tomentoſis crenatis, bracteis oblongis integerrimis. *Linn. mant.* 82. *Jacqu. hort.* 1. *p.* 29. *t.* 70.

Yellow, or Sea Stachys.

Nat. of the South of Europe.

Cult. 1714. *Philoſoph. tranſ. n.* 343. *p.* 244. *n.* 46.

Fl. July.   H. ♃.

æthiopica.   10. S. verticillis bifloris. *Linn. mant.* 82. *L'Herit. ſtirp. nov. tom.* 2. *tab.* 27.

Ethiopian Stachys.

Nat. of the Cape of Good Hope.

Introd. 1770, by Monſ. Richard.

Fl. April——July.   G. H. ♃.

hirta.   11. S. verticillis ſexfloris, caulibus proſtratis, corollis labio ſuperiore bifido divaricato reflexo. *Sp. pl.* 813.

Procumbent Stachys.

Nat. of Spain, Italy, and the Levant.

Cult. 1725, in Chelſea Garden.   *R. S. n.* 192.

Fl. June——Auguſt.   H. ♃.

recta.   12. S. verticillis ſubſpicatis, foliis cordato-ellipticis crenatis ſcabris, caulibus adſcendentibus. *Linn. mant.* 82. *Jaequ. auſtr.* 4. *p.* 31. *t.* 359.

Upright Stachys.

Nat. of the South of Europe.

Cult. 1758, by Mr. Philip Miller.

Fl. June——Auguſt.   H. ♃.

13. S.

13. S. verticillis fexfloris, foliis ovato-lanceolatis triner- *annua.*
    viis lævibus petiolatis, caule erecto. *Syft. veget.*
    536. *Jacqu. auftr.* 4. *p.* 31. *t.* 360.
    White annual Stachys.
    Nat. of Germany and France.
    Cult. 1713. *Philofoph. tranf. n.* 337. *p.* 196. *n.* 64.
    Fl. July and Auguft.   H. ⊙.

14. S. verticillis fexfloris, foliis lanceolatis bafi attenuatis *rugofa.*
    tomentofis rugofis ferratis, calycibus muticis.
    Rough Stachys.
    Nat. of the Cape of Good Hope. Mr. *Fr. Maffon.*
    Introd. 1774.
    Fl. July.   G. H. ♄.

15. S. verticillis fexfloris, foliis obtufis nudiufculis, co- *arvenfis.*
    rollis longitudine calycis, caule debili. *Syft. veget.*
    536. *Curtis lond.*
    Corn Stachys.
    Nat. of Britain.
    Fl. Auguft.   H. ⊙.

16. S. verticillis multifloris fubfpicatis, labio fuperiore *latifolia.*
    bifido: lacinulis acutis, foliis latis cordatis rugofis
    pilofis.
    Broad-leav'd Stachys.
    Nat.
    Introd. about 1775, by William Pitcairn, M. D.
    Fl. June and July.   H. ♄.

BALLOTA. *Gen. pl.* 720.

Cal. hypocrateriformis, 5-dentatus, 10-ftriatus. *Co-*
rollæ labium fuperius crenatum, concavum.

1. B. foliis cordatis indivifis ferratis, calycibus acumi- *nigra.*
   natis. *Sp. pl.* 814.

                                                  Black

Black Horehound.

Nat. of Britain.

Fl. July——November.   H. ♃.

alba.   2. B. foliis cordatis indivifis ferratis, calycibus fubtrun-
            catis. *Sp. pl.* 814.

White-flower'd black Horehound.

Nat. of England.

Fl. July.   H. ♃.

lanata.   3. B. foliis palmatis dentatis, caule lanato. *Sp. pl.* 815.

Woolly black Horehound.

Nat. of Siberia.

Cult. 1776, by Mr. James Gordon.

Fl. June——Auguft.   H. ♃.

difticha. 4. B. verticillis dimidiatis bipartitis femifpicatis. *Linn.*
             mant. 83.

Betony-leav'd black Horehound.

Nat. of the Eaft Indies.

Introd. 1783, by John Earl of Bute.

Fl. Auguft.   S. ☉.

MARRUBIUM. *Gen. pl.* 721.

Cal. hypocrateriformis, rigidus, 10-ftriatus. *Corolla*
    lab. fup. 2-fidum, lineare, rectum.

* *Calycibus 5-dentatis.*

Alyffon.   1. M. foliis cuneiformibus quinquedentatis plicatis, ver-
             ticillis involucro deftitutis. *Sp. pl.* 815.

Plaited-leav'd white Horehound.

Nat. of Spain and Italy.

Cult. before 1597, by Mr. John Gerard. *Ger. herb.*
    379. *f.* 1.

Fl. July and Auguft.   H. ♃.

                                          2. M.

2. M. foliis fubovatis lanatis fuperne emarginato-crena- *candidif-*
tis, denticulis calycinis fubulatis. *Sp. pl.* 816. *fimum.*
Woolly white Horehound.
Nat. of the Levant.
Cult. 1732, by James Sherard, M. D. *Dill. elth.*
218. *t.* 174. *f.* 214.
Fl. July——September. H. ♃.

3. M. dentibus calycinis fetaceis rectis villofis. *Sp. pl.* *fupinum.*
816.
Procumbent white Horehound.
Nat. of Spain and the South of France.
Cult. 1714, in Chelfea Garden. *Philofoph. tranf. n.* 343.
p. 241. *n.* 34.
Fl. Auguft——October. H. ♃.

** *Calycibus* 10-*dentatis.*

+ 4. M. dentibus calycinis fetaceis uncinatis. *Sp. pl.* 816. *vulgare.*
Common white Horehound.
Nat. of Britain.
Fl. June——September. H. ♃.

5. M. foliis cordatis fubrotundis emarginato-crenatis. *africa-*
Sp. pl. 816. *num.*
African white Horehound.
Nat. of the Cape of Good Hope.
Introd. 1774, by Mr. Francis Maffon.
Fl. July——September. G. H. ♃.

6. M. calycum limbis patentibus : denticulis acutis. *hifpani-*
Sp. pl. 816. *cum.*
Spanifh white Horehound.
Nat. of Spain.
Cult. 1759, by Mr. Ph. Miller. *Mill. dict. edit.* 7. *n.* 7.
Fl. July and Auguft. H. ♃.

Pseudo-dictamnus. 7. M. calycum limbis planis villosis, foliis cordatis con-
cavis, caule fruticoso. *Sp. pl.* 817.
Shrubby white Horehound.
Nat. of the Island of Candia.
Cult. 1596, by Mr. John Gerard. *Hort. Ger.*
Fl. June——August.   G. H. ♄.

acetabu-losum. 8. M. calycum limbis tubo longioribus membranaceis:
angulis majoribus rotundatis. *Sp. pl.* 817.
Saucer-leav'd white Horehound.
Nat. of the Island of Candia.
Cult. 1731, by Mr. Philip Miller. *Mill. dict. edit.* 1.
Pseudo-dictamnus 1.
Fl. June——August.   G. H. ♃.

LEONURUS. *Gen. pl.* 722.

Antheræ punctis nitidis adspersæ.

Cardiaca. 1. L. foliis caulinis lanceolatis trilobis. *Sp. pl.* 817.
Common Mother-wort.
Nat. of Britain.
Fl. July.   H. ♃.

Marru-biastrum. 2. L. foliis ovatis lanceolatisque serratis, calycibus sessili-
bus spinosis. *Sp. pl.* 817. *Jacqu. austr.* 5. *p.* 3.
t. 405.
Small-flower'd Mother-wort.
Nat. of Austria.
Cult. 1713, in Chelsea Garden. *Philosoph. transf. n.* 337.
p. 193. *n.* 56.
Fl. June——August.   H. ☉.

tataricus. 3. L. foliis tripartitis laciniatis, calycibus villosis. *Sp. pl.*
818.
Tartarian Mother-wort.

Nat. of Ruffia.
Cult. 1756, by Mr. Ph. Miller. *Mill. ic.* 53. *t.* 80.
Fl. Auguft—October.   H. ♂.

4. L. foliis tripartitis multifidis linearibus obtufiufculis. *fibiricus.*
   Sp. pl. 818.
Siberian Mother-wort.
Nat. of Siberia.
Cult. 1759, by Mr. Philip Miller. *Mill. dict. edit.* 7.
   Cardiaca 2.
Fl. June—Auguft.   H. ♂.

P H L O M I S. *Gen. pl.* 723.

Cal. angulatus. *Corollæ* lab. fuperius incumbens, com-
     preffum, villofum.

1. P. foliis fubrotundis tomentofis crenatis, involucris *fruticofa.*
   lanceolatis, caule fruticofo. *Sp. pl.* 818.
α Verbafcum latis Salviæ foliis. *Bauh. pin.* 240.
   Narrow-leav'd fhrubby Phlomis, or Jerufalem Sage.
β Phlomis latifolia capitata lutea grandiflora. *Dill. elth.*
   316. *t.* 237. *f.* 306.
   Broad-leav'd fhrubby Phlomis, or Jerufalem Sage.
Nat. of Spain and Sicily.
Cult. 1597, by Mr. John Gerard. *Ger. herb.* 625.
Fl. June and July.   H. ♄.

2. P. involucris linearibus obtufis calyce brevioribus, *purpurea.*
   foliis cordatis oblongis tomentofis, caule fuffruticofo.
   Sp. pl. 818.
Purple Phlomis.
Nat. of Portugal and Italy.
Cult. about 1661, by Mr. Edward Morgan. *Plut.*
   phyt. t. 57. *f.* 6.
Fl. June—Auguft.   G. H. ♄.

Lychni-
tis.

3. P. foliis lanceolatis tomentofis : floralibus ovatis, in-
volucris fetaceis lanatis. *Sp. pl.* 819.
Sage-leav'd Phlomis.
Nat. of the South of Europe.
Cult. 1731, by Mr. Ph. Miller. *Mill. dict. edit.* 1. *n.* 6.
Fl. June——Auguft.   G. H. ♄.

laciniata.

4. P. foliis alternatim pinnatis : foliolis laciniatis, caly-
cibus lanatis. *Sp. pl.* 819.
Jagged-leav'd Phlomis.
Nat. of the Levant.
Cult. 1731, by Mr. Ph. Miller. *Mill. dict. edit.* 1. *n.* 8.
Fl.   H. ♃.

Herba
venti.

5. P. involucris fetaceis hifpidis, foliis ovato-oblongis
fcabris, caule herbaceo. *Sp. pl.* 819.
Rough-leav'd Phlomis.
Nat. of the South of France.
Cult. 1683, by Mr. James Sutherland. *Sutherl. hort.*
edin. 223. *n.* 2.
Fl. July——September.   H. ♃.

tuberofa.

6. P. involucris hifpidis fubulatis, foliis cordatis fcabris,
caule herbaceo. *Sp. pl.* 819.
Tuberous Phlomis.
Nat. of Siberia.
Cult. 1759, by Mr. Ph. Miller. *Mill. dict. edit.* 7. *n.* 5.
Fl. June——October.   H. ♃.

zeylanica.

7. P. foliis lanceolatis fubferratis, capitulis terminalibus,
calycibus octodentatis. *Syft. veget.* 539. *Jacqu. ic.*
White Phlomis.
Nat. of the Eaft Indies.
Introd. 1777, by John Gerard Kœnig, M. D.
Fl. June——October.   S. ♂.

8. P.

8. P. foliis ovato-lanceolatis villofis, verticillis fubrotun-   *caribæa.*
dis denfiffimis, involucris fetaceis hirfutis, caule
herbaceo. *Jacqu. ic. collect.* I. *p.* 154.
Phlomis martinicenfis. *Swartz prodr.* 88.
Clinopodium martinicenfe. *Jacqu. hift.* 173. *tab.* 177.
fig. 75.
Weft Indian Phlomis.
Nat. of the Weft Indies.
Introd. 1781, by Mr. Francis Maffon.
Fl. July——September.   S. ☉.

9. P. foliis cordatis acutis ferratis fubtomentofis, caly-   *nepetifo-*
cibus fex-octodentatis: dente fupremo infimoque   *lia.*
majori, caule herbaceo. *Syft. veget.* 540.
Cat-mint-leav'd Phlomis.
Nat. of the Eaft Indies.
Introd. 1778, by Sir Jofeph Banks, Bart.
Fl. September and October.   S. ☉.

10. P. foliis lanceolatis ferratis, calycibus decagonis de-   *Leonu-*
cemdentatis muticis, caule fruticofo. *Syft. veget.*   *rus.*
540.
Narrow-leav'd Phlomis, or Lion's-tail.
Nat. of the Cape of Good Hope.
Cult. 1712, in Chelfea Garden. *Br. Muf. H. S.* 136.
fol. 55.
Fl. October——December.   G. H. ♄.

11. P. foliis ovatis obtufis fubtomentofis crenatis, calyci-   *Leonotis.*
bus feptemdentatis ariftatis, caule fruticofo. *Syft.*
veget. 541.
Dwarf fhrubby Phlomis.
Nat. of the Cape of Good Hope.
Cult. 1713, by the Dutchefs of Beaufort. *Br. Muf.*
H. S. 136. *fol.* 56.
Fl. June and July.   G. H. ♄.

MOLUCCELLA. *Gen. pl.* 724.

Cal. companulatus, ampliatus, corolla latior, fpinofus.

lævis.   1. M. calycibus campanulatis fubquinquedentatis : den-
ticulis æqualibus. *Syft. veget.* 540.
Smooth Molucca Balm.
Nat. of Syria.
Cult. 1570, by Matthias de L'Obel.   *Lobel. adv.* 221.
Fl. July and Auguft.   H. ⊙.

fpinofa.   2. M. calycibus ringentibus octodentatis. *Sp. pl.* 821.
Prickly Molucca Balm.
Nat. of the Levant.
Cult. 1739, by Mr. Philip Miller.   *Rand. chel. n.* 2.
Fl. July and Auguft.   H. ⊙.

CLINOPODIUM. *Gen. pl.* 725.

Involucrum multifetum, verticillo fubjectum.

vulgare.   1. C. capitulis fubrotundis hifpidis, bracteis fetaceis. *Sp.*
pl. 821.
Common Clinopodium, or Wild Bafil.
Nat. of Britain.
Fl. June and July.   H. ♃.

incanum.   2. C. foliis fubtus tomentofis, verticillis explanatis, brac-
teis lanceolatis. *Sp. pl.* 822.
Hoary Clinopodium.
Nat. of North America.
Cult. 1732, by James Sherard, M. D.   *Dillen. hort.*
elth. 87. *t.* 74. *f.* 85.
Fl. July——October.   H. ♃.

rugofum.   3. C. foliis rugofis, capitulis axillaribus pedunculatis explanatis radiatis. *Sp. pl.* 822.

Wrinkled

Wrinkled Clinopodium.
Nat. of Carolina.
Cult. 1690, in the Royal Garden at Hampton-court.
Catal. mff.
Fl. June and July.   S. ♃.

O R I G A N U M.   *Gen. pl.* 726.
Strobilus tetragonus, fpicatus, calyces colligens.

1. O. foliis carnofis tomentofis, fpicis nudis. *Sp. pl.* 822. *ægyptia-*
   Egyptian Marjoram.   *cum.*
   Nat. of Egypt.
   Cult. 1731, by Mr. Philip Miller. *Mill. dict. edit.* 1.
   Majorana 2.
   Fl. June——Auguft.   G. H. ♃.

2. O. foliis inferioribus tomentofis, fpicis nutantibus. *Dictam-*
   Sp. pl. 823.   *nus.*
   Dittany of Candia.
   Nat. of the Ifland of Candia.
   Cult. 1568, by Mr. Riche. *Turn. herb. part* 1, 2d *edit.*
   p. 203.
   Fl. June——Auguft.   G. H. ♄.

3. O. foliis omnibus glabris, fpicis nutantibus. *Sp. pl.* *fipyleum.*
   823.
   Dittany of Mount Sipylus.
   Nat. of the Levant.
   Cult. 1699, by Mr. Jacob Bobart. *Morif. hift.* 3.
   p. 357. *n.* 2. *f.* 11. *t.* 4. *f.* 2.
   Fl. June——September.   G. H. ♄.

4. O. fpicis tetragonis, bracteis fubrotundis maximis. *Tourne-*
   Origanum Dictamni Cretici facie, folio craffo, nunc *fortii.*
   villofo, nunc glabro. *Tourn. it.* 1. *p.* 240. *fig.*
                    X 4   Dittany

Dittany of Amorgos.
Nat. of the Ifland of Amorgos.
Introd. 1788, by John Sibthorp, M. D.
Fl. Auguft.   G. H. ♄.

heracleo-   5. O. fpicis longis pedunculatis aggregatis, bracteis lon-
ticum.   gitudine calycum. *Sp. pl.* 823.
               Winter Sweet Marjoram.
               Nat. of the South of Europe.
               Cult. 1731, by Mr. Ph. Miller. *Mill. dict. edit.* 1. *n.* 3.
               Fl. June——November.   H. ♃.

vulgare.   6. O. fpicis fubrotundis paniculatis conglomeratis, brac-
                   teis calyce longioribus ovatis. *Sp. pl.* 824. *Curtis*
                   lond.
               Common Marjoram.
               Nat. of Britain.
               Fl. June——November.   H. ♃.

Onites.   7. O. fpicis oblongis aggregatis hirfutis, foliis cordatis
                   tomentofis. *Sp. pl.* 824.
               Pot Marjoram.
               Nat. of Sicily.
               Cult. 1759. *Mill. dict. edit.* 7. *n.* 13.
               Fl. July——November.   H. ♃.

Majora-   8. O. foliis ovalibus obtufis, fpicis fubrotundis compac-
na.   tis pubefcentibus. *Sp. pl.* 825.
               Sweet, or Knotted Marjoram.
               Nat.
               Cult. 1597. *Ger. herb.* 538. *f.* 1.
               Fl. June and July.   G. H. ♂.

                                    THYMUS.

THYMUS. *Gen. pl.* 727.

Calycis bilabiati faux villis clausa.

1. T. floribus capitatis, caulibus repentibus, foliis pla- *Serpyl-*
nis obtusis basi ciliatis. *Syst. veget.* 541. *Curtis* *lum.*
lond.

α Serpyllum vulgare minus. *Bauh. pin.* 220.
Common smooth Mother of Thyme.

β Serpyllum foliis citri odore. *Bauh. pin.* 220.
Lemon Thyme.

γ Serpyllum villosum fruticosius floribus dilute rubentibus. *Raj. syn.* 231.
Hoary Mother of Thyme.

δ Serpyllum angustifolium hirsutum. *Bauh. pin.* 220.
Hairy Mother of Thyme.
Nat. of Britain.
Fl. June——August. H. ♄ .

2. T. floribus axillaribus solitariis pedunculatis, foliis *filiformis.*
cordatis acutis integerrimis petiolatis, caulibus fili-
formibus decumbentibus.
Small-leav'd Thyme.
Nat. of the Balearic Islands.
Introd. 1770, by Mr. William Malcolm.
Fl. June and July. G. H. ♄ .

3. T. erectus, foliis revolutis ovatis, floribus verticillato- *vulgaris.*
spicatis. *Sp. pl.* 825.

α Thymus vulgaris, folio tenuiore. *Bauh. pin.* 219.
Narrow-leav'd Garden Thyme.

β Thymus vulgaris, folio latiore. *Bauh. pin.* 219.
Broad-leav'd Garden Thyme.
Nat. of France, Spain, and Italy.
Cult.

Cult. 1596, by Mr. John Gerard.   *Hort. Ger.*

Fl. May——Auguſt.   H. ♄.

Zygis.   4. T. floribus verticillato-ſpicatis, caule ſuffruticoſo erecto, foliis linearibus baſi ciliatis.   *Sp. pl.* 826.

Linear-leav'd Thyme.

Nat. of Spain.

Introd. 1786, by William Pitcairn, M.D.

Fl. Auguſt.   H. ♄.

Acinos.   5. T. floribus verticillatis, pedunculis unifloris, caulibus
erectis ſubramoſis, foliis acutis ſerratis.   *Syſt. veget.*
542.   *Curtis lond.*

Corn Thyme, or Baſil.

Nat. of Britain.

Fl. June——Auguſt.   H. ☉.

patavi-   6. T. floribus verticillatis : fauce inflata calyce longiore,
nus.   foliis ovatis ſerratis, caulibus ſuffruticoſis.

Thymus patavinus.   *Jacqu. obſ.* 4. *p.* 7. *tab.* 87.

Clinopodium perenne Pulegii odore.   *Bocc. muſ.* 60.
tab. 45. *fig. B.*

Great-flower'd Thyme.

Nat.

Introd. 1776, by Joſeph Nicholas de Jacquin, M. D.

Fl. June——Auguſt.   H. ♂.

alpinus.   7. T. verticillis ſexfloris, foliis obtuſiuſculis concavis
ſubſerratis.   *Sp. pl.* 826.   *Jacqu. auſtr.* 1. *p.* 60.
t. 97.

Alpine Thyme.

Nat. of Auſtria and Switzerland.

Cult. 1731, by Mr. Philip Miller.   *Mill. dict. edit.* 1.
Acinos 2.

Fl. June—— September.   H. ♃.

8. T.

8. T. floribus verticillatis, calycibus lanuginofis : denti- *Mafti-*
   bus fetaceis villofis. *Syfl. veget.* 542.   *china.*
   Maftick Thyme.
   Nat. of Spain.
   Cult. 1596, by Mr. John Gerard. *Hort. Ger.*
   Fl. July——September.   G. H. ♄.

9. T. capitulis terminalibus, caule erecto, foliis lanceola- *virgini-*
   tis. *Syfl. veget.* 542.   *cus.*
   Satureja virginiana. *Sp. pl.* 793.
   Virginian, or Savory Thyme.
   Nat. of North America.
   Cult. 1739, by Mr. Philip Miller. *Mill. dict. vol.* 2.
   Addenda. Clinopodium 1.
   Fl. July.   H. ♃.

   M E L I S S A. *Gen. pl.* 728.

   Cal. aridus, fupra planiufculus : lab. fuperiore fubfafti-
   giato. *Corollæ* lab. fuper. fubfornicatum, 2-fidum ;
   lab. inf. lobo medio cordato.

1. M. racemis axillaribus verticillatis : pedicellis fim- *officinalis,*
   plicibus. *Sp. pl.* 827.
   α Meliffa hortenfis. *Bauh. pin.* 229.
   Common Balm.
   β Meliffa *romana*, floribus verticillatis feffilibus, foliis
     hirfutis. *Mill. dict.*
   Roman Balm.
   Nat. of the South of Europe.
   Cult. 1596, by Mr. John Gerard. *Hort. Ger.*
   Fl. June ——October.   H. ♃.

2. M. pedunculis axillaribus dichotomis longitudine flo- *grandi-*
   rum. *Sp. pl.* 827.   *flora.*
   Great flower'd Balm.
                                          Nat.

Nat. of Italy.
Cult. 1640. *Park. theat.* 37. *f.* 2.
Fl. June——September.   H. ♃.

Calamin-   3. M. pedunculis axillaribus dichotomis longitudine fo-
tha.   liorum. *Sp. pl.* 827.
              Mountain Balm, or Calamint.
              Nat. of England.
              Fl. Auguſt.   H. ♃.

Nepeta.   4. M. pedunculis axillaribus dichotomis folio longiori-
                  bus, caule adſcendente hirſuto. *Syſt. veget.* 542.
              Field Balm, or Calamint.
              Nat. of England.
              Fl. July——October.   H. ♃.

cretica.   5. M. racemis terminalibus, pedunculis ſolitariis bre-
                  viſſimis. *Sp. pl.* 828.
              Cretan Balm.
              Nat. of the South of Europe.
              Cult. 1731, by Mr. Philip Miller. *Mill. dict. edit.* 1.
                  Calamintha 4.
              Fl. June and July.   G. H. ♄.

fruticoſa. 6. M. ramis attenuatis virgatis, foliis ſubtus tomentoſis,
                  caule fruticoſo. *Sp. pl.* 828.
              Shrubby Balm.
              Nat. of Spain.
              Cult. 1752, by Mr. Philip Miller. *Mill. dict. edit.* 6.
                  Calamintha 5.
              Fl. July——September.   G. H. ♄.

DRACO-

DRACOCEPHALUM. *Gen. pl.* 729.

Corollæ faux inflata : lab. fup. concavum.

* *Spicata.*

1. D. floribus fpicatis confertis, foliis lineari-lanceolatis   *virginia-*
    ferratis.   *num.*
    Dracocephalum virginianum.   *Sp. pl.* 828.
    Virginian Dragon's Head.
    Nat. of Canada and Virginia.
    Cult. 1683, by Mr. James Sutherland.   *Sutherl. hort.*
      edin. 104. *n.* 2.
    Fl. July——September.   H. ♃.

2. D. floribus fpicatis remotis, foliis obovato-lanceolatis   *denticu-*
    fuperne denticulatis.   *latum.*
    Denticulated Dragon's Head.
    Nat. of Carolina.
    Introd. 1787, by Meffrs. Watfon.
    Fl. Auguft.   H. ♃.

3. D. floribus fpicatis, foliis compofitis.   *Sp. pl.* 829.   *canari-*
    Canary Dragon's Head, or Balm of Gilead.   *eafe.*
    Nat. of the Canary Iflands.
    Cult. 1697, by the Dutchefs of Beaufort.   *Br. Muf.*
      Sloan. mff. 3343.
    Fl. July——September.   G. H. 3.

4. D. floribus fubfpicatis, foliis caulinis ovato-oblongis   *peregri-*
    incifis, bracteis lineari-lanceolatis denticulato-fpino-   *num.*
    fis.   *Syft. veget.* 543.   *L'Herit. ftirp. nov. tori* 2.
    tab. 28.
    Prickly-leav'd Dragon's Head.
    Nat. of Siberia.
                                                   Cult.

Cult. before 1699, by Mr. Jacob Bobart. *Morif. hift.* 3.
p. 364. *n.* 9.
Fl. July and Auguſt.   H. ♃.

auſtria-
cum.

5. D. floribus ſpicatis, foliis bracteiſque linearibus parti-
    tis ſpinoſis. *Sp. pl.* 829. *Jacqu. ic. collect.* 1.
    p. 119.
Auſtrian Dragon's Head.
Nat. of Auſtria and Hungary.
Introd. 1788, by Joſeph Nicholas de Jacquin, M. D.
Fl.   H. ♃.

Ruyfchi-
ana.

6. D. floribus ſpicatis, foliis bracteiſque lanceolatis indi-
    viſis muticis. *Sp. pl.* 830.
Hyſſop-leav'd Dragon's Head.
Nat. of Norway, Sweden, and Siberia.
Cult. 1758, by Mr. Philip Miller.
Fl. June and July.   H. ♃.

** *Verticillata.*

ſibiricum.

7. D. floribus ſubverticillatis, pedunculis bifidis ſecun-
    dis, foliis cordato-oblongis acuminatis nudis. *Sp.*
    pl. 830.
Siberian Dragon's Head.
Nat. of Siberia.
Cult. 1760, by Mr. James Gordon.
Fl. June——Auguſt.   H. ♃.

Molda-
vica.

8. D. floribus verticillatis, bracteis lanceolatis: ſerratu-
    ris capillaceis. *Sp. pl.* 830.
Moldavian Dragon's Head, or Balm.
Nat. of Moldavia.
Cult. 1596, by Mr. John Gerard. *Hort. Ger.*
Fl. July and Auguſt.   H. ☉.

9. D.

9. D. floribus verticillatis, bracteis oblongis : ferraturis *canescens.*
spinofis, foliis fubtomentofis. *Sp. pl.* 831.
Hoary Dragon's Head.
Nat. of the Levant.
Cult. 1712. *Philofoph. tranf. n.* 333. *p.* 416. *n.* 58.
Fl. July and Auguft. H. ⊙.

10. D. floribus verticillatis, bracteis orbiculatis ferrato- *peltatum.*
ciliatis. *Sp. pl.* 831.
Willow-leav'd Dragon's Head.
Nat. of the Levant.
Cult. 1731. *Mill. dict. edit.* 1. Moldavica 6.
Fl. July and Auguft. H. ⊙.

11. D. foliis crenatis : radicalibus cordatis ; caulinis or- *grandiflo-*
biculatis feffilibus, bracteis acuminato-dentatis. *rum.*
Linn. fuppl. 274.
Great-flower'd Dragon's Head.
Nat. of Siberia.
Introd. 1783, by Mr. John Bell.
Fl. July. H. ♃.

12. D. floribus verticillatis, bracteis oblongis ovatis in- *nutans.*
tegerrimis, corollis majufoulis nutantibus. *Sp. pl.*
831.
Nodding Dragon's Head.
Nat. of Siberia.
Cult. 1768, by Mr. Philip Miller. *Mill. dict. edit.* 8.
Fl. July and Auguft. H. ♃.

13. D. floribus verticillatis, bracteis oblongis integerri- *thymiflo-*
mis, corollis vix calyce majoribus. *Sp. pl.* 831. *rum.*
Small-flower'd Dragon's Head.
Nat. of Siberia.

Cult.

Cult. 1759, by Mr. Ph. Miller.   *Mill. dict. edit.* 7. *n.* 7.
Fl. June——September.   H. ☉.

M E L I T T I S.   *Gen. pl.* 731.

Cal. tubo corollæ amplior.   *Cor.* lab. fup. planum ;
        lab. inf. crenatum.   *Antheræ* cruciatæ.

Meliffo-   1. MELITTIS.   *Sp. pl.* 832.   *Jacqu. auftr.* 1. *p.* 18.
phyllum.   *t.* 26.
α foliis ellipticis.
Common Baftard Balm.
β foliis ovatis obcordatis.
Alpine Baftard Balm.
Nat. α. of England ; β. of Switzerland.
Fl. May and June.   H. ♃.

O C Y M U M.   *Gen. pl.* 732.

Calyx lab. fuperiore orbiculato ; inferiore quadrifido.
        Corollæ refupinatæ alterum lab. 4-fidum ; alterum
        indivifum.   *Filamenta* exteriora bafi proceffum
        emittentia.

gratiffi-   1. O. caule fuffruticofo, foliis lanceolato-ovatis, race-
mum.   mis teretibus.   *Syft. veget.* 545.
Shrubby Bafil.
Nat. of the Eaft Indies.
Cult. 1752, by Mr. Philip Miller.   *Mill. dict. edit.* 6.
        n. 17.
Fl. July.   S. ♄.

Bafili-   2. O. foliis ovatis glabris, calycibus ciliatis.   *Sp. pl.*
cum.   833.
Common Sweet Bafil.

                                                        Nat.

Nat. of India and Perſia.
Cult. 1596, by Mr. John Gerard. *Hort. Ger.*
Fl. July and Auguſt. G. H. ☉.

3. O. foliis ovatis integerrimis. *Sp. pl.* 833. *minimum.*
Buſh Baſil.
Nat. of the Eaſt Indies.
Cult. 1596, by Mr. John Gerard. *Hort. Ger.*
Fl. July and Auguſt. S. ☉.

4. O. foliis oblongiuſculis obtuſis ſerratis undulatis, caule *ſanctum.*
hirto, bracteis cordatis. *Syſt. veget.* 545.
Purple-ſtalked Baſil, or Sacred-herb.
Nat. of the Eaſt Indies.
Introd. 1758, by Hugh Duke of Northumberland.
Fl. September. S. ☉.

5. O. foliis ovato-oblongis ſerratis, bracteis cordatis re- *tenuiflo-*
flexis concavis, ſpicis filiformibus. *Sp. pl.* 833. *rum.*
Slender-ſpiked Baſil.
Nat. of the Eaſt Indies.
Cult. 1708, by the Dutcheſs of Beaufort. *Br. Muſ.*
H. S. 133. *fol.* 22.
Fl. July and Auguſt. S. ♂.

6. O. corollis quadrifidis, racemis aphyllis apice nutan- *polyſta-*
tibus. *Linn. mant.* 567. *chyon.*
Many-ſpiked Baſil.
Nat. of the Eaſt Indies.
Introd. 1783, by John Earl of Bute.
Fl. July and Auguſt. S. ☉.

7. O. foliis lineari-lanceolatis ſerratis. *Sp. pl.* 834. *menthoi-*
Mint-leav'd Baſil. *des.*
Nat. of the Eaſt Indies.
VOL. II. Y *Introd.*

Introd. 1783, by John Earl of Bute.
Fl. July.   S. ⊙.

molle.   8. O. foliis ovatis cordatis acutis ferratis rugofis: finubus claufis, bracteis fubrotundo-cuneiformibus.
Heart-leav'd Bafil.
Nat. of the Eaft Indies. *John Gerard Kœnig*, M.D.
Introd. 1781, by Sir Jofeph Banks, Bart.
Fl. September and October.   S. ⊙.
DESCR. Tota planta pubefcens, fuaveolens. *Caulis*
craffus, obtufe tetragonus. *Folia* late ovata, acuta,
ferrata : ferraturis obtufis inæqualibus ; rugofa,
mollia, profunde cordata: lobis approximatis, inflexis. *Petioli* foliis vix breviores. *Corollæ* e vio-
laceo-albidæ: *Tubus* longitudine calycis ; *Labium*
fuperius orbiculatum, fornicatum ; *inferius* quadri-
lobum, feu potius trilobum, lobo intermedio bifido.
Filamenta fimplicia.

PLECTRANTHUS. *L'Herit. ftirp. nov.*

Calyx lacinia fumma majore.   *Corolla* refupinata, ringens.   *Nectarium* calcaratum, fupinum.

frutico-   1. P. nectario calcarato, racemis compofitis, pedunculis
fus.   tripartitis, caule fruticofo lævigato. *L'Herit. ftirp.*
nov. p. 85. *tab.* 41.
Shrubby Plectranthus.
Nat. of the Cape of Good Hope. Mr. *Fr. Maffon.*
Introd. 1774.
Fl. June——September.   G. H. ♃.

puncta-   2. P. nectario gibbofo, floribus fpicatis, caule herbaceo
tus.   hirto. *L'Herit. ftirp. nov. p.* 87. *tab.* 42.
Ocymum punctatum. *Linn. fuppl.* 275.
§   Dotted

Dotted Plectranthus.
Nat. of Africa. *James Bruce*, Efq.
Introd. 1775.
Fl. January——May. G. H. ♂.

TRICHOSTEMA. *Gen. pl.* 733.

Calycis labium fuperius trifidum; inferius bifidum, bre-
viffimum. *Stamina* longa, incurva.

1. T. ftaminibus longiffimis exfertis. *Sp. pl.* 834. *dichoto-*
Marjoram-leav'd Trichoftema. *ma.*
Nat. of Virginia and Penfylvania.
Cult. 1759, by Mr. Ph. Miller. *Mill. dict. edit.* 7. *n.* 1.
Fl. June and July. H. ☉.

SCUTELLARIA. *Gen. pl.* 734.

Calyx ore integro: poft florefcentiam claufo, operculato.

1. S. foliis incifis fubtus tomentofis, fpicis rotundato- *orientalis.*
tetragonis. *Sp. pl.* 834.
Yellow-flower'd Skull-cap.
Nat. of Barbary and the Levant.
Cult. 1729, by Mr. Philip Miller. *R. S. n.* 362.
Fl. July and Auguft. H. ♃.

2. S. foliis fubcordatis ferratis rugofis opacis, fpicis fe- *albida.*
cundis, bracteis ovatis. *Linn. mant.* 248.
Hairy Skull-cap.
Nat. of the Levant.
Introd. 1780, by Monf. Thouin.
Fl. June and July. H. ♃.

3. S. foliis cordatis incifo-ferratis crenatis, fpicis imbri- *alpina.*
catis rotundato-tetragonis. *Sp. pl.* 834.
Alpine

Alpine Skull-cap.
Nat. of Switzerland.
Cult. 1752, by Mr. Ph. Miller. *Mill. dict. edit.* 6. *n.* 3.
Fl. June——October.   H. ♃.

lupulina. 4. S. foliis cordatis incifo-ferratis acutis glabris, fpicis
     imbricatis rotundato-tetragonis. *Sp. pl.* 835.
    Great-flower'd Skull-cap.
    Nat. of Siberia.
    Cult. 1739, by Mr. Ph. Miller. *Rand. chel.* Caffida 5.
    Fl. June——September.   H. ♃.

lateriflo- 5. S. foliis lævibus carina fcabris, racemis lateralibus fo-
ra.   liofis. *Sp. pl.* 835.
    Virginian Skull-cap.
    Nat. of Canada and Virginia.
    Cult. 1752, by Mr. Ph. Miller. *Mill. dict. edit.* 6. *n.* 5.
    Fl. July——September   H. ♃.

galericu- 6. S. foliis cordato-lanceolatis crenatis; floribus axillari-
lata.   bus. *Sp. pl.* 835. *Curtis lond.*
    Common Skull-cap, or Hooded Willow-herb.
    Nat. of Britain.
    Fl. June——September.   H. ♃.

minor. 7. S. foliis cordato-ovatis fubintegerrimis, floribus axil-
    laribus. *Sp. pl.* 835. *Curtis lond.*
    Leaft Skull-cap, or Hooded Willow-herb.
    Nat. of Britain.
    Fl. July and Auguft.   H. ♃.

peregri- 8. S. foliis fubcordatis ferratis, fpicis elongatis fecundis.
na.   *Sp. pl.* 836.
    Florentine Skull-cap.
    Nat. of Italy.

                Cult.

Cult. 1683, by Mr. James Sutherland. *Sutherl. hort.*
edin. 182. *n.* 4.
Fl. June——October. H. ♃.

9. S. foliis cordato-oblongis acuminatis ferratis, fpicis *altiffima.*
fubnudis. *Sp. pl.* 836.
Tall Skull-cap.
Nat. of the Levant.
Cult. 1731, by Mr. Philip Miller. *Mill. dict. edit.* 1.
Caffida 3.
Fl. July and Auguft. H. ♃.

P R U N E L L A. *Gen. pl.* 735.
Filamenta bifurca: altero apice antherifera. *Stigma*
bifidum.

1. P. foliis omnibus ovato-oblongis ferratis petiolatis. *vulgaris.*
Syft. veget. 547. *Curtis lond.*
α floribus cæruleis.
Common Self-heal.
β floribus albis.
White-flower'd Self-heal.
γ Prunella grandiflora. *Jacqu. auftr.* 4. *p.* 40. *t.* 377.
Great-flower'd Self-heal.
Nat. α. β. of Britain; γ. of Auftria.
Fl. July and Auguft. H. ♃.

2. P. foliis ovato-oblongis petiolatis: fupremis quatuor *laciniata.*
lanceolatis dentatis. *Sp. pl.* 837. *Jacqu. auftr.* 4.
p. 41. *t.* 378.
Jagged-leav'd Self-heal.
Nat. of Germany.
Cult. 1713. *Philofoph. tranf. n.* 337. *p.* 42. *n.* 34.
Fl. July——September. H. ♃.

Y 3 3. P.

byſſopifo-   3. P. foliis lanceolatis integerrimis feſſilibus, caule
lia.   erecto. *Syſt. veget.* 547.
                   Hyſſop-leav'd Self-heal.
                   Nat. of the South of France.
                   Cult. 1731, by Mr. Philip Miller. *Mill. dict. edit.* 1,
                      Brune.la 6.
                   Fl. July——October.   H. ♃,

C L E O N I A.   *Gen. pl.* 736.

Filamenta bifurca : apice altero antherifera. *Stigma*
quadrifidum.

luſitani-   1. Cleonia.   *Sp. pl.* 837.
ca.   Sweet-ſcented Cleonia.
                   Nat. of Spain and Portugal.
                   Cult. 1756, by Mr. Ph. Miller. *Mill. ic.* 47. *t.* 70. *f.* 1,
                   Fl. June and July.   H. ☉.

P R A S I U M.   *Gen. pl.* 737.

Baccæ 4, monoſpermæ.

majus.   1. P. foliis ovato-oblongis ſerratis.   *Sp. pl.* 838.
                   Great Spaniſh Hedge-nettle.
                   Nat. of Spain and Italy.
                   Cult. 1729, by Mr. Philip Miller.   *R. S. n.* 367.
                   Fl. June——Auguſt.   G. H. ♄.

ANGIOSPERMIA.

BARTSIA. *Gen. pl.* 739.

Cal. 2-lobus, emarginatus, coloratus. *Corolla* minus ipfo calyce colorata: labio fuperiore longiore. *Capfula* 2-locularis.

1. B. foliis alternis lanceolatis integerrimis: floralibus   *pallida.*
ovatis dentatis. *Sp. pl.* 839.
Pale-flower'd Bartfia.
Nat. of Siberia and Hudfon's-bay.
Introd. 1782, by the Hudfon's-bay Company.
Fl. June——September.   H. ♃.

RHINANTHUS. *Gen. pl.* 740.

Cal. 4-fidus, ventricofus. *Capfula* 2-locularis, obtufa, compreffa.

1. R. corollarum labio fuperiore compreffo breviore.   *Crifta*
Sp. pl. 840. *Curtis lond.*   *galli.*
Yellow Rattle.
Nat. of Britain.
Fl. July and Auguft.   H. ☉.

EUPHRASIA. *Gen. pl.* 741.

Cal. 4-fidus, cylindricus. *Capf.* 2-locularis, ovato-oblonga. *Antheræ* inferiores altero lobo bafi fpinofæ.

1. E. foliis ovatis lineatis argute dentatis. *Sp. pl.* 841.   *officinalis.*
Curtis lond.
Common Eye-bright.

Y 4   *Nat.*

Nat. of Britain.

Fl. July——September. H. ☉,

Odon-
tites.

2. E. foliis linearibus : omnibus ferratis. *Sp. pl.* 841,
   Curtis lond.

Red Eye-bright.

Nat. of Britain.

Fl. July——September. H. ☉,

MELAMPYRUM. *Gen. pl.* 742.

Cal. 4-fidus. *Corollæ* lab. fup. compreffum, margine
replicato. *Capf.* 2-locularis, obliqua, hińc dehifcens. *Semina* 2, gibba.

crifta-
tum.

1. M. fpicis quadrangularibus: bra&eis cordatis com-
   pa&is denticulatis imbricatis. *Sp. pl.* 842.

Crefted Cow-wheat.

Nat. of Britain.

Fl. July and Auguft. H. ☉.

pratenfe.

2. M. floribus fecundis lateralibus : foliorum conjugati-
   onibus remotis, corollis claufis. *Syft. veget.* 550,

Meadow Cow-wheat.

Nat. of England.

Fl. July and Auguft. H. ☉.

PEDICULARIS. *Gen. pl.* 746.

Calyx 5-fidus. *Capfula* 2-locularis, mucronata, obliqua. *Semina* tunicata.

paluftris.

1. P. caule ramofo, calycibus criftatis callofo-pun&atis,
   corollis labio obliquis. *Syft. veget.* 551.

Marfh Loufe-wort.

Nat. of Britain.

Fl. June. H. ☉.

2. P,

2. P. caule ramofo, calycibus oblongis angulatis lævi- *fylvatica.*
bus, corollis labio cordato. *Sp. pl.* 845.
Common Loufe-wort.
Nat. of Britain.
Fl. May and June.   H. ☉.

3. P. caule fimplici, foliis pinnatis retro imbricatis. *Sp.* *flammea.*
pl. 846.
Upright Loufe-wort.
Nat. of Switzerland.
Introd. 1775, by the Doctors Pitcairn and Fothergill.
Fl. July.   H. ♃.

4. P. caule fimplici, fpica foliofa, corollis galea acuta *comofa.*
emarginata, calycibus quinquedentatis. *Syft. veget.*
552.
Spiked Loufe-wort.
Nat. of the Alps of Italy.
Introd. 1775, by the Doctors Pitcairn and Fothergill.
Fl.   H. ♃.

G E R A R D I A. *Gen. pl.* 747.

Cal. 5-fidus. *Corolla* 2-labiata: lab. inf. 3-partito:
lobis emarginatis: medio 2-partito. *Capf.* 2-locu-
laris, dehifcens.

1. G. foliis linearibus. *Sp. pl.* 848.   *purpurea.*
Purple Gerardia.
Nat. of North America.
Introd. 1772, by Samuel Martin, M. D.
Fl. July and Auguft.   H. ☉.

C H E L O N E.

CHELONE. *Gen. pl.* 748.

Cal. 5-phyllus. *Cor.* ringens, ventricofa. *Rudimentum* filamenti quinti fimplex. *Capfula* 2-locularis.

glabra. 1. C. foliis petiolatis lanceolatis ferratis : fummis oppofitis. *Syft. veget.* 553.
White Chelone.
Nat. of Virginia and Canada.
Cult. 1730, by Mr. Philip Miller. *R. S. n.* 420.
Fl. Auguft——Oƈtober.   H. ♃.

obliqua. 2. C. foliis petiolatis lanceolatis ferratis oppofitis. *Syft. veget.* 553.
Red Chelone.
Nat. of North America.
Cult. 1752, by Mr. Philip Miller. *Mill. diƈt. edit.* 6. Appendix.
Fl. Auguft——Oƈtober.   H. ♃.

CYRILLA. *L'Herit. ftirp. nov.*

Cal. fuperus, 5-phyllus. *Cor.* declinata, infundibulif. *Limbus* planus, 5-partitus, fubæqualis. *Rudimentum* filamenti quinti. *Capf.* femibilocularis.

pulchella. 1. CYRILLA. *L'Herit. ftirp. nov. tab.* 71.
Gefneria pulchella. *Swartz prodr.* 90.
Columnea ereƈta. *De Lamarck encycl.* 2. *p.* 66.
Buchneria coccinea. *Scop. infubr.* 2. *p.* 10. *tab.* 5.
Achimenes minor ereƈta fimplex, foliis crenatis ovatis oppofitis vel ternatis, floribus petiolatis fingularibus ad alas. *Browne jam.* 271. *t.* 30. *f.* 1.
Scarlet-flower'd Cyrilla.
Nat. of Jamaica.

Introd.

Introd. 1778, by Mr. William Forfyth.
Fl. Auguft——Oâober.   S. ♃.

G L O X I N I A.   *L'Herit. ſtirp. nov.*

Cal. ſuperus, 5-phyllus.   *Cor.* campanulata, *limbo* obliquo.   *Filam.* cum rudimento quinti, receptaculo inſerta.

1. GLOXINIA.   *L'Herit. ſtirp. nov. p.* 149.   *maculata.*
Martynia perennis. *Sp. pl.* 862.   *Medic. bot. beobacht.*
    1783. *p.* 238.
Spotted Gloxinia.
Nat. of South America.
Cult. 1739, by Mr. Philip Miller. *Rand. chel.* Martynia 2.

G E S N E R I A.   *Gen. pl.* 749.

Cal. 5-fidus, germini inſidens.   *Corolla* incurva recurvaque.   *Capſ.* infera, 2-locularis.

1. G. foliis ovato-lanceolatis crenatis hirſutis, pedunculis lateralibus longiſſimis corymbiferis. *Sp. pl.* 851.   *tomentoſa.*
Woolly Geſneria.
Nat. of South America.
Cult. 1759, by Mr. Ph. Miller. *Mill. dict. edit.* 7. *n.* 1.
Fl. moſt part of the Year.   S. ♄.

A N T I R R H I N U M.   *Gen. pl.* 750.

Cal. 5-phyllus.   *Corollæ* baſis deorſum prominens, nectarifera.   *Capſ.* 2-locularis.

1. A. foliis cordatis quinquelobis alternis, caulibus procumbentibus. *Sp. pl.* 851.   *Curtis lond.*   *Cymbalaria.*
Ivy-leav'd Toad-flax.

                                        Nat.

Nat. of England.

Fl. June——November.   H. ♃.

Elatine.   2. A. foliis haftatis . alternis, caulibus procumbentibus.
       Sp. pl. 851. *Curtis lond.*
       Sharp-pointed Toad-flax, or Fluellin.
       Nat. of England.
       Fl. Auguft——November.   H. ☉.

fpurium.   3. A. foliis ovatis alternis, caulibus procumbentibus.
       Sp. pl. 851. *Curtis lond.*
       Round-leav'd Toad-flax, or Fluellin.
       Nat. of England.
       Fl. Auguft.   H. ☉.

cirrho-   4. A. foliis haftatis alternis, caulibus patulis, petiolis
fum.   paffim cirrhefcentibus. *Linn. mant.* 249. *Jacqu.*
       hort. 1. *p.* 35. *t.* 82.
       Tendrill'd Toad-flax.
       Nat. of Egypt.
       Introd. 1777, by Jofeph Nicholas de Jacquin, M. D.
       Fl. July.   H. ☉.

ægyptia-   5. A. foliis haftatis alternis, caule erecto ramofiffimo,
cum.   pedunculis rigefcentibus. *Syft. veget.* 555.
       Egyptian Toad-flax.
       Nat. of Egypt.
       Introd. 1771, by Monf. Richard.
       Fl. July.   H. ☉.

triphyl-   6. A. foliis ternis ovatis. *Sp. pl.* 852.
lum.   Three-leav'd Toad-flax.
       Nat. of Sicily.
       Cult. 1640. *Park. theat.* 457. *n.* 5.
       Fl. June——September.   H. ☉.

                                        7. A.

7. A. foliis quaternis linearibus, caule florifero erecto   *purpure-*
spicato. *Sp. pl.* 853.   *um.*
Purple Toad-flax.
Nat. of the South of Europe.
Cult. 1648, in Oxford Garden. *Hort. oxon. edit.* 1. *p.* 5.
Fl. July——September.   H. ♃.

8. A. foliis lineari-lanceolatis: inferioribus ternis, caule   *verficolor.*
erecto fpicato. *Jacqu. ic. mifcell.* 2. *p.* 336.
Spiked-flower'd Toad-flax.
Nat.
Introd. 1777, by Monf. Thouin.
Fl. July——September.   H. ☉.

9. A. foliis linearibus confertis: inferne quaternis, ca-   *repens.*
lycibus capfulam æquantibus. *Syſt. veget.* 555.
Creeping Toad-flax.
Nat. of Britain.
Fl. July——October.   H. ♃.

10. A. foliis fubulatis canaliculatis carnofis: inferioribus   *fparteum.*
ternis, caule paniculato corollifque glaberrimis. *Sp.*
pl. 854.
Branching Toad-flax.
Nat. of Spain.
Introd. 1772, by Monf. Richard.
Fl. June——October.   H. ♂.

11. A. foliis linearibus glabris: inferioribus quaternis,   *bipuncta-*
caule erecto paniculato, floribus fpicato-capitatis.   *tum.*
Sp. pl. 853.
Dotted-flower'd Toad-flax.
Nat. of Spain and Italy.
Introd. 1777, by Monf. Thouin.
Fl. June——Auguft.   H. ☉.
12. A.

trifte.　12. A. foliis linearibus fparfis : inferioribus oppofitis, nec-
tariis fubulatis, floribus fubfeffilibus. *Syft. veget.* 555.
Black-flower'd Toad-flax.
Nat. of Spain.
Introd. 1727, by Sir Charles Wager. *Mill. dict. edit.* 1.
Linaria 5.
Fl. July and Auguft.　　　　　　　　　　　　　　G. H. ♃.

fupinum.　13. A. foliis fubquaternis linearibus, caule diffufo, floribus
racemofis, calcari recto. *Syft. veget.* 556.
Procumbent Toad-flax.
Nat. of Spain.
Cult. 1728, by Mr. Philip Miller. *R. S. n.* 333.
Fl. July.　　　　　　　　　　　　　　　　　　　H. ☉.

arvenfe.　14. A. foliis fublinearibus : inferioribus quaternis, calyci-
bus pilofo-vifcidis, floribus fpicatis, caule erecto.
Sp. pl. 855.
α Linaria arvenfis cærulea. *Bauh. pin.* 213.
Blue Corn Toad-flax.
β Linaria pumila, foliolis carnofis, flofculis minimis fla-
vis. *Bauh. pin.* 213.
Yellow Corn Toad-flax.
Nat. of England.
Fl. July and Auguft.　　　　　　　　　　　　　H. ☉.

pelifferia-　15. A. foliis caulinis linearibus alternis; radicalibus lan-
num.　　　ceolatis ternis, floribus corymbofis. *Sp. pl.* 855.
Violet-colour'd Toad-flax.
Nat. of France and Italy.
Cult. 1640. *Park. theat.* 459. *f.* 9.
Fl. June and July.　　　　　　　　　　　　　　H. ☉.

vifcofum.　16. A. foliis radicalibus quaternis lanceolatis, caulinis linearibus alternis, calycibus villofis cauli approxi-
matis. *Sp. pl.* 855.
　　　　　　　　　　　　　　　　　　　　　　Clammy

Clammy Snap-dragon.
Nat. of Spain.
Introd. 1786, by Monf. Thouin.
Fl. July. H. ☉.

17. A. foliis quinis linearibus carnofis, floribus capitatis. *multi-*
    Syft. veget. 556. *caule.*
    Many-ftalked Toad-flax.
    Nat. of Sicily and the Levant.
    Cult. 1731, by Mr. Philip Miller. *Mill. dict. edit.* 1.
    Linaria 6.
    Fl. May——July. H. ☉.

18. A. foliis quaternis lineari lanceolatis glaucis, caule *alpinum.*
    diffufo, floribus racemofis : calcari recto. *Syft.*
    veget. 556. *Jacqu. auftr.* 1. *p.* 36. *t.* 58.
    Alpine Toad-flax.
    Nat. of Auftria and Switzerland.
    Cult. 1570, by Mr. Hugh Morgan. *Lobel. adv.* 176.
    Fl. July——November. H. ♂.

19. A. foliis oppofitis ovato-oblongis ferratis, caule erec- *bicorne.*
    to, floribus racemofis, capfulis bicornibus. *Sp. pl.*
    856.
    Horned Toad-flax.
    Nat. of the Cape of Good Hope.
    Introd. 1774, by Mr. Francis Maffon.
    Fl. July and Auguft. G. H. ☉.

20. A. foliis oppofitis ovatis ferratis petiolatis, pedun- *macro-*
    culis axillaribus unifloris, capfulis compreffis cari- *carpum.*
    natis truncatis.
    Large-fruited Toad-flax.
    Nat. of the Cape of Good Hope. Mr. *Fr. Maffon.*
    Introd. 1787.
    Fl. March. G. H. ♃.

                                    21. A.

villosum.   21. A. foliis omnibus oppositis ovatis villosis, caulibus simplicibus, floribus oppositis lateralibus. *Sp. pl.* 852.
Villous Snap-dragon.
Nat. of Spain.
Introd. 1786, by Sir Francis Drake, Bärt.
Fl. July and August.   G. H. ♃.

origani-folium.   22. A. foliis plerisque oppositis oblongis, floribus alternis. *Sp. pl.* 852.
Marjoram-leav'd Snap-dragon.
Nat. of the South of Europe.
Introd. 1785, by Messrs. Lee and Kennedy.
Fl. most part of the Summer.   G. H. ♃.

minus.   23. A. foliis plerisque alternis lanceolatis obtusis, caule ramosissimo diffuso. *Sp. pl.* 852. *Curtis lond.*
Lesser Toad-flax.
Nat. of England.
Fl. June——November.   H. ☉.

hirtum.   24. A. foliis lanceolatis hirtis alternis, floribus spicatis: foliolo calycino supremo maximo. *Sp. pl.* 857. *Jacqu. ic. miscell.* 2. *p.* 334.
Hairy Toad-flax.
Nat. of Spain.
Introd. 1777, by Mr. John Hyacinth de Magellan.
Fl. June——September.   H. ☉.

genistifo-lium.   25. A. foliis lanceolatis acuminatis, panicula virgata flexuosa. *Sp. pl.* 858. *Jacqu. austr.* 3. *p.* 25. *t.* 244.
Broom-leav'd Toad-flax.
Nat. of Austria and Siberia.

         Cult.

Cult. 1732, by James Sherard, M. D. *Dill. elth.* 201.
t. 164. *f.* 200.
Fl. July and Auguft.　　　　　　　　　H. ♃.

26. A. foliis linearibus alternis, caulé paniculato virga-　*junceum.*
to, floribus racemofis. *Sp. pl.* 858.
Rufh-ftalk'd Toad-flax.
Nat. of Spain.
Introd. 1780, by Monf. Thouin.
Fl. July.　　　　　　　　　　　　　　H. ☉.

27. A. foliis lineari-filiformibus fucculentis fparfis con-　*monfpef-*
fertis, caule erecto, calcaribus calyce brevioribus.　*fulanum.*
Antirrhinum monfpeffulanum. *Sp. pl.* 854.
Montpelier Toad-flax.
Nat. of France.
Cult. 1748, by Mr. Ph. Miller. *Mill. dict. edit.* 5.
Linaria 16.
Fl. June——Auguft.　　　　　　　　　H. ♃.

28. A. foliis lanceolato-linearibus confertis, caule erecto,　*Linaria.*
fpicis terminalibus feffilibus: floribus imbricatis.
Sp. pl. 858. *Curtis lond.*
Common Yellow Toad-flax.
Nat. of Britain.
Fl. July——September.　　　　　　　　H. ♃.

29. A. foliis lineari-lanceolatis alternis, floribus race-　*chale-*
mofis, calycibus corolla longioribus, caule erecto.　*penfe.*
Sp. pl. 859.
White-flower'd Toad-flax.
Nat. of the Levant.
Cult. 1680, in Oxford Garden. *Morif. hift.* 2.
p. 502. *n.* 24. *f.* 5. *t.* 35. *f.* 9.
Fl. June and July.　　　　　　　　　　H. ☉.
Vol. II.　　　　　　Z　　　　　　30. A.

majus.   30. A. corollis ecaudatis, floribus fpicatis, calycibus rotundatis. *Sp. pl.* 859.

α Antirrhinum majus, rotundiore folio. *Bauh. pin.* 211.
Common Snapdragon.

β Antirrhinum majus alterum, folio longiore. *Bauh.
pin.* 211.
Long-leav'd Snapdragon.

γ floribus ruberrimis : palato aureo-punctato.
Scarlet-flower'd Snapdragon.

δ floribus plenis.
Double-flower'd Snapdragon.

Nat. of England.

Fl. June——Auguft.   H. ♂ .

Oronti-   31. A. corollis ecaudatis, floribus fubfpicatis, calycibus
um.   corolla longioribus. *Syft. veget.* 557. *Curtis lond.*
Small Toad-flax, or Calf's-fnout.

Nat. of England.

Fl. July——September.   H. ☉.

Afarina.   32. A. foliis oppofitis cordatis crenatis, corollis ecaudatis, caulibus procumbentibus. *Sp. pl.* 860.
Heart-leav'd Toad-flax.

Nat. of Italy.

Cult. 1699, by Mr. Jacob Bobart. *Morif. hift.* 3.
p. 432. *n.* 1. *f.* 11. *t.* 21. *f.* 1.

Fl. July.   H. ♃.

molle.   33. A. foliis oppofitis ovatis tomentofis, corollis ecaudatis, caulibus procumbentibus. *Sp. pl.* 860.
Woolly-leav'd Toad-flax, or Snapdragon.

Nat. of Spain.

Cult. 1748, by Mr. Philip Miller. *Mill. dict. edit.* 5.
n. 8.

Fl. July——October.   G. H. ♄ .

34. A.

34. A. foliis radicalibus lingulatis dentatis lineatis; cauli- *bellidifo-*
nis partitis integerrimis. *Syft. veget.* 558. *lium:*
Daify-leav'd Toad-flax.
Nat. of France:
Cult. 1629. *Park. parad.* 267. *f.* 3.
Fl. June——Auguft. H. ♂.

M A R T Y N I A. *Gen. pl.* 753.

Cal. 5-fidus. *Cor.* ringens. *Capf.* lignofa, corticatá,
roftro hamato, 4-locularis, bivalvis.

1. M. caule ramofo, foliis integerrimis cordatis: finubus *Probofci-*
dilatatis. *dea.*
Martynia Probofcidea. *Gloxin obf. bot. p.* 7 & 14.
Martynia caule ramofo, foliis cordato-ovatis pilofis.
Mill. ic. 191. *tab.* 286.
Martynia Louifiana. *Mill. dict.*
Martynia annua villofa. *Kretzfchmar monogr. cum fig.*
Probofcidea. *Schmid. ic.* 49. *tab.* 12, 13.
Probofcidea Juffievii. *Medic. bot. beobacht.* 1783.
p. 20.
Hairy Martynia.
Nat. of America.
Cult. 1759, by Mr. Philip Miller. *Mill. ic. loc. cit.*
Fl. June and Auguft. S. ☉.

2. M. caule fimplici, foliis fubrotundis repandis, tubo *longiflora:*
corollæ bafi gibbofo complanato.
Martynia longiflora. *Syft. veget.* 559. *Meerb. ic.* 7.
Martynia capenfis. *Gloxin obf. bot. p.* 13.
Long-flower'd Martynia.
Nat. of the Cape of Good Hope.
Introd. 1781, by the Countefs of Strathmore.
Fl. July and Auguft. G. H. ☉.

Z 2 BRUN-

BRUNFELSIA. *Gen. pl.* 260.

Cal. 5-dentatus, anguſtus.   *Corollæ* tubus longiſſimus.
Capſ. unilocularis, polyſperma : *conceptaculo* carnoſo
maximo.

america-   1. BRUNFELSIA. *Sp. pl.* 276.
na.   α foliis ellipticis, laciniis limbi tubo ter brevioribus.
           Oval-leav'd Brunfelſia.
        β. foliis lanceolato-oblongis, laciniis limbi tubo dimidio
           brevioribus.
           Spear-leav'd Brunfelſia.
           Nat. of the Weſt Indies.
           Cult. 1739, by Mr. Philip Miller.   *Rand. chel.*
           Fl. June and July.   S. ♄.

SCROPHULARIA. *Gen. pl.* 756.

Cal. 5-fidus.   *Cor.* ſubgloboſa, reſupinata.   *Capſula*
2-locularis.

marilan-   1. S. foliis cordatis ſerratis acutis baſi rotundatis, caule
dica.   obtuſangulo.   *Sp. pl.* 863.
           Maryland Figwort.
           Nat. of North America.
           Cult. 1759, by Mr. Philip Miller.   *Mill. dict. edit.* 7.
           n. 17.
           Fl. May——July.   H. ♃.

nodoſa.   2. S. foliis cordatis trinervatis, caule obtuſangulo.   *Syſt.*
           veget. 560.
           Knobby-rooted Figwort.
           Nat. of Britain.
           Fl. May——July.   H. ♃.

3. S.

3. S. foliis cordatis petiolatis decurrentibus obtufis, caule *aquatica.*
    membranis angulato, racemis terminalibus. *Sp. pl.*
    864. *Curtis lond.*
    Water Figwort.
    Nat. of Britain,
    Fl. May——July.   H. ♃.

4. S. foliis cordatis fubtus tomentofis bafi appendiculatis, *auricula-*
    racemis terminalibus. *Syft. veget.* 560.   *ta.*
    Ear'd-leav'd Figwort.
    Nat. of Spain.
    Introd. 1772, by Monf. Richard.
    Fl. July.   H. ♃.

5. S. foliis cordatis duplicato-ferratis pubefcentibus, pa-   *Scorodo-*
    niculis terminalibus trichotomis foliis interftinctis.   *nia.*
    Scrophularia Scorodonia. *Sp. pl.* 864.
    Balm-leav'd Figwort.
    Nat. of England.
    Fl. May——July.   H. ♃.

6. S. foliis oblongo-lanceolatis cordatis duplicato-ferratis *glabrata.*
    glabris, paniculis racemofis terminalibus trichoto-
    mis, caule fuffruticofo.
    Spear-leav'd Figwort.
    Nat. of the Canary Iflands. Mr. *Francis Maffon.*
    Introd. 1779.
    Fl. April and May.   G. H. ♂.

7. S. foliis cordatis oblongis dentatis: dentibus integer-   *betonici-*
    rimis: bafeos profundioribus. *Syft. veget.* 560.   *folia.*
    Petony-leav'd Figwort.
    Nat. of Spain.
    Cult. 739, by Mr. Ph. Miller. *Mill. dict. vol.* 2. *n.* 3.
    Fl. July——Auguft.   H. ♃.

                    Z 3   8. S.

orientalis.   8. S. foliis lanceolatis ferratis petiolatis : caulinis ter-
nis ; rameis oppofitis.   *Sp. pl.* 864.
Hemp-leav'd Figwort.
Nat. of the Levant.
Cult. 1712.   *Philofoph. tranfaƈt. n,* 333. *p*. 416.
n. 60.
Fl. July and Auguft.   H. ♃.

frutef-   9. S. foliis fubcarnofis feffilibus lævibus apice recurvis.
çens.   *Syft. veget.* 560.
Shrubby Figwort.
Nat. of Portugal.
Cult. 1768, by Mr. Ph. Miller.   *Mill. diƈt. edit.* 8.
Fl. June——Auguft.   G. H. ♄.

vernalis.   10. S. foliis cordatis pubefcentibus duplicato-ferratis,
paniculis axillaribus dichotomis.
Scrophularia vernalis.   *Sp. pl.* 864.
Yellow Figwort.
Nat. of England.
Fl. March——May.   H. ♂

arguta.   11. S. foliis cordatis glabris duplicato-ferratis, paniculis
axillaribus dichotomis, capfulis acuminatis.
Slender upright Figwort.
Nat. of Madeira and Teneriffe.   Mr. *Fr. Maffon.*
Introd. 1778.
Fl. May and June.   G. H. ☉.
OBS. Differt a Scr. vernali caule foliifque glabris, et
floribus minoribus rubris.

fambuci-   12. S. foliis interrupte pinnatis cordatis inæqualibus, r⁻
folia.   cemo terminali : pedunculis axillaribus gennis
dichotomis.   *Sp. pl.* 865.
Elder-leav'd Figwort.
§   *Nat.*

Nat. of Spain and Portugal.
Cult. before 1629, by Mr. John Parkinſon, *Park. theat.* 611. *f.* 8. conf. *Alp. exot.* 203.
Fl. July——September. H. ♃.

13. S. foliis pinnatis, racemo terminali nudo, pedunculis *canina.* bifidis. *Sp. pl.* 865.
Cut-leav'd Figwort.
Nat. of the South of Europe.
Cult. 1683, by Mr. James Sutherland. *Sutherl. hort. edin.* 313. *n.* 4.
Fl. June——Auguſt. H. ♃.

14. S. foliis pinnatis ſubinterruptis: foliolis ſubquinis *mellifera.* oblongis, floribus axillaribus, corollæ fundo melli-fero. *L'Herit. ſtirp. nov. tom.* 2. *tab.* 31.
Barbary Figwort.
Nat. of Barbary.
Introd. 1786, by Monſ. Thouin.
Fl. July and Auguſt. G. H. ♃.

15. S. foliis inferioribus bipinnatis ſubcarnoſis glaberri- *lucida.* mis, racemis bipartitis. *Syſt. veget.* 561.
Shining-leav'd Figwort.
Nat. of the Levant.
Cult. before 1680, by **Robert** Moriſon, M. D. *Moriſ. hiſt.* 2. *p.* 483. *n.* 7.
Fl. June——Auguſt. H. ♃.

16. S. foliis cordatis lineatis lucidis, pedunculis axillari- *peregri-* bus bifloris, caule ſexangulari. *Syſt. veget.* 561. *na.*
Nettle-leav'd Figwort.
Nat. of Italy.
Cult. 1683, by Mr. James Sutherland. *Sutherl. hort. edin.* 313. *n.* 3.
Fl. June——Auguſt. H. ☉.

Z 4 CELSIA.

CELSIA. *Gen. pl.* 757.

Cal. 5-partitus.   *Cor.* rotata.   *Filamenta* barbata,
Capf. 2-locularis.

orientalis.   1. C. foliis bipinnatis.   *Sp. pl.* 866.
Oriental Celfia.
Nat. of the Levant.
Cult. 1739, in Chelfea Garden.   *Rand. chel.*   Ver-
bafcum 8.
Fl. July and Auguft.   H. ⊙,

Arcturus.   2. C. foliis radicalibus lyrato-pinnatis, pedunculis flore
longioribus. *Syft. veget.* 561. *Jacqu. hort.* 2. *p.* 53.
t. 117.
Verbafcum Arcturus.   *Sp. pl.* 254.
Scollop-leav'd Celfia.
Nat. of the Ifland of Candia.
Introd. about 1780.
Fl. July——September.   G. H. ♂ .

cretica.   3. C. foliis radicalibus lyratis ; caulinis fubcordatis am-
plexicaulibus, floribus fubfeffilibus.   *Linn. fuppl.*
281.
Great-flower'd Celfia.
Nat. of the Eaft Indies.
Introd. 1776, by Monf. Thouin.
Fl. July.   G. H. ♂ .

DIGITALIS. *Gen. pl.* 758.

Cal. 5-partitus.   *Cor.* campanulata, 5-fida, ventricofa.
Capf. ovata, 2-locularis.

purpurea.   1. D. calycinis foliolis ovatis acutis, corollis obtufis :
labio fuperiore integro. *Sp. pl.* 866. *Curtis lond.*
α Digitalis purp rea, folio afpero.   *Bauh. pin.* 243.
Purple Fox-glove.
                                              β Digitalis

β Digitalis alba, folio afpero. *Bauh. pin.* 244.
White Fox-glove.
Nat. of Britain.
Fl. June——September. H. ♂.

2. D. foliis decurrentibus. *Sp. pl.* 867. *Thapfi.*
Spanifh Fox-glove.
Nat. of Spain.
Cult. 1759, by Mr. Ph. Miller. *Mill. dict. edit.* 7. *n.* 2.
Fl. June——Auguft. H. ♃.

3. D. calycinis foliolis lanceolatis, corollis acutis : labio *lutea.*
fuperiore bifido. *Sp. pl.* 867. *Jacqu. hort.* 2.
p. 47. *t.* 105.
Small yellow Fox-glove.
Nat. of France and Italy.
Cult. 1629. *Park. parad.* 382. *n.* 7.
Fl. July and Auguft. H. ♃.

4. D. calycinis foliolis lanceolatis, corollis galeâ emar- *ambigua.*
ginata, foliis fubtus pubefcentibus. *Linn. fuppl.* 282.
Digitalis ochroleuca. *Jacqu. auftr.* 1. *p.* 36. *t.* 57.
Greater yellow Fox-glox.
Nat. of the Alps of Switzerland.
Cult. 1596, by Mr. John Gerard. *Hort. Ger.*
Fl. July and Auguft. H. ♃.

5. D. calycinis foliolis ovatis obtufis patentibus, corollæ *ferrugi-*
labio inferiore barbato. *Syft. veget.* 562. *nea.*
Iron-colour'd Fox-glove.
Nat. of Italy.
Cult. 1597. *Ger. herb.* 647.
Fl. July and Auguft. H. ♃.

6. D. foliis lineari-lanceolatis integerrimis bafi adnatis. *obfcura.*
Syft. veget. 562. *Jacqu. hort.* 1. *p.* 40. *t.* 91.
Willow-

Willow-leav'd Fox-glove.
Nat. of Spain.
Introd. about 1778.
Fl. July and Auguſt.   H. ♃.

canarien-   7. D. calycinis foliolis lanceolatis, corollis bilabiatis acu-
ſis.   tis, caule fruticoſo. *Sp. pl.* 868.
Canary ſhrubby Fox-glove.
Nat. of the Canary Iſlands.
Cult. 1698, by the Dutcheſs of Beaufort. *Br. Muſ.*
Sloan. mſſ. 3358. fol. 20.
Fl. June and July.   G. H. ♄.

Sceptrum.   8. D. calycinis foliolis ſubulatis, bracteis linearibus flo-
ribus longioribus, corollis obtuſis, foliis ellipticis
ſerratis, caule fruticoſo.
Digitalis Sceptrum. *Linn. ſuppl.* 282. *L'Herit. ſert.*
angl. tab. 24.
Madeira ſhrubby Fox-glove.
Nat. of Madeira. Mr. *Francis Maſſon.*
Introd. 1777.
Fl. July and Auguſt.   G. H. ♄.

BIGNONIA. *Gen. pl.* 759.

Cal. 5-fidus, cyathiformis. *Cor.* fauce campanulata,
5-fida, ſubtus ventricoſa. *Siliqua* 2-locularis. *Sem,*
membranaceo-alata.

Catalpa.   1. B. foliis ſimplicibus cordatis, caudice erecto, ſemini-
bus membranaceo-alatis.
Bignonia Catalpa. *Sp. pl.* 868. (excluſis ſynonymis
Jacquini, Brownei, Plumerii, & Rheedii).
Common Catalpa.
Nat. of Carolina.

Introd.

Introd. about 1726, by Mr. Mark Catefby, *Hort. angl.* 13. *n.* 4.

Fl. July and Auguſt. H. ♄.

2. B. foliis ſimplicibus oblongis acuminatis, caudice erec- *longiſſi-*
to, ſeminibus lanatis. *ma.*

Bignonia longiſſima. *Jacqu. hiſt.* 182. *tab.* 176. *fig.*
78. *Swartz prodr.* 91.

Bignonia Quercus. *De Lamarck encycl.* 1. *p.* 417.

Bignonia arborea, foliis ovatis verticillato-ternatis,
ſiliqua gracili longiſſima. *Brown. jam.* 264.

Bignonia arbor, folio ſingulari undulato, ſiliquis longiſſimis et anguſtiſſimis. *Plum. ic.* 47. *t.* 57.

Wave-leav'd Trumpet-flower.

Nat. of the Weſt Indies.

Introd. 1777, by Thomas Clark, M. D.

Fl. S. ♄.

3. B. foliis ſimplicibus lanceolatis, caule volubili. *Sp.* *ſempervi-*
pl. 869. *rens.*

Yellow ſweet-ſcented Trumpet-flower.

Nat. of North America.

Cult. 1640, by Mr. John Parkinſon. *Park. theat.*
1465. *n.* 5.

Fl. June and July. G. H. ♄.

4. B. foliis conjugatis cirrhofis : foliolis cordato-lanceo- *capreola-*
latis ; foliis imis ſimplicibus. *Sp. pl.* 870. *ta.*

Four-leav'd Trumpet-flower.

Nat. of North America.

Cult. 1730. *Hort. angl.* 13. *n.* 3.

Fl. June. H. ♄.

5. B. foliis digitatis : foliolis integerrimis obovatis. *Sp.* *penta-*
pl. 870. *phylla.*

Hairy

Hairy five-leav'd Trumpet-flower.

Nat. of Jamaica.

Introd. before 1733, by William Houftoun, M. D.
Mill. dict. edit. 8.

Fl.   S. ♄.

Leucoxy-   6. B. foliis digitatis: foliolis integerrimis ovatis acumi-
lon.   natis. *Sp. pl.* 870.

Smooth five-leav'd Trumpet-flower.

Nat. of the Weft Indies.

Cult. 1759, by Mr. Ph. Miller. *Mill. dict. edit.* 7. *n.* 10.

Fl. June.   S. ♄.

radicans.   7. B. foliis pinnatis : foliolis incifis, caule geniculis ra-
                    dicatis. *Sp. pl.* 871.

major.   α Pfeudo-Gelfeminum filiquofum. *Riv. mon.* 101.

Great Afh-leav'd Trumpet-flower.

minor.   β Bignonia fraxini foliis, coccineo flore minore. *Catefb.
              car.* 1. *p.* 65. *t.* 65.

Small Afh-leav'd Trumpet-flower.

Nat. of North America.

Cult. 1640.   *Park. theat.* 385. *n.* 6.

Fl. July and Auguft.   H. ♄.

flans.   8. B. foliis pinnatis :. foliolis ferratis, caule erecto firmo,
                    floribus racemofis. *Sp. pl.* 871.

Branching-flower'd Trumpet-flower.

Nat. of America.

Cult. 1739, by Mr. Philip Miller. *Rand. chel. n.* 3.

Fl. Auguft.   S. ♄.

indica.   9. B. foliis bipinnatis : foliolis integerrimis ovatis acu-
                    minatis. *Syft. veget.* 564.

Indian Trumpet-flower.

Nat. of India.

Introd.

Introd. 1775, by Daniel Charles Solander, LL.D.
Fl.   S. ♄.

CITHAREXYLUM. *Gen. pl.* 760.

Cal 5-dentatus, campanulatus. *Cor.* infundibuliformi-
rotata : laciniis fupra villofis, æqualibus. *Bacca*
2-fperma. *Sem.* 2-locularia.

1. C. ramis teretibus, calycibus truncatis. *Sp. pl.* 872.   *cauda-*
Oval-leav'd Fiddle-wood.   *tum.*
Nat. of Jamaica.
Introd. 1763, by Mr. John Bufh.
Fl.   S. ♄.

2. C. ramis tetragonis. *Syft. veget.* 564. *Jacqu. hort.* 1.   *quadran-*
p. 8. *t.* 22.   *gulare.*
Square-ftalk'd Fiddle-wood.
Nat. of Jamaica.
Cult. 1759, by Mr. Philip Miller. *Mill. dict. edit.* 7.
n. 1.
Fl.   S. ♄.

3. C. foliis villofis. *Jacqu. ic. collect.* 1. *p.* 72.   *villofum.*
Citharexylon fubferratum. *Swartz prodr.* 91.
Hairy-leav'd Fiddle-wood.
Nat. of the Ifland of St. Domingo.
Introd. 1784, by Mr. John Græfer.
Fl.   S. ♄.

HALLERIA. *Gen. pl.* 761.

Cal. 3-fidus. *Cor.* 4-fida. *Filamenta* corolla longiora.
Bacca infera, 2-locularis.

1. HALLERIA. *Sp. pl.* 872.   *lucida.*

African

African Fly Honey-fuckle.

Nat. of the Cape of Good Hope.

Cult. 1752, by Mr. Philip Miller.   *Mill. dict. edit.* 6.

Fl. June——Auguft.   G. H. ♃.

CRESCENTIA. *Gen. pl.* 762.

Cal. 2-partitus, æqualis. *Cor.* gibba. *Bacca* pedicellata, 1-locularis, polyfperma. *Semina* 2-locularia.

Cujete. 1. C. foliis cuneato-lanceolatis. *Syft. veget.* 564.
Calabafh Tree.

Nat. of Jamaica.

Introd. 1690, by Mr. Bentick. *Br. Muf. Sloan. mff.*
3370.

Fl.   S. ♃.

LANTANA. *Gen. pl.* 765.

Cal. 4-dentatus obfolete. *Stigma* uncinato-refractum.
Drupa nucleo 2-loculari.

trifolia. 1. L. foliis ternis quaternifve ellipticis fupra rugofis
fubtus villofis, caule inermi, fpicis oblongis imbricatis.

Lantana trifolia. *Sp. pl.* 873.

Three-leav'd Lantana.

Nat. of the Weft Indies.

Introd. before 1733, by William Houftoun, M.D.
Mill. dict. edit. 8.

Fl. June——September.   S. ♃.

annua. 2. L. foliis oppofitis, caule inermi herbaceo, fpicis oblongis. *Syft. veget.* 566.

Annual Lantana.

Nat. of South America.

Introd.

Introd. before 1733, by William Houftoun, M.D.
 Mill. dict. edit. 8.
Fl. July.   S. ☉.

3. L. foliis oppofitis, caule inermi ramofo, floribus capi-   *Camara.*
 tato-umbellatis aphyllis. *Sp. pl.* 874.
Various-colour'd Lantana.
Nat. of the Weft Indies.
Cult. 1691, in the Royal Garden at Hampton-court.
 Pluk. alm. 385. *t.* 114. *f.* 4.
Fl. April——September.   •   S. ♄.

4. L. foliis oppofitis ternifque ellipticis rugofis, caule in-   *odorata.*
 ermi, capitulis fquarrofis, bracteis lanceolatis, pedun-
 culis folio brevioribus.
Lantana odorata. *Syft. veget.* 566.
Sweet-fcented Lantana.
Nat. of the Weft Indies.
Cult. 1758, by Mr. Philip Miller.
Fl. May——November.   S. ♄.

5. L. foliis oppofitis ovalibus rugofis, caule inermi, ca-   *recta.*
 pitulis fquarrofis : bracteis oblongis, pedunculis fo-
 lio longioribus.
Upright Lantana.
Nat. of Jamaica.
Cult. 1758, by Mr. Philip Miller.
Fl. June——Auguft.   S. ♄.

6. L. foliis oppofitis ternifque rhombeo-ovatis obtufis   *involu-*
 rugofis tomentofis, caule inermi, capitulis fquarro-   *crata.*
 fis : bracteis ovatis.
Lantana involucrata. *Sp. pl.* 874.
Round-leav'd Lantana.
Nat. of the Weft Indies.

                                               Cult.

Cult. 1690, in the Royal Garden at Hampton-court.
Catal. mff.
Fl. May——July.   S. ♄.

meliffæfo- 7. L. foliis oppofitis ovato-oblongis villofis mollibus,
lia.   caule aculeato, fpicis hemifphæricis, bracteis tubo
                 dimidio brevioribus.
             Lantana flava. *Medicus act. palat. vol.* 3. *phyf. p.* 225.
             Camara Meliffæ folio, flore flavo. *Dillen. elth.* 66.
                 t. 57. *f.* 66.
             Balm-leav'd Lantana.
             Nat. of the Weft Indies.
             Cult. 1732, by James Sherard, M. D. *Dill. elth. loc. cit.*
             Fl. July——September.   S. ♄.

fcabrida. 8. L. foliis oppofitis ovato-ellipticis fcabris, caule aculea-
                 to, fpicis hemifphæricis, bracteis tubo dimidio bre-
                 vioribus lanceolatis acutis.
             Rough Lantana.
             Nat. of the Weft Indies.   Mr. *Gilbert Alexander.*
             Introd. 1774.
             Fl. September.   S. ♄.

aculeata. 9. L. foliis oppofitis ovatis fubcordatis fubtus molliufcu-
                 lis, caule aculeato, capitulorum bracteis lineari-
                 cuneiformibus.
             Lantana aculeata. *Sp. pl.* 874.
             Prickly Lantana.
             Nat. of the Weft Indies.
             Cult. 1692, in the Royal Garden at Hampton-court.
                 Pluk. alm. 385. *t.* 233. *f.* 5.
             Fl. April——November.   S. ♄.

africana. 10. L. foliis alternis feffilibus, floribus folitariis. *Sp. pl.*
                 875.

                                                        Ilex-

Ilex-leav'd Lantana.

Nat. of the Cape of Good Hope.

Cult. 1731, by Mr. Philip Miller. *Mill. dict. edit.* 1.
Jasminum 10.

Fl. February——November. G. H. ♄.

CORNUTIA. *Gen. pl.* 766.

Cal. 5-dentatus. *Stamina* corolla longiora. *Stylus*
longissimus. *Bacca* monosperma.

1. CORNUTIA. *Sp. pl.* 875.   *pyramid-*
Hoary-leav'd Cornutia.   *ata.*
Nat. of the West Indies.
Introd. before 1733, by William Houstoun, M.D:
Mill. dict. edit. 8.
Fl.   S. ♄.

CAPRARIA. *Gen. pl.* 768.

Cal. 5-partitus. *Cor.* campanulata, 5-fida, acuta.
Caps. 2-valvis, 2-locularis, polysperma.

1. C. foliis alternis, floribus geminis. *Sp. pl.* 875.   *biflora,*
Two-flower'd Capraria, or Sweet-weed.
Nat. of South America.
Cult. 1759, by Mr. Philip Miller. *Mill. dict. edit.* 7.
Fl. July and August.   S. ♄.

2. C. foliis oppositis oblongis acutis argute serrulatis læ- *lucida.*
vibus, petiolis alatis, pedunculis trifloris.
Shining Capraria.
Nat. of the Cape of Good Hope. Mr. *Fr. Masson.*
Introd. 1774.
Fl. April and May.   G. H. ♂.
DESCR. *Planta* glabra. *Caules* tetragoni. *Folia* sesqui-
uncialia: *petioli* foliis ter breviores. *Pedunculi* axil-
lares, oppositi, petiolis paulo longiores, tetragoni,
VOL. II.   A a   triflori.

triflori. *Pedicelli* exteriores interdum triflori. *Bracteæ* subulatæ, longitudine pedicellorum. *Calycis* laciniæ subulatæ, trilineares. *Corolla* hypocrateriformis : *tubus* cylindricus, pallide purpurascens, calyce paulo longior, basi extus gibbus, supra gibbum parum recurvus, tandem erectus, latere exteriore parum elongatus, unde *limbus* omnino horizontalis ; *laciniæ* ovatæ, obtusæ, æquales, e rubro purpurascentes, macula atro-purpurea prope faucem : *faux* pilosa. *Antheræ* oblongæ, compressæ. *Germen* subrotundum. *Stylus* staminibus brevior. *Stigma* magnum, convexum, obliquum.

lanceola-   3. C. foliis oppositis lineari-lanceolatis integerrimis, ra-
ta.   cemis compositis terminalibus.
        Capraria lanceolata.   *Linn. suppl.* 284.
        Willow-leav'd Capraria.
        Nat. of the Cape of Good Hope.   Mr. *Fr. Masson.*
        Introd. 1774.
        Fl.   G. H. ♄ .

undulata.   4. C. foliis oppositis ovato-oblongis integerrimis undu-
        latis : supremis subcordatis verticillatis, racemis
        spiciformibus.
        Capraria undulata.   *Linn. suppl.* 284.   *L'Herit. sert.*
        angl. tab. 25.
        Wave-leav'd Capraria.
        Nat. of the Cape of Good Hope.   Mr. *Fr. Masson.*
        Introd. 1774.
        Fl. March——July.   G. H. ♄ .

humilis.   5. C. pubescens, foliis oppositis ternisve ovatis serratis
        petiolatis, pedunculis axillaribus petiolo brevioribus.
        Dwarf Capraria.
        Nat. of the East Indies.   *John Gerard Kœnig,* M. D.
                                       Introd.

Introd. 1781, by Sir Joſeph Banks, Bart.
Fl. July and Auguſt, S. ☉.

S E L A G O. *Gen. pl.* 769.

Cal. 5-fidus. *Cor.* tubus capillaris; limbus ſubæqualis.
Sem. 1 ſ. 2.

1. S. corymbo multiplici, floribus disjunctis, foliis fili-　　*corymbo-*
   formibus faſciculatis. *Sp. pl.* 876.　　　　　　　　　　　*ſa.*
   Fine-leav'd Selago.
   Nat. of the Cape of Good Hope.
   Cult. 1759, by Mr. Philip Miller. *Mill. dict. edit.* 7.
   Fl. July and Auguſt.　　　　　　　　　　　G. H. ♄.

2. S. ſpicis corymboſis, foliis linearibus denticulatis. *Sp.*　*ſpuria.*
   pl. 877.
   Linear-leav'd Selago.
   Nat. of the Cape of Good Hope.
   Introd. 1786, by Mr. Francis Maſſon.
   Fl. July.　　　　　　　　　　　　　　　　G. H. ♂.

3. S. corymbo multiplici, foliis obovatis glabris ſerratis.　*faſcicula-*
   Linn. mant. 250.　　　　　　　　　　　　　　　　　*ta.*
   Cluſter-flower'd Selago.
   Nat. of the Cape of Good Hope.
   Introd. 1774, by Mr. Francis Maſſon.
   Fl. June.　　　　　　　　　　　　　　　　G. H. ♂.

4. S. ſpicis ſtrobilinis ovatis terminalibus, foliis ſparſis　*ovata.*
   linearibus, caule fruticoſo. *L'Herit. ſtirp. nov.*
   tom. 2. *tab.* 33.
   Lippia ovata. *Linn. mant.* 89.
   Oval-headed Selago.
   Nat. of the Cape of Good Hope.

　　　　　　　　　　　A a 2　　　　　　　*Introd.*

Introd. 1774, by Mr. Francis Maſſon.
Fl. June and July.   G. H. ♄.

M A N U L E A.   *Linn. mant.* 12.

Cal. 5-partitus. *Cor.* limbo 5-partito, ſubulato : laciniis ſuperioribus 4 magis connexis. *Capſ.* 2-locu-
laris, polyſperma.

tomentoſa.   1. M. foliis tomentoſis, caulibus folioſis, pedunculis
multifloris. *Syſt. veget.* 569.
Selago tomentoſa. *Sp. pl.* 877.
Woolly Manulea.
Nat. of the Cape of Good Hope.
Introd. 1774, by Mr. Francis Maſſon.
Fl. May——November.   G. H. ♂.

H E B E N S T R E T I A. *Gen. pl.* 770. *mant.* 142.

Cal. emarginatus, ſubtus fiſſus. *Cor.* 1-labiata : lab.
adſcendente, 4-fido. *Capſ.* 2-ſperma. *Stamina*
margini limbi corollæ inſerta.

dentata.   1. H. foliis linearibus dentatis, ſpicis lævibus. *Syſt. veget.*
570.
Dentated Hebenſtretia.
Nat. of the Cape of Good Hope.
Introd. 1770, by Monſ. Richard.
Fl. February——November.   G. H. ♂.

cordata.   2. H. foliis ſubcarnoſis cordatis ſeſſilibus. *Syſt. veget.*
570.
Heart-leav'd Hebenſtretia.
Nat. of the Cape of Good Hope.
Introd. 1774, by Mr. Francis Maſſon.
Fl.   G. H. ♄.

E R I N U S.

E R I N U S. *Gen. pl.* 771.

Cal. 5-phyllus. *Cor.* limbus 5-fidus, æqualis. *Capf.* 2-locularis.

1. E. floribus racemofis, foliis fpathulatis. *Syfl. veget.* 570.

Alpine Erinus.

N . of the Pyrenees and Switzerland.

Cult. 1759, by Mr. Ph. Miller. *Mill. dict. edit.* 7. *n.* 1.

Fl. March and April. H. ♃.

alpinus.

2. E. foliis lanceolato-oblongis dentatis, limbi laciniis integris.

Selago Lychnidea. *Sp. pl.* 877. *Berg. cap.* 158.

α limbo corollæ fordide purpurafcente.

Dark-flower'd Erinus.

β limbo corollæ fordide flavefcente.

Yellow-flower'd Erinus.

Nat. of the Cape of Good Hope.

Introd. 1776, by Meffrs. Kennedy and Lee.

Fl. May and June. G. H. ♃.

Obs. Erinus Lychnidea *Linn. fuppl.* 287. non eft Selago Lychnidea *Sp. pl.* fed Erinus capenfis *Linn. mant.* 252. in Syftemate vegetabilium omiffus.

fragrans.

B U C H N E R A. *Gen. pl.* 772.

Cal. obfolete 5-dentatus. *Corollæ* limbus 5-fidus, æqualis: lobis cordatis. *Capf.* 2-locularis.

1. B. foliis lineari-lanceolatis laxe dentatis fubglutinofis, floribus pedunculatis, caule fruticofo. *L'Herit. ftirp. nov. tom.* 2. *tab.* 34.

Clammy Buchnera.

vifcofa.

A a 3 *Nat.*

Nat. of the Cape of Good Hope. Mr. *Fr. Maſſon,*
Introd. 1774.
Fl. moſt part of the Summer.   G. H. ♄.

BROWALLIA. *Gen. pl.* 773.

Cal. 5-dentatus. *Cor.* limbus 5-fidus, æqualis, patens :
    umbilico clauſo Antheris duabus majoribus, *Capſ,*
    1-locularis.

demiſſa.   1. B. pedunculis unifloris. *Sp. pl.* 879.
    Spreading Browallia.
    Nat. of South America.
    Introd. 1735, by Mr. Robert Millar. *Mill, dict. edit.* 8,
    Fl. July——September.   S. ☉.

elata,   2. B. pedunculis unifloris multiflorifque. *Sp. pl.* 880,
        Curtis magaz. 34.
    Upright Browallia.
    Nat. of Peru.
    Cult. 1768, by Mr. Philip Miller. *Mill, dict. edit.* 8.
    Fl. July——September.   S. ☉.

LINNÆA. *Gen. pl.* 774.

Cal. duplex : fructus 2-phyllus ; floris 5-partitus, ſupe-
    rus. *Corolla* campanulata, *Bacca* ſicca, trilocularis,

borealis.   1. L. floribus geminatis. *Sp. pl.* 880.
    Two-flower'd Linnæa.
    Nat. of the North of Europe, Aſia, and America.
    Introd. 1762, by Mr. Andrew Kallſtrœm,
    Fl. May and June.   H. ♄.

SIBTHORPIA. *Gen. pl. 775.*

Cal. 5-partitus. *Cor.* 5-partita, æqualis. *Stam.* pa-
ribus remotis. *Capf.* compreffa, orbicularis, 2-
locularis : diffepimento tranfverfo.

1. S. foliis reniformibus fubpeltatis crenatis. *Sp. pl.*   *europæa.*
880.
Cornifh Sibthorpia, or Money-wort.
Nat. of England.
Fl. Auguft.   H. ♃.

LIMOSELLA. *Gen. pl. 776.*

Cal. 5-fidus. *Cor.* 5-fida, æqualis. *Stam.* per paria
approximata. *Capf.* 1-locularis, 2-valvis, polyfperma.

1. L. foliis lanceolatis. *Syft. veget. 572.*   *aquatica.*
Baftard Plantain, or Mud-wort.
Nat. of Britain.
Fl. Auguft and September.   H. ♃.

DODARTIA. *Gen. pl. 780.*

Cal. 5-dentatus. *Cor.* labium inf. duplo longius.
     Capf. 2-locularis, globofa.

1. D. foliis linearibus integerrimis glabris. *Sp. pl.* 883. *orientalis.*
Oriental Dodartia.
Nat. of the Levant.
Cult. 1739, by Mr. Ph. Miller. *Mill. dict. vol. 2. n. 1.*
Fl. July.   H. ♃.

SESAMUM. *Gen. pl. 782.*

Cal. 5-partitus. *Cor.* campanulata, 5-fida : lobo in-fimo majore. *Rudimentum* Filamenti quinti. *Stig-ma* lanceolatum. *Capf.* 4-locularis.

orientale.   1. S. foliis ovato-oblongis integris. *Sp. pl.* 883.
        Oriental Sefamum, or Oily-grain.
        Nat. of the Eaft Indies.
        Cult. 1731, by Mr. Ph. Miller. *Mill. dict. edit.* 1. *n.* 1.
        Fl. July.   S. ⊙,

indicum.   2. S. foliis inferioribus trifidis. *Sp. pl.* 884.
        Indian Sefamum, or Oily-grain.
        Nat. of India.
        Cult. 1731, by Mr. Ph. Miller. *Mill. dict. edit.* 1. *n.* 2.
        Fl. July.   S. ⊙,

PENTSTEMON.

Cal. 5-phyllus. *Cor.* bilabiata, ventricofa. *Rudimen-tum* Filamenti quinti fuperne barbatum. *Capf.* bilocularis.

pubefcens.   1. P. caule pubefcente, filamento fterili ab apice infra medium barbato.
        Chelone Pentftemon. *Sp. pl.* 850.

latifolia.   α foliis ovato-oblongis.
        Dracocephalus hirfutus Lyfimachiæ foliis latioribus ferratis. *Morif. hift.* 3. *p.* 417. *f.* 11. *t.* 21. *f.* 3.
        Broad-leav'd hairy Pentftemon.

angufti-folia.   β foliis lanceolatis.
        Dracocephalus latifolius glaber, Lyfimachiæ luteæ foliis. *Morif. hift.* 3. *p.* 417. *f.* 11. *t.* 21. *f.* 2. (De-fcriptio mala e fpecimine ficco).

                                   Narrow-

Narrow-leav'd hairy Pentſtemon.
Nat. of North America.
Cult. 1758, by Mr. Ph. Miller. *Mill. ic.* 168. *t.* 252.
Fl. Auguſt and September.　　　　　H. ♃.

2. P. caule glabro, filamento ſterili ſuperne barbato,　　*lævigata.*
Chelone Pentſtemon. *J. F. Miller ic.* 4.
Chelone foliis inferioribus ovato-acuminatis petiolatis
　integerrimis : ſuperioribus amplexicaulibus lanceo-
　latis dentatis : corollis patentibus bilabiatis. *Ar-*
　duin. animadv. 1. *p.* 14. *t.* 5.
Digitalis perfoliata glabra, flore violaceo minore.
　Moriſ. hiſt. 2. *p.* 479. *ſ.* 5. *t.* 8. *ſ.* 6.
Smooth Pentſtemon.
Nat. of North America.
Cult. 1776, by John Fothergill, M. D.
Fl. Auguſt and September.　　　　　H. ♃.

　　　M I M U L U S. *Gen. pl.* 783.

Cal. 4-dentatus, priſmaticus. *Cor.* ringens : labio ſu-
　periore lateribus replicato. *Capſ.* 2-locularis, po-
　lyſperma.

1. M. erectus, foliis oblongis linearibus ſeſſilibus. *Sp.*　*ringens.*
　pl. 884.
Oblong-leav'd Monkey-flower.
Nat. of Virginia and Canada.
Cult. 1759, by Mr. Philip Miller. *Mill. dict. edit.* 7.
Fl. July and Auguſt.　　　　　H. ♃.

2. M. erectus, foliis ovatis petiolatis, caule tetragono　*alatus.*
　alato.
Oval-leav'd Monkey-flower.
Nat. of North America.
　　　　　　　　　　　　　　　Introd.

Introd. 1783, by Mr. William Malcolm.
Fl. July and Auguſt.   H: ♃.

R U E L L I A.   *Gen. pl.* 784.

Cal. 5-partitus. *Cor.* fubcampanulata. *Stamina* per
paria approximata. *Capſ.* dentibus elaſticis diſ-
filiens.

Blechum.   1. R. foliis ovatis integerrimis, ſpicis ovatis : braĉteis
interioribus geminis, floribus binis feſſilibus. *Sp. pl.*
884.
Thick-ſpiked Ruellia.
Nat. of South America.
Introd. 1780, by Monſ. Thouin.
Fl.   S. ♃.

ſtrepens.   2. R. foliis petiolatis, pedunculis trifloris brevibus. *Syſt.*
veget. 575.
Verticil'd Ruellia.
Nat. of Virginia and Carolina.
Cult. 1726, by James Sherard, M.D.   *Dill. elth.* 330.
t. 249. *f.* 321.
Fl. July and Auguſt.   S. ♃.

clandeſti-   3. R. foliis petiolatis, pedunculis longis ſubdiviſis nudis.
na.   *Sp. pl.* 885.
Three-flower'd Ruellia.
Nat. of Barbadoes.
Cult. 1728, by James Sherard, M.D.   *Dill. elth.* 328.
t. 248. *f.* 320.
Fl. July and Auguſt.   S. ♃.

biflora.   4. R. floribus geminis feſſilibus. *Sp. pl.* 886.
Two-flower'd Ruellia.

Nat.

Nat. of Carolina.
Introd. 1765, by Mr. John Cree.
Fl.   G. H. ♃.

BARLERIA. *Gen. pl.* 785.
Cal. 4-partitus. *Stamina* 2 longe minora. *Capf.*
4-angularis, bilocularis ? 2-valvis, elaftica abfque
unguibus. *Sem.* 2.

1. B. fpinis verticillorum fenis, foliis enfiformibus lon-   *longifolia.*
giffimis fcabris. *Sp. pl.* 887.
Long-leav'd Barleria.
Nat. of the Eaft Indies.
Introd. 1781, by Sir Jofeph Banks, Bart.
Fl. July——September.   S. ♂.

DURANTA. *Gen. pl.* 786.
Cal. 5-fidus, fuperus. *Bacca* 4-fperma. *Sem.* 2-
locularia.

1. D. calycibus fructefcentibus contortis. *Sp. pl.* 888.   *Plumi-*
Smooth Duranta.   *eri.*
Nat. of South America.
Introd. before 1733, by William Houftoun, M.D.
Mill. dict. vol. 2. Caftorea 1.
Fl. October.   S. ♄.

2. D. calycibus fructefcentibus erectis. *Sp. pl.* 888.   *Ellifia.*
Jacqu. hort. 3. *p.* 51. *t.* 99.
Prickly Duranta.
Nat. of the Weft Indies.
Cult. 1759, by Mr. Philip Miller. *Mill. dict. edit.* 7.
Addenda. Ellifia.
Fl. Auguft.   S. ♄.
VOLKA-

VOLKAMERIA. *Gen. pl.* 788.

Cal. 5-fidus. *Cor.* laciniis fecundis. *Bacca* 2-fperma,
Semina 2-locularia.

aculeata. 1. V. fpinis petiolorum rudimentis. *Sp. pl.* 889.
Prickly Volkameria.
Nat. of the Weft Indies.
Cult. 1739, by Mr. Philip Miller. *Rand. chel.* Liguf-
trum 5.
Fl. Auguft——October.   S. ♄.

inermis. 2. V. ramis inermibus. *Sp. pl.* 889.
α foliis ovatis.
Oval-leav'd fmooth Volkameria.
β foliis lanceolato-oblongis.
Long-leav'd fmooth Volkameria.
Nat. of the Eaft Indies.
Cult. 1692, in the Royal Garden at Hampton-court.
Pluk. phyt. t. 211. *f.* 4.
Fl. Auguft——November.   S. ♄.

CLERODENDRUM. *Gen. pl.* 789.

Cal. 5-fidus, campanulatus. *Cor.* tubo filiformi; limbo
5-partito, æquali. *Stam.* longiffima, intra lacinias
maxime hiantes. *Bacca* 1-fperma.

fortuna- 1. C. foliis lanceolatis integerrimis. *Syft. veget.* 578.
tum. Intire-leav'd Clerodendrum.
Nat. of the Eaft Indies.
Introd. 1784, by Meffrs. Kennedy and Lee.
Fl.   S. ♄.

VITEX,

V I T E X. *Gen. pl.* 790.

Cal. 5-dentatus. *Cor.* limbus 6-fidus. *Bacca* 4-fperma.

1. V. foliis digitatis ferratis, fpicis verticillatis. *Sp. pl.*   *Agnus*
   890.   *Caftus.*

α Vitex foliis anguftioribus cannabis modo difpofitis.   angufti-
   Bauh. pin. 475.   folia.
   Narrow-leav'd Chafte-tree.

β Vitex latiore folio. *Bauh. pin.* 475.   latifolia.
   Broad-leav'd Chafte-tree.
   Nat. of Sicily.
   Cult. 1570. *Lobel. adv.* 423.
   Fl. September.   H. ♄.

2. V. foliis ternatis quinatifque integerrimis, paniculis   *trifolia.*
   dichotomis. *Sp. pl.* 890.
   Three-leav'd Chafte-tree.
   Nat. of the Eaft Indies.
   Cult. 1759, by Mr. Ph. Miller. *Mill. dict. edit.* 7. *n.* 3.
   Fl.   S. ♄.

3. V. foliis quinatis ternatifque ferratis, floribus racemo-   *Negundo.*
   fo paniculatis. *Sp. pl.* 890.
   Five-leav'd Chafte-tree.
   Nat. of the Eaft Indies.
   Cult. 1697, by the Dutchefs of Beaufort. *Br. Muf.*
   Sloan. mff. 3357. *fol.* 71.
   Fl. July and Auguft.   G. H. ♄.

B O N T I A. *Gen. pl.* 791.

Cal. 5-partitus. *Cor.* bilabiata : labio inferiore 3-par-
   tito, revoluto. *Drupa* ovata, monofperma, apice
   obliquo.

1. B. foliis alternis, pedunculis unifloris. *Sp. pl.* 890.   *daph-*
                          Barbadoes   *noides.*

Barbadoes Wild Olive.
Nat. of the Weſt Indies.
Introd. 1690, by Mr. Bentick.   *Br. Muſ. Sloan. mſſ.*
337o.
Fl. June.   S. ♄.

C O L U M N E A.   *Gen. pl.* 792.

Cal. 5-phyllus.   *Cor.* ringens : lab. ſup. 3-partito :
   intermedia fornicata ; ſupra baſin gibba.   *Antheræ*
   connexæ.   *Capſ.* 1-locularis.

hirſuta.   1. C. foliis ovatis acuminatis ſerratis ſuperne hirtis, fo-
       liolis calycinis denticulatis lanceolatis, corolliſque
       hirſutis, labio ſuperiori bifido.   *Swartz prodr.* 94.
       Achimenes major herbacea ſubhirſuta oblique aſſur-
       gens, foliis ovatis crenatis oppoſitis, alternis mino-
       ribus ; floribus geminatis ad alas alternas.   *Brown.*
       jam. 270. *t.* 30. *f.* 3.
       Hairy Columnea.
       Nat. of Jamaica.
       Introd. 1780, by the Marquis of Rockingham.
       Fl. November.   S. ♄.

A C A N T H U S.   *Gen. pl.* 793.

Cal. bifolius, 2-fidus.   *Cor.* 1-labiata, deflexa, 3-fida.
         Capſula 2-locularis.

mollis.   1. A. foliis ſinuatis inermibus.   *Sp. pl.* 891.
     α Acanthus ſativus vel mollis Virgilii.   *Bauh. pin.* 383.
       Smooth Bear's-breech.
     β Acanthus *nigra*, foliis ſinuatis inermibus glabris lu-
       cidè virens.   *Mill. diſt.*
       Portugueſe Bear's-breech.
       Nat. α. of Italy and Sicily ; and β. of Portugal.
       §   *Cult.*

Cult. 1551, in Sion Garden. *Turn. herb. part* 1.
fign. Bj.
Fl. July——September. H. ♃.

2. A. foliis pinnatifidis fpinofis. *S̨. pl.* 891. *fpinofus.*
Prickly Bear's-breech.
Nat. of Italy.
Cult. 1629. *Park. parad.* 331. *f.* 2.
Fl. July and September. H. ♃.

P E D A L I U M. *Gen. pl.* 794.

Cal. 5-partitus. *Cor.* fubringens : limbo 5-fido. *Nux*
fuberofa, tetragona, angulis fpinofa, 2 -locularis.
Sem. bina.

1. PEDALIUM. *Sp. pl.* 892. *Murex.*
Prickly-fruited Pedalium.
Nat. of the Eaft Indies.
Introd. 1778, by Sir Jofeph Banks, Bart.
Fl. Auguft and September. S. ☉.

M E L I A N T H U S. *Gen. pl.* 795.

Cal. 5-phyllus : folio inferiore gibbo. *Petala* 4 ; nectario infra infima. *Capf.* 4-locularis.

1. M. ftipulis folitariis petiolo adnatis. *Sp. pl.* 892. *major.*
Great Honey-flower.
Nat. of the Cape of Good Hope.
Introd. 1690, by Mr. Bentick. *Br. Muf. Sloan. mff.*
3370.
Fl. May——July. G. H. ♄.

2. M.

minor.   2. M. ſtipulis geminis diſtinctis. *Sp. pl.* 892.
Small Honey-flower.
Nat. of the Cape of Good Hope.
Cult. 1708, by the Dutcheſs of Beaufort. *Br. Muſ.*
   H. S. 137. *fol.* 53.
Fl. Auguſt.   G. H. ♄.

Claſſis

Classis XV.

TETRADYNAMIA

SILICULOSA.

MYAGRUM. *Gen. pl.* 796.

Silicula ftylo conico terminata; loculo fubmonofpermo.

1. M. filiculis biarticulatis monofpermis, foliis extrorfum   *perenne.*
   finuatis denticulatis. *Syft. veget.* 583.
   Perennial Gold of Pleafure.
   Nat. of Germany.
   Cult. 1739, by Mr. Philip Miller. *Rand. chel.* Rapif-
   trum 1.
   Fl. July.   H. ♃.

2. M. filiculis obcordatis fubfeffilibus, foliis amplexi-   *perfolia-*
   caulibus. *Sp. pl.* 893.   *tum.*
   Perfoliate Gold of Pleafure.
   Nat. of France and Switzerland.
   Cult. 1658, in Oxford Garden. *Hort. oxon. edit.* 2.
   p. 113.
   Fl. June and July.   H. ☉.

3. M. filiculis obovatis pedunculatis polyfpermis. *Sp. pl.*   *fativum.*
   894.
   Cultivated Gold of Pleafure.
   Nat. of Britain.
   Fl. May and June.   H. ☉.

                4. M.

panicula-
tum.

4. M. filiculis lentiformibus orbiculatis punctato-rugofis.
　　Sp. pl. 894.
　　Panicl'd Gold of Pleafure.
　　Nat. of Europe.
　　Introd. 1787, by Mr. Zier.
　　Fl. July and Auguft.　　　　　　　　　　　　H. ⊙.

faxatile.

5. M. filiculis lentiformibus obovatis glabris, foliis petio-
　　latis oblongis ferratis fcabris, caule paniculato. *Syft.*
　　veget. 584. *Jacqu. auftr.* 2. *p.* 17. *t.* 128.
　　Rock Gold of Pleafure.
　　Nat. of the Alps of Auftria and Switzerland.
　　Introd. 1775, by the Doctors Pitcairn and Fothergill.
　　Fl. June and July.　　　　　　　　　　　　H. ♃.

V E L L A.　　*Gen. pl.* 797.

Silicula diffepimento valvulis duplo majore, extus ovato.

annua.

1. V. foliis pinnatifidis, filiculis pendulis. *Sp. pl.* 895.
　　Annual Vella, or Crefs-rocket.
　　Nat. of England.
　　Fl. June.　　　　　　　　　　　　　　　H. ⊙.

Pfeudo-
Cytifus.

2. V. foliis integris obovatis ciliatis, filiculis erectis. *Sp.*
　　pl. 895.
　　Shrubby Vella.
　　Nat. of Spain.
　　Cult. 1759, by Mr. Ph. Miller. *Mill. dict. edit.* 7. *n.* 2.
　　Fl. April and May.　　　　　　　　　　　H. ♄.

ANASTA-

ANASTATICA. *Gen. pl.* 798.

Silicula retufa, margine coronata valvulis diffepimento
duplo longioribus; *Stylo* intermedio mucronato,
obliquo : loculis 1-fpermis.

1. A. foliis obtufis, fpicis axillaribus breviffimis, filiculis *hiero-*
ungulatis fpinofis. *Sp. pl.* 895. *Jacqu. hort.* 1. *chuntica.*
p. 23. *t.* 58.
Common Anaftatica, or Rofe of Jericho.
Nat. of the Levant.
Cult. 1656, by Mr. John Tradefcant, Jun. *Muf.*
Trad. 163.
Fl. June.   H. ☉.

2. A. foliis acutis, fpicis folio longioribus, filiculis ovatis *fyriaca.*
roftratis. *Sp. pl.* 895. *Jacqu. auftr.* 1. *p.* 7. *t.* 6.
Syrian Anaftatica.
Nat. of Auftria and the Levant.
Introd. 1778, by Monf. Thouin.
Fl. July and Auguft.   H. ☉.

SUBULARIA. *Gen. pl.* 799.

Silicula integra, ovata : valvis ovatis, concavis, diffepimento contrariis. *Stylus* filicula brevior.

1. SUBULARIA. *Sp. pl.* 896.   *aquatica.*
Awl-wort.
Nat. of Wales and Scotland.
Fl. July.   H. ☉.

DRABA.

D R A B A.   *Gen. pl.* 800.

Silicula integra, ovali-oblonga : valvis planiufculis,
   diffepimento parallelis.   *Styllus* nullus.

aizoides.   1. D. fcapo nudo fimplici, foliis enfiformibus carinatis
              ciliatis.   *Syft. veget.* 585.   *Jacqu. auftr.* 2. *p.* 55.
              t. 192.
              Hairy-leav'd Alpine Whitlow-grafs.
              Nat. of the Alps of Europe.
              Cult. 1731, by Mr. Philip Miller.   *Mill. dict. edit.* 1.
              Alyffon 3.
              Fl. February——April.   H. ♃.

verna.   2. D. fcapis nudis, foliis fubferratis.   *Syft. veget.* 585.
              Curtis lond.
              Common Whitlow-grafs.
              Nat. of Britain.
              Fl. April.   H. ☉.

muralis.   3. D. caule ramofo, foliis ovatis feffilibus dentatis.   *Syft.*
              veget. 585.
              Wall Whitlow-grafs.
              Nat. of England.
              Fl. May.   H. ☉.

incana.   4. D. foliis caulinis numerofis incanis, filiculis oblongis
              obliquis fubfeffilibus.   *Syft. veget.* 585.
              Hoary Whitlow-grafs.
              Nat. of Britain.
              Fl. May.   H. ♂.

LEPIDIUM.

L E P I D I U M. *Gen. pl.* 801.

Silicula emarginata, cordata, polyfperma : valvulis carinatis, contrariis.

1. L. foliis caulinis pinnato-multifidis, ramiferis cordatis *perfolia-*
amplexicaulibus integris. *Sp. pl.* 897. *Jacqu.* *tum.*
auftr. 4..*p.* 24. *t.* 346.
Various-leav'd Pepper-wort.
Nat. of Auftria and the Levant.
Cult. 1640. *Park. theat.* 849. *n.* 3.
Fl. July.   H. ☉.

2. L. foliis pinnatis integerrimis, fcapo fubradicato, fili- *alpinum.*
culis lanceolatis mucronatis. *Sp. pl.* 898. *Jacqu.*
auftr. 2. *p.* 23. *t.* 137.
Alpine Pepper-wort.
Nat. of the Alps of Germany, Switzerland, and Italy.
Introd. 1775, by the Doctors Pitcairn and Fothergill.
Fl. April —— June.   H. ♃.

3. L. foliis pinnatis integerrimis, petalis emarginatis ca- *petræum.*
lyce minoribus. *Sp. pl.* 899. *Jacqu. auftr.* 2.
p. 19. *t.* 131.
Rock Pepper-wort.
Nat. of Britain.
Fl. April and May.   H. ☉.

4. L. floribus tetradynamis, foliis oblongis multifidis. *fativum.*
Sp. pl. 899.
α Nafturtium hortenfe vulgatum. *Bauh. pin.* 103.
Garden, or Common Crefs.
β Nafturtium hortenfe crifpum. *Bauh. pin.* 104. *prodr.*
44. *t.* 43.
Curl'd Crefs.

Nat.

Nat.
Cult. 1562.   *Turn. herb. part* 2. *fol.* 64 *verſo.*
Fl. June and July.   H. ⊙.

latifoli-   5. L. foliis ovato-lanceolatis integris ſerratis,   *Sp. pl.*
um.   899.
             Broad-leav'd Pepper-wort, or Dittander.
             Nat. of Britain.
             Fl. June and July.   H. ♃.

olerace-   6. L. foliis elliptico-oblongis acutis ſerratis, floribus te-
um.   trandris.   *Forſt. pl. eſcul. p.* 69. *fl. auſtr. p.* 46.
             Lepidium bidentatum.   *Montin in nov. act. acad. nat.*
                curioſ. 6. *p.* 324. *tab.* 5 *a.*
             Notch'd-leav'd Pepper-wort.
             Nat. of New Zealand.
             Introd. 1779, by Jonas Dryander, M. A.
             Fl. September.   H. ⊙.

ſubula-   7. L. foliis ſubulatis indiviſis ſparſis, caule ſuffruticoſo.
tum.   *Sp. pl.* 899.
             Awl-leav'd Pepper-wort.
             Nat. of Spain.
             Cult. 1739, by Mr. Ph. Miller. *Mill. dict. vol.* 2. *n.* 5.
             Fl. July and Auguſt.   G. H. ♄.

didymum.   8. L. floribus diandris, foliis pinnatifidis, fructibus didy-
                mis.
             Lepidium didymum.   *Linn. mant.* 92.
             Lepidium anglicum.   *Hudſ. angl.* 280.
             Procumbent Pepper-wort.
             Nat. of England.
             Fl. June and Auguſt.   H. ⊙.

                                                         9. L.

9. L. floribus diandris apetalis, foliis radicalibus den- *ruderale.*
tato-pinnatis ; ramiferis linearibus integerrimis.
Sp. pl. 900.
Narrow-leav'd Pepper-wort.
Nat. of England.
Fl. June and July. H. ⊙.

10. L. floribus fubtriandris tetrapetalis, foliis linearibus *virgini-*
pinnatis. *Sp. pl.* 900. *cum.*
Virginian Pepper-wort.
Nat. of Virginia and Jamaica.
Cult. 1713. *Philofoph. tranf. n.* 337. *p.* 200. *n.* 82.
Fl. June and July. H. ⊙.

11. L. foliis pinnatifidis, caule ramofiffimo, filiculis ova- *divarica-*
tis fubemarginatis. *tum.*
Divaricated Pepper-wort.
Nat. of the Cape of Good Hope. Mr. *Fr. Maffon.*
Introd. 1774.
Fl. May——Auguft. G. H. ♄.

12. L. floribus diandris tetrapetalis, foliis inferioribus *Iberis.*
lanceolatis ferratis; fuperioribus linearibus inte-
gerrimis. *Sp. pl.* 900.
Bufhy Pepper-wort.
Nat. of Germany, France, and Italy.
Cult. 1683, by Mr. James Sutherland. *Sutherl. hort.*
edin. 170.
Fl. Auguft and September. H. ♃.

T H L A S P I. *Gen. pl.* 802.

Silicula emarginata, obcordata, polyfperma : valvulis
navicularibus, marginato-carinatis.

1. T. filiculis orbiculatis, foliis oblongis dentatis gla- *arvenfe.*
bris. *Sp. pl.* 901.

Field

Field Baſtard-creſs.
Nat. of Britain.
Fl. June and July.   H. ☉.

ſaxatile.   2. T. ſiliculis ſubrotundis, foliis lanceolato-linearibus carnoſis obtuſis. *Sp. pl.* 901. *Jacqu. auſtr.* 3. *p.* 21 *t.* 236.
Rock Baſtard-creſs.
Nat. of the South of Europe.
Cult. 1748, by Mr. Philip Miller.   *Mill. dict. edit.* 5. *n.* 13.
Fl. July.   H. ☉.

hirtum.   3. T. ſiliculis ſubrotundis piloſis, foliis caulinis ſagittatis villoſis. *Sp. pl.* 901.
Hairy Baſtard-creſs.
Nat. of Wales.
Fl. July.   H. ♂.

campeſtre.   4. T. ſiliculis ſubrotundis, foliis ſagittatis dentatis incanis. *Sp. pl.* 902. *Curtis lond.*
Wild Baſtard-creſs, or Mithridate-muſtard.
Nat. of Britain.
Fl. June and July.   H. ♂.

monta-num.   5. T. ſiliculis obcordatis, foliis glabris : radicalibus carnoſis obovatis integerrimis ; caulinis amplexicaulibus, corollis calyce majoribus. *Sp. pl.* 902. *Jacqu. auſtr.* 3. *p.* 22. *t.* 237.
Mountain Baſtard-creſs.
Nat. of England.
Fl. July.   H. ♂.

perfolia-tum.   6. T. ſiliculis obcordatis, foliis caulinis cordatis glabris ſubdentatis, petalis longitudine calycis, caule ramoſo. *Sp. pl.* 902. *Jacqu. auſtr.* 4. *p.* 19. *t.* 337.
                            Perfoliate

Perfoliate Baftard-crefs.
Nat. of France, Switzerland, and Germany.
Cult. 1748, by Mr. Philip Miller. *Mill. dict. edit.* 5.
n. 11.
Fl. June and July. H. ♂.

7. T. filiculis obcordatis, foliis fubdentatis: caulinis *alpeftre.*
amplexicaulibus, petalis longitudine calycis, caule
fimplici. *Sp. pl.* 903.
Dwarf Baftard-crefs.
Nat. of England.
Fl. May and June. H. ☉.

8. T. filiculis obcordatis, foliis radicalibus pinnatifidis. *Burfa*
Sp. pl. 903. *Curtis lond.* *Paftoris.*
Common Baftard-crefs, or Shepherd's-purfe.
Nat. of Britain.
Fl. March——September. H. ☉.

9. T. glaberrimum, caule fulcato, foliis fagittatis lan- *cerato-*
ceolatis fubferratis, filiculis bilobat s. *Linn. fuppl.* *carpon.*
295.
Thlafpi ceratocarpon. *Murray in nov. comm. gotting.* 5.
p. 26. *tab.* 1. *Scop. infubr.* 1. *p.* 10. *tab.* 4.
Lepidium ceratocarpon. *Pallas it.* 2. *p.* 740. *tab. U.*
Siberian Baftard-crefs.
Nat. of Siberia.
Introd. 1779, by Jonas Dryander, M. A.
Fl. July. H. ☉.

C O C H L E A R I A. *Gen. pl.* 803.
Silicula emarginata, turgida, fcabra : valvulis gibbis,
obtufis.

1. C. foliis radicalibus cordato-fubrotundis; caulinis *officinalis.*
oblongis fubfinuatis. *Syft. veget.* 588.
Common

Common Scurvy-grafs.
Nat. of Britain.
Fl. April and May.   H. ♃.

danica.   2. C. foliis haftato-angulatis : omnibus deltoidibus. *Syft.*
            veget. 588.
            Danifh Scurvy-grafs.
            Nat. of Britain.
            Fl. May.   H. ♂.

anglica. 3. C. foliis omnibus ovato-lanceolatis. *Syft. veget.* 588.
            Sea Scurvy-grafs.
            Nat. of Britain.
            Fl. May.   H. ♂.

groenlan- 4. C. foliis reniformibus integris carnofis. *Syft. veget.*
dica.   588.
            Greenland Scurvy-grafs.
            Nat. of Britain.
            Fl. April and May.   H. ♂.

Corono-   5. C. foliis pinnatifidis, caule depreffo. *Syft. veget.* 588.
pus.   Wild Scurvy-grafs, or Swine's-crefs.
            Nat. of Britain.
            Fl. June——Auguft.   H. ☉.

Armora-   6. C. foliis radicalibus lanceolatis crenatis ; caulinis in-
cia.   cifis. *Sp. pl.* 904.
            Horfe-radifh.
            Nat. of England.
            Fl. May.   H. ♃.

glaftifo- 7. C. foliis caulinis cordato-fagittatis amplexicaulibus.
lia.   *Syft. veget.* 588.
            Woad-leav'd Scurvy-grafs.

                                                                  Nat.

Nat. of Germany.

Cult. 1683, by Mr. James Sutherland. *Sutherl. hort.*
edin. 191. *n.* 3.

Fl. May——July.　　　　　　　　　　H. ♂ .

I B E R I S.　*Gen. pl.* 804.

Cor. irregularis : petalis 2 exterioribus majoribus.
Silicula polyfperma, emarginata.

1. I. frutefcens, foliis cuneiformibus integerrimis obtufis.　*femper-*
　Sp. pl. 904.　　　　　　　　　　　　　　　　*florens.*
　Broad-leav'd evergreen Candy-tuft.
　Nat. of Perfia and Sicily.
　Cult. 1680, in Oxford Garden. *Morif. hift.* 2. *p.* 296.
　n. 23.
　Fl. moft part of the Year.　　　　　G. H. ♄ .

2. I. frutefcens, foliis linearibus acutis integerrimis.　*Sp.*　*fempervi-*
　pl. 905.　　　　　　　　　　　　　　　　　　　*rens.*
　Narrow-leav'd evergreen Candy-tuft.
　Nat. of the Ifland of Candia.
　Cult. 1739, in Chelfea Garden. *Rand. chel. n.* 10.
　Fl. April——June.　　　　　　　　　H. ♄ .

3. I. frutefcens, foliis apice dentatis.　*Sp. pl.* 905.　　*gibralta-*
　Gibraltar Candy-tuft.　　　　　　　　　　　*rica.*
　Nat. of Spain.
　Cult. 1732, by James Sherard, M. D. *Dill. elth.* 382.
　t. 287. *f.* 371.
　Fl. May.　　　　　　　　　　　　G. H. ♄ .

4. I. herbacea, foliis ovatis : caulinis amplexicaulibus　*rotundi-*
　lævibus fuccofis. *Sp. pl.* 905.　　　　　　　*folia.*
　Round-leav'd Candy-tuft.

　§　　　　　　　　　　　　　　*Nat.*

Nat. of Switzerland.
Cult. 1748, by Mr. Philip Miller. *Mill. dict. edit.* 5.
Thlafpi 14.
Fl. May——July. H. ♃.

umbella- 5. I. herbacea, foliis lanceolatis acuminatis : inferioribus
ta. ferratis ; fuperioribus integerrimis. *Sp. pl.* 906.
Purple Candy-tuft.
Nat. of the South of Europe.
Cult. 1596, by Mr. John Gerard. *Hort. Ger.*
Fl. June and July. H. ☉.

amara. 6. I. herbacea, foliis lanceolatis acutis fubdentatis, flori-
bus racemofis. *Sp. pl.* 906.
White Candy-tuft.
Nat. of England.
Fl. June and July. H. ☉.

linifolia. 7. I. herbacea, foliis linearibus integerrimis ; caulinis
ferratis, caule paniculato, corymbis hemifphæricis.
Syft. veget. 589.
Flax-leav'd Candy-tuft.
Nat. of Spain and Portugal.
Cult 1759, by Mr. Ph. Miller. *Mill. dict. edit.* 7. *n.* 8.
Fl. July. H. ☉.

nudicau- 8. I. herbacea, foliis finuatis, caule nudo fimplici. *Sp.*
lis. *pl.* 907.
Naked-ftalked Candy-tuft, or Rock-crefs.
Nat. of Britain.
Fl. May. H. ☉.

pinnata. 9. I. herbacea, foliis pinnatifidis. *Syft. veget.* 589.
Winged Candy-tuft.
Nat. of the South of Europe.

Cult.

Cult. 1640. *Park. theat.* 840. *f.* 4.
Fl. June——Auguft. H. ☉.

A L Y S S U M. *Gen. pl.* 805.

Filamenta quædam íntrorfum denticulo notata. *Silicula*
emarginata.

* *Suffruticofa.*

1. A. caulibus fuffruticofis diffufis, foliis lineari-lanceo- *halimifo-*
latis integerrimis villofiufculis, ftaminibus fimplici- *lium.*
bus, filiculis fubrotundis integris.
Alyffum halimifolium. *Sp. pl.* 907.
Sweet-fcented Mad-wort, or Alyffon.
Nat. of the South of Europe.
Cult. 1722, in Cheкea Garden. *R. S. n.* 44.
Fl. June——November. H. ♄.

2. A. caulibus frutefcentibus paniculatis, foliis lanceola- *faxatile.*
tis molliffimis repandis. *Syft. veget.* 590.
Shrubby Mad-wort.
Nat. of the Ifland of Candia.
Cult. 1731, by Mr. Philip Miller. *Mill. dict. edit.* 1.
Alyffon 1.
Fl. April and May. H. ♄.

* * *Herbacea.*

3. A. caule erecto, foliis lanceolatis incanis integerrimis, *incanum.*
floribus corymbofis, petalis bifidis. *Syft. veget.* 590.
Hoary Mad-wort.
Nat. of the North of Europe.
Cult. 1640, by Mr. John Parkinfon. *Park. theat.* 847.
f. 5.
Fl. July——September. H. ♂.

4. A.

calyci-
num.

4. A. caulibus herbaceis, ftaminibus omnibus dentatis, calycibus perfiftentibus. *Sp. pl.* 908. *Jacqu. auftr.* 4. *p.* 20. *t.* 338.
Small-flower'd Mad-wort.
Nat. of Auftria and France.
Cult. 1768, by Mr. Philip Miller. *Mill. dict. edit.* 8.
Fl. July and Auguft.   H. ⊙.

monta-
num.

5. A. caulibus herbaceis diffufis, foliis fublanceolatis punctato-echinatis. *Syft. veget.* 590. *Jacqu. auftr.* 1. *p.* 24. *t.* 37.
Mountain Mad-wort.
Nat. of Germany and Switzerland.
Cult. 1759, by Mr. Ph. Miller. *Mill. dict. edit.* 7. *n.* 4.
Fl. July.   H. ♃.

campef-
tre.

6. A. caule herbaceo, ftaminibus ftipatis pari fetarum, calycibus deciduis. *Sp. pl.* 909.
Thyme-leav'd Mad-wort.
Nat. of France.
Cult. 1768, by Mr. Philip Miller. *Mill. dict. edit.* 8.
Fl. July.   H. ⊙.

clypea-
tum.

7. A. caule erecto herbaceo, filiculis feffilibus ovalibus compreffo-planis, petalis acuminatis linearibus. *Syft. veget.* 590.
Upright Mad-wort.
Nat. of the South of Europe.
Cult. 1596, by Mr. John Gerard. *Hort. Ger.*
Fl. June.   H. ⊙.

*** *Siliculis inflatis f. calycibus oblongis claufis.*

finuatum.

8. A. caule herbaceo, foliis lanceolato-deltoidibus, filiculis inflatis. *Syft. veget.* 590.
Sinuated Mad-wort.

Nat.

Nat. of Spain.
Cult. 1680. *Moriſ. hiſt.* 2. *p.* 247. *n.* 6. *ſ.* 3. *t.* 9. *f.* 6.
Fl. April——June. H. ♂.

9. A. caule fruticoſo, foliis lanceolatis ſubdentatis tomen- *creticum.*
toſis, ſiliculis inflatis globoſis. *Linn. mant.* 92.
Cretan Mad-wort.
Nat. of the Iſland of Candia.
Cult. 1739, by Mr. Philip Miller. *Mill. dict. vol.* 2.
Alyſſoides 3.
Fl. May——Auguſt. H. ♃.

10. A. caule herbaceo erecto, foliis lævibus lanceolatis *utricula-*
integerrimis, ſiliculis inflatis. *Syſt. veget.* 591. *tum.*
Bladder Mad-wort.
Nat. of the Levant.
Cult. 1739, by Mr. Philip Miller. *Rand. chel.* Alyſ-
foides 1.
Fl. April——June. H. ♃.

11. A. caulibus ſuffruteſcentibus proſtratis, foliis lanceo- *deltoide-*
lato-deltoidibus, ſiliculis hirtis. *Sp. pl.* 908. *um.*
Purple Mad-wort.
Nat. of the Levant.
Cult. 1739, by Mr. Philip Miller. *Rand. chel.* Alyſ-
fon 10.
Fl. March——May. H. ♃.

CLYPEOLA. *Gen. pl.* 807.

Silicula emarginata, orbiculata, compreſſo-plana, decidua.

1. C. ſiliculis unilocularibus monoſpermis. *Sp. pl.* 910. *Jonthlaſ-*
Annual Treacle Muſtard. *pi.*
Nat. of France and Italy.
Cult.

Cult. 1739, by Mr. Philip Miller.   *Mill. dict. vol.* 2.
Jonthlafpi 2.
Fl. May——July.   H. ☉.

P E L T A R I A.   *Gen. pl.* 806.

Silicula integra, fuborbiculata, compreffo-plana, non
dehifcens.

alliacea.   1. PELTARIA. *Sp. pl.* 910. *Jacqu. auftr.* 2. *p.* 14. *t.* 123.
Garlick-fcented Peltaria.
Nat. of Auftria.
Cult. 1768, by Mr. Philip Miller.   *Mill. dict. edit.* 8.
Fl. May.   H. ♃.

B I S C U T E L L A.   *Gen. pl.* 808.

Silicula compreffo-plana, rotundata, fupra infraque
biloba.   *Cal.* foliola bafi gibba.

auricula-   1. B. calycibus nectario utrinque gibbis, filiculis in fty-
ta.   lum coëuntibus.   *Sp. pl.* 911.
Ear-podded Bifcutella.
Nat. of France and Italy.
Cult. 1683, by Mr. James Sutherland.   *Sutherl. hort.*
edin. 334. *n.* 5.
Fl. June and July.   H. ☉.

apula.   2. B. filiculis fcabris, foliis lanceolatis feffilibus ferratis.
Linn. mant. 254.
Spear-leav'd Bifcutella.
Nat. of Italy.
Cult. 1759.   *Mill. dict. edit.* 7. *n.* 3.
Fl. June and July.   H. ☉.

lævigata. 3. B. filiculis glabris, foliis lanceolatis ferratis.   *Linn.*
mant. 255.   *Jacqu. auftr.* 4. *p.* 20. *t.* 339.
Smooth

Smooth Biſcutella.

Nat. of Auſtria and Italy.

Introd. 1777, by Monſ. Thouin.

Fl. June and July.   H. ☉.

4. B. ſiliculis ſcabriuſculis, foliis lanceolatis tomentoſis.   *ſempervi-*
    Linn. mant. 255.   *rens.*
    Shrubby Biſcutella.
    Nat. of Spain.
    Introd. 1784, by Meſſrs. Lee and Kennedy.
    Fl. June and July.   G. H. ♃.

L U N A R I A.   *Gen. pl.* 809.

Silicula integra, elliptica, compreſſo-plana, pedicellata :
    valvis diſſepimento æqualibus, parallelis, planis.
    Cal. foliolis ſaccatis.

1. L. foliis alternis.   *Sp. pl.* 911.   *rediviva.*
    Perennial Honeſty.
    Nat. of Auſtria and Hungary.
    Cult. 1597. *Ger. herb.* 378.
    Fl. May and June.   H. ♃.

2. L. foliis oppoſitis.   *Sp. pl.* 911.   *annua.*
    Annual Honeſty.
    Nat. of Germany.
    Cult. 1597. *Ger. herb.* 377.
    Fl. May and June.   H. ☉.

S I L I Q U O S A.

RICOTIA. *Gen. pl.* 810.

Siliqua unilocularis, oblonga, compreffa: valvulis planis.

ægyptia-   1. RICOTIA. *Sp. pl.* 912.
ca.
     Egyptian Ricotia.
     Nat. of Egypt.
     Cult. 1757, by Mr. Ph. Miller. ·*Mill. ic.* 113. *t.* 169.
     Fl. June and July.   H. ⊙.

DENTARIA. *Gen. pl.* 811.

Siliqua elaftice diffiliens valvulis revolutis. *Stigma*
     emarginatum. *Cal.* longitudinaliter connivens.

bulbifera.   1. D. foliis inferioribus pinnatis; fummis fimplicibus.
     Sp. pl. 912.
     Bulbiferus Tooth-wort, or Coral-wort.
     Nat. of England.
     Fl. April and May.   H. ♃.

pinnata.   2. D. foliis omnibus pinnatis.
     Dentaria pinnata. *De Lamarck encycl.* 2. *p.* 268.
     Dentaria pentaphyllos *α.* *Sp. pl.* 912.
     Dentaria foliis pinnatis et digitatis. *Ger. prov.* 356.
     Dentaria heptaphyllos. *Bauh. pin.* 322. *Garid. prov.*
       152. *t.* 28.
     Seven-leav'd Tooth-wort.
     Nat. of Switzerland and the South of France.
     Cult. 1683, by Mr. James Sutherland. *Sutherl. hort.*
       edin. 102. *n.* 4.
     Fl. May and June.   H. ♃.

                     3. D.

3. D. foliis digitatis.

    penta-
    phyllos:

Dentaria pentaphyllos β. *Sp. pl.* 912.

Dentaria digitata. *De Lamarck encycl.* 2. *p.* 268.

Dentaria foliis omnibus quinato-digitatis. *Ger. prov.*
356.

Dentaria pentaphyllos, foliis mollioribus. *Garid. prov.*
152. *t.* 29.

Five-leav'd Tooth-wort.

Nat. of Switzerland and the South of France.

Cult. 1656, by Mr. John Tradefcant. *Trad. muf.* 109.

Fl. May and June.   H. ♃.

CARDAMINE. *Gen. pl.* 812.

Siliqua elaftice diffiliens valvulis revolutis. *Stigma*
integrum. *Cal.* fubhians.

1. C. foliis fimplicibus fubcordatis. *Sp. pl.* 913.   *afarifolia.*
Heart-leav'd Lady's Smock.
Nat. of Italy.
Introd. 1779, by Anthony Chamler, Efq.
Fl.   H. ♃.

2. C. foliis ternatis obtufis, caule fubnudo. *Sp. pl.* 913. *trifolia.*
   Jacqu. auftr. 1. *p.* 18. *t.* 27.
Three-leav'd Lady's Smock.
Nat. of Lapland, Auftria, and Switzerland.
Cult. 1629, by Mr. John Parkinfon. *Park. parad:*
389. *n.* 2.
Fl. March and April.   H. ♃.

3. C. foliis pinnatis incifis ftipulatis, floribus apetalis. *impati-*
   Sp. pl. 914.   *ens.*
Impatient Lady's Smock.
Nat. of Britain.
Fl. April.   H. ☉.

parviflo-
ra.

4. C. foliis pinnatis exftipulatis : foliolis lanceolatis ob-
    tufis, floribus corollatis. *Syft. veget.* 594.
    Small-flower'd Lady's Smock.
    Nat. of England.
    Fl. March——May.   H. ☉.

hirfuta.

5. C. foliis pinnatis, floribus tetrandris. *Sp. pl.* 915.
    Curtis lond.
    Hairy Lady's Smock.
    Nat. of Britain.
    Fl. April——Auguft.   H. ☉.

pratenfis.

6. C. foliis pinnatis : foliolis radicalibus fubrotundis ;
    caulinis lanceolatis. *Sp. pl.* 915. *Curtis lond.*
 α floribus fimplicibus.
    Common Lady's Smock.
 β floribus plenis.
    Double Lady's Smock.
    Nat. of Britain.
    Fl. April and May.   H. ♃.

amara.

7. C. foliis pinnatis, axillis ftoloniferis. *Sp. pl.* 915.
    Curtis lond.
    Bitter Lady's Smock, or Crefs.
    Nat. of Britain.
    Fl. April and May.   H. ♃.

SISYMBRIUM. *Gen. pl.* 813.

Siliqua dehifcens valvulis rectiufculis. *Calyx* patens.
Corolla patens.

* *Siliquis declinatis brevibus.*

Naftur-
tium.

1. S. filiquis declinatis, foliis pinnatis : foliolis fubcorda-
    tis. *Sp. pl.* 916.

                                     Common

Common Water-crefs.
Nat. of Britain.
Fl. May and June.   H. ♂.

2. S. filiquis declinatis, foliis pinnatis : foliolis lanceola- *fylveftre.*
tis ferratis. *Sp. pl.* 916. *Curtis lond.*
Water Rocket.
Nat. of Britain.
Fl. June.   H. ♃.

3. S. filiquis declinatis oblongo-ovatis, foliis pinnatifidis *amphibi-*
ferratis. *Sp. pl.* 917.   *um.*
Water-radifh.
Nat. of Britain.
Fl. June and July.   H. ♃.

4. S. radice annua, foliis pinnatifidis dentato-ferratis, fi- *terreftre.*
liquis fœcundis. *Curtis lond.*
Annual Water-radifh.
Nat. of England.
Fl. June——September.   H. ☉.

5. S. filiquis fubovatis, foliis inferioribus lyratis ; fupe- *pyrenai-*
rioribus bipinnatifidis amplexicaulibus, ftylis filifor- *cum.*
mibus. *Sp. pl.* 916.
Pyrenean Sifymbrium.
Nat. of the Alps of Switzerland.
Introd. 1775, by Monf. Thouin.
Fl. May and June.   H. ♃.

6. S. foliis pinnatis : foliolis lanceolatis incifo-ferratis : *tanaceti-*
extimis confluentibus. *Syft. veget.* 594.   *folium.*
Tanfey-leav'd Sifymbrium.
Nat. of Italy.

Cult.

Cult. 1731, by Mr. Philip Miller. *Mill. dict. edit.* 1.
Eruca 4.
Fl.   H. ♃.

fagitta- 7. S. pubefcens, filiquis declinatis fubcylindraceis re-
tum.   curvis, foliis obovato-oblongis fagittatis dentatis.
Arrow-leav'd Sifymbrium.
Nat. of Siberia.
Introd. 1780, by Peter Simon Pallas, M.D.
Fl. May and June.   H. ♃.

** *Siliquis feffilibus axillaribus.*

fupinum. 8. S. filiquis axillaribus fubfeffilibus folitariis, foliis den-
tato-finuatis. *Sp. pl.* 917.
Dwarf Sifymbrium.
Nat. of the South of Europe.
Introd. 1778, by Monf. Thouin.
Fl. June and July.   H. ☉.

polycera- 9. S. filiquis axillaribus feffilibus fubulatis aggregatis,
tium.   foliis repando-dentatis. *Sp. pl.* 918. *Jacqu. hort.* 1.
p. 34. *t.* 79.
Dandelion-leav'd Sifymbrium.
Nat. of France and Italy.
Cult. 1633, by Mr. John Parkinfon. *Ger. emac.* 254.
f. 2.
Fl. June and July.   H. ☉.

*** *Caule nudo.*

monenfe. 10. S. acaule, foliis pinnato-dentatis fubpilofis, fcapis læ-
vibus. *Syft. veget.* 595.
Procumbent Sifymbrium.
Nat. of Britain.
Fl. June——Auguft.   H. ♂.

§   11. S.

11. S. caule fubnudo ramofo, foliis radicalibus runcinatis   *Barrelie-*
    dentatis hifpidis. *Sp. pl.* 919.   *ri.*
    Small Sifymbrium.
    Nat. of Spain and Italy.
    Introd. 1770, by Monf. Richard.
    Fl. July.   H. ⊙.

          **** *Foliis pinnatis.*

12. S. filiquis fcabris, foliis pinnatifidis: pinnis lineari-   *afperum.*
    lanceolatis fubdentatis, corollis calyce longioribus.
    Sp. pl. 920.
    Rough-podded Sifymbrium.
    Nat. of the South of France.
    Introd. 1778, by Monf. Thouin.
    Fl. May and June.   H. ⊙.

13. S. foliis fupradecompofitis tomentofis, petalis calyce   *millefoli-*
    majoribus.   *um.*
    Sinapis millefolia. *Jacqu. ic. collect.* 1. *p.* 41.
    Milfoil-leav'd Sifymbrium.
    Nat. of the Canary Iflands.   Mr. *Francis Maſſon.*
    Introd. 1779.
    Fl. May——September.   G. H. ♄.

14. S. foliis fupradecompofitis: lacinulis linearibus, pe-   *Sophia.*
    talis calyce minoribus.
    Sifymbrium Sophia. *Sp. pl.* 920.
    Flix-weed.
    Nat. of Britain.
    Fl. July.   H. ⊙.

15. S. foliis runcinatis flaccidis : foliolis fublinearibus   *altiſſi-*
    integerrimis, pedunculis laxis. *Sp. pl.* 920.   *mum.*
    Tall Sifymbrium.
    Nat. of Siberia.
                        C c 4   *Cult.*

Cult. 1768, by Mr. Philip Miller. *Mill. dict. edit.* 8.
Fl. Auguft.   H. ⊙.

Irio.   16. S. foliis runcinatis dentatis nudis, caule lævi, filiquis
           erectis. *Sp. pl.* 921. *Jacqu. auftr.* 4. *p.* 11. *t.* 322.
           Curtis lond.
           Broad-leav'd Hedge-muftard.
           Nat. of England.
           Fl. May——Auguft.   H. ⊙.

Lœfelii.   17. S. foliis runcinatis acutis hirtis, caule hifpido. *Syft.*
            veget. 596. *Jacqu. auftr.* 4. *p.* 12. *t.* 324.
            Hairy Sifymbrium.
            Nat. of Auftria and Pruffia.
            Introd. 1787, by Mr. Zier.
            Fl. Auguft.   H. ⊙.

orientale. 18. S. foliis runcinatis tomentofis, caule lævi. *Sp. pl.*
             921.
             Oriental Sifymbrium.
             Nat. of the Levant.
             Introd. 1775, by Chevalier Murray.
             Fl. July.   H. ⊙.

pannoni-   19. S. foliis caulinis fuperioribus glabris pinnatis : pinnis
cum.   linearibus integerrimis : extima elongata. *Jacqu.*
                ic. collect. 1. *p.* 70.
                Hungarian Sifymbrium.
                Nat. of Hungary.
                Introd. 1787, by Mr. Zier.
                Fl. Auguft.   H. ⊙.

***** *Foliis integris.*

ftrictiffi- 20. S. foliis lanceolatis dentato-ferratis caulinis. *Sp. pl.*
mum.   922. *Jacqu. auftr.* 2. *p.* 56. *t.* 194.

                                                            Spear-

Spear-leav'd Sifymbrium.
Nat. of Switzerland and Italy.
Cult. 1658, in Oxford Garden. *Hort. oxon. edit.* 2.
   p. 56.
Fl. June——Auguſt.   H. ♃.

E R Y S I M U M. *Gen. pl.* 814.

Siliqua columnaris, exaɗe tetraëdra.   *Cal.* clauſus.

1. E. ſiliquis ſpicæ adpreſſis, foliis runcinatis. *Sp. pl.* *officinale.*
   922. *Curtis lond.*
Common Hedge-muſtard,
Nat. of Britain.
Fl. May.   H. ☉.

2. E. foliis lyratis : extimo ſubrotundo. *Sp. pl.* 922. *Barba-*
α floribus ſimplicibus.   *rea.*
Common Winter Hedge-muſtard, or Creſs.
β floribus plenis.
Double Winter Hedge-muſtard, or Creſs.
Nat. of Britain.
Fl. May.   H. ♃.

3. E. foliis cordatis. *Sp. pl.* 922. *Curtis lond.*   *Alliaria.*
Stinking Hedge-muſtard, or Sauce-alone.
Nat. of Britain.
Fl. May.   H. ♃.

4. E. foliis lanceolatis deňtatis, racemis oppoſitifoliis, *repan-*
   ſiliquis raçemoſis ſubſeſſilibus, corollis minutis. *Sp.* *dum.*
   pl. 923. *Jacqu. auſtr.* 1. *p.* 16. *t.* 22.
Small-flower'd Hedge-muſtard.
Nat. of Spain and Auſtria.
Introd. 1772, by Monſ. Richard.
Fl. May and June.   H. ☉.
                       5. E.

cheiran-
thoides.

5. E. foliis lanceolatis integerrimis, ſiliquis patulis. *Syſt.*
    veget. 597. *Jacqu. auſtr.* 1. *p.* 16. *t.* 23.
    Treacle Hedge-muſtard, or Worm-ſeed.
    Nat. of Britain.
    Fl. July and Auguſt.   H. ☉.

bicorne.

6. E. foliis lanceolatis piloſis, ſiliquis apice bicornibus.
    Horned Hedge-muſtard.
    Nat. of the Canary Iſlands.   Mr. *Francis Maſſon.*
    Introd. 1779.
    Fl. Auguſt and September.   G. H. ☉.
    DESCR. *Flores* parvi, flavi. *Calyx* flaveſcens, extus
      piloſus. *Siliquæ* approximatæ, adpreſſæ, ex anci-
      piti tetragonæ, piloſæ, tri- vel quadrilineares, apice
      bicornes. *Stylus* perſiſtens, capillaris, longitudine
      cornuum. *Stigma* capitatum, integrum.

CHEIRANTHUS. *Gen. pl.* 815.

Germen utrinque denticulo glandulato. *Cal.* clauſus :
    foliolis duobus baſi gibbis. *Semina* plana.

eryſimoi-
des.

1. C. foliis lanceolatis dentatis nudis, caule erecto, ſiliquis
    tetragonis. *Syſt. veg.* 597. *Jacqu. auſtr.* 1. *p.* 48. *t.* 74.
    Wild Stock.
    Nat. of England.
    Fl. May.   H. ♂.

alpinus.

2. C. foliis lineari-cuneiformibus integris denticulatiſve
    ſubpiloſis : pilis adpreſſis.
    Cheiranthus alpinus. *Linn. mant.* 93. *Jacqu. auſtr.* 1.
      p. 48. *t.* 75.
    Alpine Stock.
    Nat. of the South of Europe.
    Cult. 1731, by Mr. Philip Miller. *Mill. dict. edit.* 1.
      Leucojum 19.
    Fl. May——July.   H. ♃.
                                                    3. C.

3. C. foliis filiformibus integerrimis fubfericeis, caule *tenuifo-* frutefcente ramofo.   *lius.*
Narrow-leav'd fhrubby Stock.
Nat. of Madeira.   Mr. *Francis Maſſon.*
Introd. 1777.
Fl. May and June.   G. H. ♄.

4. C. foliis lanceolatis acuminatis argute ferratis, caule *mutabilis,* frutefcente, filiquis pedunculatis.
Broad-leav'd fhrubby Stock.
Nat. of Madeira.   Mr. *Francis Maſſon.*
Introd. 1777.
Fl. March——May.   G. H. ♄.

5. C. foliis lanceolatis acutis glabris integerrimis, ramis *Cheiri.* angulatis, caule fruticofo.
Cheiranthus Cheiri. *Sp. pl.* 924.
α Leucojum luteum vulgare. *Bauh. pin.* 202.
   Wild Wall-flower.
β Leucojum luteum, magno flore. *Bauh. pin.* 202.
   Single Garden Wall-flower.
γ Leucojum luteum, pleno flore, majus. *Bauh. pin.* 202.
   Double Garden Wall-flower.
Nat. of Britain.
Fl. April——July.   H. ♂.

6. C. foliis ellipticis obtufis nudis fcabriufculis, caule *mariti-* diffufo fcabro. *Syſt. veget.* 597.   *mus.*
Dwarf annual Stock Gillyflower.
Nat. of the South of Europe.
Cult. 1739. *Mill. dict. edit.* 1. Hefperis 8.
Fl. May and June.   H. ☉.

7. C. foliis lanceolatis integerrimis obtufis incanis, fili- *incanus.* quis apice truncatis compreſſis, caule fuffruticofo.
Sp. pl. 924.

                     α Leucojum

α Leucojum incanum majus. *Bauh. pin.* 200.
Queen's Stock Gillyflower.

β Cheiranthus *coccineus*, foliis lanceolatis undatis, caule
erecto indiviso. *Mill. dict.*
Brompton Stock Gillyflower.

γ Cheiranthus *albus*, foliis lanceolatis integerrimis obtufis incanis, ramis floriferis axillaribus, caule suffruticofo. *Mill. dict.*
White Stock Gillyflower.

δ Cheiranthus *glabrus*, foliis lanceolatis acutis petiolatis
viridibus, caule suffruticofo.. *Mill. dict.*
Wall-flower-leav'd Stock Gillyflower.
Nat. of the South of Europe.
Cult. 1597. *Ger. herb.* 372.
Fl. June——November.   H. δ.

feneftra-
lis.   8. C. foliis conferto-capitatis recurvatis undulatis, caule
indivifo. *Sp. pl.* 924. *Jacqu. hort.* 2. *p.* 84. *t.* 179.
Clufter-leav'd Stock, or Gillyflower.
Nat.
Cult. 1759, by Mr. Ph. Miller. *Mill. dict. edit.* 7. *n.* 10.
Fl. July and Auguft.   G. H. δ.

annuus. 9. C. foliis lanceolatis fubdentatis obtufis incanis, filiquis
cylindricis apice acutis, caule herbaceo. *Sp. pl.* 925.
Ten-week Stock Gillyflower.
Nat. of the South of Europe.
Cult. 1731, by Mr. Philip Miller. *Mill. dict. edit.* 1.
Leucojum 14.
Fl. June——Auguft.   H. ☉.

littoreus. 10. C. foliis lanceolatis fubdentatis fubtomentofis fubcarnofis, petalis emarginatis, filiquis tomentofis. *Sp.*
pl. 925.
Small Sea Stock.
                                                       Nat.

Nat. of the South of Europe.

Cult. 1683, by Mr. James Sutherland. *Sutherl. hort. edin.* 193. *n.* 5.

Fl. June——November. G. H. ♄.

11. C. foliis linearibus subsinuatis, floribus sessilibus: pe-   *tristis.*
talis undatis, caule suffruticoso. *Sp. pl.* 925.

Dark-flower'd Stock.

Nat. of the South of Europe.

Cult. 1768, by Mr. Philip Miller. *Mill. dict. edit.* 8.

Fl. May——July. H. ♂.

12. C. foliis tomentosis obtusis subsinuatis: rameis inte-   *sinuatus.*
gris, siliquis muricatis. *Sp. pl.* 926.

Greater Sea Stock.

Nat. of England.

Fl. May and June. H. ♂.

H E L I O P H I L A. *Gen. pl.* 816.

Nectaria 2, recurvata versus calycis basin vesicularem.

1. H. foliis linearibus pinnatifidis. *Sp. pl.* 927.   *coronopi-*
Buck's-horn Heliophila.   *folia.*

Nat. of the Cape of Good Hope.

Cult. 1768, by Mr. Philip Miller. *Mill. dict. edit.* 8.

Fl. June——October. G. H. ♄.

2. H. foliis spathulatis integerrimis pubescentibus, sili-   *incana.*
quis villosis.

Heliophila. *Burmann. in nov. act. upsal.* 1. *p.* 94. *t.* 7.

Hoary Heliophila.

Nat. of the Cape of Good Hope.

Introd. 1774, by Mr. Francis Masson.

Fl. May and August. G. H. ♄.

3. H.

filiformis.   3. H. foliis fubulatis filiformibus glabris, filiquis pen-
dulis, ramis divaricatis.   *Linn. fuppl.* 296.
Divaricated Heliophila.
Nat. of the Cape of Good Hope.
Introd. 1786, by John Sibthorp, M.D.
Fl. July and Auguft.   G. H. ☉.

H E S P E R I S.   *Gen. pl.* 817.

Petala oblique flexa. *Glandula* intra ftamina breviora.
Siliqua ftri&a: *Stigma* bafi bifurca apice conni-
vente.   *Cál.* claufus.

triftis.   1. H. caule hifpido ramofo patente.   *Sp. pl.* 927.   *Jacqu.*
auftr. 2. *p.* 1. *t.* 102.
Night-fmelling Rocket.
Nat. of Auftria and Hungary.
Cult. 1739, by Mr. Ph. Miller.   *Rand. chel. n.* 5.
Fl. April——June.   H. ♂.

matrona-   2. H. caule fimplici erecto, foliis ovato-lanceolatis den-
lis.   ticulatis, petalis mucrone emarginatis.   *Sp. pl.* 929.
α flore fimplici.
Single Garden Rocket, or Dame's Violet.
β flore pleno albo.
Double white Garden Rocket.
γ flore pleno purpureo.
Double purple Garden Rocket.
Nat. of Italy.
Cult. 1597.   *Ger. herb.* 376.
Fl. May——Auguft.   H. ♃.

africana.   3. H. caule ramofiffimo diffufo, foliis petiolatis lanceolatis
acute dentatis fcabris, filiquis feffilibus.   *Sp. pl.*
928.

                                                        African

African Rocket.
Nat. of Africa.
Cult. 1759, by Mr. Ph. Miller. *Mill. dict. edit.* 7. *n.* 8.
Fl. June and July. H. ☉.

4. H. caule erecto ramofo, foliis cordatis amplexicaulibus   *verna.*
    ferratis villofis. *Sp. pl.* 928.
Early-flowering Rocket.
Nat. of the South of France.
Cult. 1739, by Mr. Ph. Miller. *Rand. chel.* Turritis 5.
Fl. May. H. ☉.

ARABIS. *Gen. pl.* 818.

Glandulæ nectariferæ 4, fingulæ intra calycis foliola,
    fquamæ inftar reflexæ.

1. A. foliis amplexicaulibus dentatis. *Sp. pl.* 928.   *alpina.*
Alpine Wall-crefs.
Nat. of Switzerland.
Cult. 1658, in Oxford Garden. *Hort. oxon. edit.* 2.
    p. 55.
Fl. March——May. H. ♃.

2. A. foliis petiolatis lanceolatis integerrimis. *Sp. pl.*   *thaliana.*
    929. *Curtis lond.*
Common Wall-crefs.
Nat. of Britain.
Fl. April and May. H. ☉.

3. A. foliis fubdentatis: radicalibus obovatis; caulinis   *bellidifo-*
    lanceolatis. *Syft. veget.* 599. *Jacqu. auftr.* 3. *p.* 44.   *lia.*
    t. 280.
Daify-leav'd Wall-crefs.
Nat. of Auftria and Switzerland.

                                        Introd.

Introd. 1773, by John Earl of Bute.
Fl. May and June.   H. ♃.

hispida.   4. A. foliis cuneiformibus fublyratis hifpidis : caulinis
femiamplexicaulibus lanceolatis, filiquis ftrictis an-
cipitibus.
Arabis hifpida.   *Linn. fuppl.* 298.
Arabis ftricta.   *Hudf. angl.* 292.
Rough Wall-crefs.
Nat. of England.
Fl. May.   H. ♂.

pendula.   5. A. foliis amplexicaulibus, filiquis ancipitibus lineari-
bus, calycibus fubpilofis. *Sp. pl.* 930. *Jacqu. hort.* 3.
p. 20. *t.* 34.
Pendulous Wall-crefs.
Nat. of Siberia.
Cult. 1759, by Mr. Ph. Miller. *Mill. dict. edit.* 7. *n.* 3.
Fl. May and June.   H. ☉.

Turrita.   6. A. foliis amplexicaulibus, filiquis decurvis planis li-
nearibus, calycibus fubrugofis. *Sp. pl.* 930. *Jacqu.*
auftr. 1. *p.* 10. *t.* 11.
Tower Wall-crefs.
Nat. of England.
Fl. April and May.   H. ☉.

TURRITIS.   *Gen. pl.* 819.

Siliqua longiffima, angulata.   *Cal.* connivens, erectus.
Cor. erecta.

glabra.   1. T. foliis radicalibus dentatis hifpidis, caulinis integerrimis amplexicaulibus glabris. *Sp. pl.* 930. *Curtis*
lond.

Smooth

Smooth Tower-muftard.
Nat. of England.
Fl. May.   H. ♂.

2. T. foliis omnibus hifpidis, caulinis amplexicaulibus. *hirfuta.*
   Sp. pl. 930. *Jacqu. ic. colleEt.* 1. *p.* 70.
   Hairy Tower-muftard.
   Nat. of Britain.
   Fl. June.   H. ♂.

BRASSICA. *Gen. pl.* 820.

Cal. ereĉtus, connivens. *Sem.* globofa. *Glandula* inter ftamina breviora et piftillum, interque longiora et calycem.

1. B. foliis cordatis amplexicaulibus glabris : radicalibus *orien-* fcabris integerrimis, filiquis tetragonis. *Sp. pl.* 931. *talis.* *Jacq. auftr.* 3. *p.* 45. *t.* 282.
   Perfoliate Cabbage.
   Nat. of England.
   Fl. June.   H. ☉.

2. B. radice cauleque tenui, foliis caulinis uniformibus *campef-* cordatis feffilibus. *Sp. pl.* 931. *tris.*
   Field Cabbage.
   Nat. of England.
   Fl. June.   H. ☉.

3. B. radice caulefcente fufiformi. *Sp. pl.* 931. *Napus.*
   Wild Cabbage, Rape, or Navew.
   Nat. of Britain.
   Fl. May.   H. ♂.

4. B. radice caulefcente orbiculari depreffa carnofa. *Sp.* *Rapa.* *pl.* 931.
   VoL. II.   D d   Turnep.

Turnep.
Nat. of England.
Fl. April. H. ♂.

oleracea. 5. B. radice caulefcente tereti carnofa. *Sp. pl.* 932.
capitata. α Braffica capitata alba. *Bauh. pin.* 111.
   White Cabbage.
rubra. β Braffica capitata rubra. *Bauh. pin.* 111.
   Red Cabbage.
fabauda. γ Braffica alba crifpa. *Bauh. pin.* 111.
   Savoy Cabbage.
fabellica. δ Braffica fimbriata. *Bauh. pin.* 112.
   Borecole.
botrytis. ε Braffica cauliflora. *Bauh. pin.* 111.
   Cauliflower.
   ζ Braffica italica broccoli dicta. *Mill. dict.*
   Broccoli.
Napo- η Braffica radice napiformi. *Tourn. inft.* 219.
braffica. Turnep-rooted Cabbage.
   Nat. of the Englifh fea-coafts.
   Fl. April——June. H. ♂.

chinenfis. 6. B. foliis ovalibus fubintegerrimis : floralibus amplexi-
   caulibus lanceolatis, calycibus ungue petalorum
   longioribus. *Sp. pl.* 932.
   Chinefe cabbage.
   Nat. of China.
   Introd. 1770, by Monf. Richard.
   Fl. July. H. ☉.

muralis. 7. B. foliis lanceolatis finuato-ferratis læviufculis, caule
   erecto glabro. *Hudf. angl.* 290. *Curtis lond.*
   Wall Cabbage, or wild Rocket.
   Nat. of England.
   Fl. May——July. H. ♃.

SINAPIS.

S I N A P I S. *Gen. pl.* 821.

Cal. patens. *Cor.* ungues recti. *Glandula* inter stamina breviora et pistillum, interque longiora et calycem.

1. S. siliquis multangulis torofo-turgidis lævibus rostro   *arvenfis.*
ancipite longioribus. *Syst. veget.* 602. *Curtis lond.*
Wild Mustard, or Charlock.
Nat. of Britain.
Fl. June and July.   H. ☉.

2. S. siliquis retrorfum hifpidis apice fubtetragonis com-   *orientalis.*
preffis. *Sp. pl.* 933.
Oriental Mustard.
Nat. of the Levant.
Introd. 1778, by Monf. Thouin.
Fl. June and July.   H. ☉.

3. S. siliquis hifpidis: rostro obliquo longiffimo enfi-   *alba.*
formi. *Syst. veget.* 602. *Curtis lond.*
White Mustard.
Nat. of Britain.
Fl. June and July.   H. ☉.

4. S. siliquis glabris racemo adpreffis. *Syst. veget.* 602.   *nigra.*
Common black Mustard.
Nat. of Britain.
Fl. May and June.   H. ○.

5. S. siliquis lævibus fubarticulatis patulis, foliis lyrato-   *chinenfis.*
runcinatis fubhirtis. *Linn. mant.* 95.
Chinefe Mustard.
Nat. of China.
Introd. 1782, by Monf. Thouin.
Fl. July.   H. ☉.

juncea.   6. S. ramis fasciculatis, foliis summis lanceolatis integer-
rimis. *Sp. pl.* 934. *Jacqu. hort.* 2. *p.* 80. *t.* 171.
Fine-leav'd Mustard.
Nat. of China.
Cult. 1731, by Mr. Philip Miller. *Mill. dict. edit.* 1.
n. 3.
Fl. June and July.   H. ☉.

incana.   7. S. siliquis racemo adpressis lævibus, foliis inferioribus
lyratis scabris ; summis lanceolatis, caule scabro.
Sp. pl. 934. *Jacqu. hort.* 2. *p.* 79. *t.* 169.
Hoary Mustard.
Nat. of France, Spain, and Portugal.
Introd. 1778, by Monsf. Thouin.
Fl. July.   H. ♂ .

frutes-   8. S. siliquis linearibus lævibus, foliis inferioribus ob-
cens.   longis dentatis ; superioribus lanceolatis integris,
caule glabro frutescente.
Shrubby Mustard.
Nat. of Madeira.   Mr. *Francis Masson.*
Introd. 1777.
Fl. December——June.   G. H. ♄ .

lævigata.   9. S. siliquis lævibus patulis, foliis lyratis glabris : sum-
mis lanceolatis, caule lævi. *Sp. pl.* 934.
Smooth Mustard.
Nat. of Spain and Portugal.
Introd. 1770, by Monsf. Richard.
Fl. June.   H. ♂ .

R A P H A N U S.   *Gen. pl.* 822.

Cal. clausus.   *Siliqua* torosa, subarticulata, teres.
Glandulæ melliferæ 2 inter stamina breviora et
piftillum,

piftillum, totidem inter ftamina longiora et calycem.

1. R. filiquis teretibus torofis bilocularibus. *Sp. pl.* 935. *fativus.*
α Raphanus minor oblongus. *Bauh. pin.* 96.
Common Garden Radifh.
β Raphanus major orbicularis, vel rotundus. *Bauh.*
pin. 96.
Turnep Radifh.
γ Raphanus niger. *Bauh. pin.* 96.
Black Spanifh Radifh.
Nat. of China.
Cult. 1597. *Ger. herb.* 183.
Fl. May and June.   H. ☉.

2. R. filiquis teretibus articulatis lævibus unilocularibus.   *Rapha-*
     Sp. pl. 935. *Curtis lond.*   *niftrum.*
Wild Radifh.
Nat. of Britain.
Fl. June and July.   H. ☉.

3. R. filiquis fubulatis articulatis bilocularibus foliifque   *tenellus.*
     glabris lanceolatis dentatis: infimis pinnatifidis.
Raphanus tenellus. *Pall. it.* 3. *p.* 741. *tab.* L. *f.* 3.
Small Radifh.
Nat. of Siberia.
Introd. 1780, by Peter Simon Pallas, M. D.
Fl. June and July.   H. ☉.

B U N I A S. *Gen. pl.* 823.

Silicula decidua, tetraëdra, angulis inæqualibus acuminatis muricata.

1. B. filiculis tetragonis: angulis bicriftatis. *Sp. pl.*   *Erucago.*
   935. *Jacqu. auftr.* 4. *p.* 21. *t.* 340.
Prickly-podded Bunias.

Nat.

Nat. of Auſtria and the South of France.
Cult. 1640.   *Park. theat.* 821. *f.* 3.
Fl. June and July.   H. ⊙.

orientalis. 2. B. filiculis ovatis gibbis verrucoſis.   *Sp. pl.* 936.
Oriental Bunias.
Nat. of Ruſſia.
Cult. 1739, by Mr. Ph. Miller. *Rand. chel.* Crambe 2.
Fl. May —— July.   H. ♃.

Cakile. 3. B. filiculis ovatis lævibus ancipitibus.   *Sp. pl.* 936.
Sea Bunias.
Nat. of Britain.
Fl. June and July.   H. ⊙.

ægyptia- 4. B. filiculis tetragonis undique verrucoſo-muricatis,
ca.   foliis runcinatis. *Syſt. veget.* 603. *Jacqu. hort.* 2.
              p. 68. *t.* 145.
Egyptian Bunias.
Nat. of Egypt.
Introd. 1787, by Mr. Zier.
Fl. Auguſt.   H. ⊙.

balearica. 5. B. filiculis hiſpidis, foliis pinnatis : foliolis ſubdenta-
               tis. *Syſt. veget.* 603. *Jacqu. hort.* 2. *p.* 68. *t.* 144.
Minorca Bunias.
Nat. of Minorca.
Introd. 1781, by P. M. A. Brouſſonet, M. D.
Fl. June and July.   H. ⊙.

I S A T I S.   *Gen. pl.* 824.

Siliqua lanceolata, unilocularis, 1-ſperma, decidua,
          bivalvis : valvulis navicularibus.

                                                      1. I.

1. I. foliis radicalibus crenatis; caulinis sagittatis, sili-   *tinctoria.*
culis oblongis. *Sp. pl.* 936.
Common Dyers Woad.
Nat. of England.
Fl. May and June.   H. ♂.

2. I. foliis radicalibus crenatis; caulinis sagittatis, pe-   *lusitani-*
dunculis subtomentosis. *Sp. pl.* 936.   *ca.*
Portugal Woad.
Nat. of Spain and the Levant.
Cult. 1739, by Mr. Philip Miller. *Rand. chel. n.* 3.
Fl. June and July.   H. ⊙.

C R A M B E. *Gen. pl.* 825.

Filamenta 4 longiora apice bifurca: altero antherifero.
Bacca sicca, globosa, decidua.

1. C. foliis cauleque glabris. *Sp. pl.* 937.   *mariti-*
Sea Colewort.   *ma.*
Nat. of Britain.
Fl. May and June.   H. ♃.

2. C. foliis cauleque scabris. *Sp. pl.* 937.   *hispani-*
Spanish Colewort.   *ca.*
Nat. of Spain.
Cult. 1759, by Mr. Ph. Miller. *Mill. dict. edit.* 7. *n.* 4.
Fl. June and July.   H. ⊙.

3. C. fruticosa, foliis ovatis pinnatifidis serratis canis, ra-   *fruticosa.*
cemis in panicula effusa dichotoma. *Linn. suppl.* 299.
Shrubby Colewort.
Nat. of Madeira. Mr. *Francis Masson.*
Introd. 1777.
Fl. Most part of the Year.   G. H. ♄.

ſtrigoſa.   4. C. fruteſcens, folis baſi inæqualibus biauritis ſtrigoſis.
            L'Herit. ſtirp. nov. p. 151. *tab.* 72.
            Myagrum arboreſcens.   *Jacqu. ic. collect.* 1. *p.* 39.
            Rough-leav'd ſhrubby Colewort.
            Nat. of the Canary Iſlands.   Mr. *Fr. Maſſon.*
            Introd. 1779.
            Fl. May and June.   G. H. ♄.

                    C L E O M E,   *Gen. pl.* 826.

            Glandulæ nectariferæ 3, ad ſingulum ſinum calycis
               ſingulæ, excepto infimo.   *Petala* omnia adſcenden-
               tia.   *Siliqua* unilocularis, bivalvis.

hepta-   1. C. floribus gynandris, foliis ſubſeptenatis, caule acu-
phylla.   leato.   *Sp. pl.* 937.
            Seven-leav'd Cleome.
            Nat. of both Indies.
            Cult. 1748, by Mr. Philip Miller.   *Mill. dict. edit.* 5,
               Sinapiſtrum 2.
            Fl. June and July.   S. ☉.

penta-   2. C. floribus gynandris, foliis quinatis, caule inermi.
phylla.   *Sp. pl.* 938.   *Jacqu. hort.* 1. *p.* 9. *t.* 24.
            Five-leav'd Cleome.
            Nat. of both Indies.
            Cult. 1640, by Mr. John Parkinſon.   *Park. theat.*
               397. *f.* 3.
            Fl. June and July.   S. ☉.

triphylla. 3. C. floribus gynandris, foliis ternatis, caule inermi.
            Sp. pl. 938.
            Three-leav'd Cleome.
            Nat. of the Weſt Indies.

                                                      Introd.

Introd. 1730, by William Houftoun, M. D. *Mill.*
dict. edit. 8.
Fl. June and July.   S. ☉.

4. C. floribus dodecandris, foliis quinatis ternatifque.   *vifcofa.*
    Sp. pl. 938.
    Vifcous Cleome.
    Nat. of Ceylon.
    Cult. 1739, by Mr. Philip Miller. *Rand. chel.* Si-
        napiftrum 5.
    Fl. June and July.   S. ☉.

5. C. floribus hexandris, foliis feptenatis, caule inermi. *gigantea.*
    Syft. veget. 605.
    Gigantic Cleome.
    Nat. of South America.
    Introd. 1774, by John Fothergill, M. D.
    Fl. June and July.   S. ♄.

6. C. floribus hexandris, foliis feptenatis quinatifque,   *fpinofa.*
    caule fpinofo. *Sp. pl.* 939.
    Prickly Cleome.
    Nat. of the Weft Indies.
    Introd. 1731, by William Houftoun, M. D. *Mill.*
        dict. edit. 8.
    Fl. July.   S. ☉.

7. C. floribus hexandris, foliis ternatis : foliolis ovali- *ornitho-*
    lanceolatis. *Sp. pl.* 940.   *podioides.*
    Bird's-foot Cleome.
    Nat. of the Levant.
    Cult. 1732, by James Sherard, M. D, *Dill. elth.*
        359. *t.* 266. *f.* 345.
    Fl. June and July.   S. ☉.

8. C.

violacea.   8. C. floribus hexandris, foliis ternatis folitariifque : fo-
       liolis lineari-lanceolatis integerrimis. *Sp. pl.* 940.
       Violet-colour'd Cleome.
       Nat. of Portugal.
       Introd. 1776, by Monf. Thouin.
       Fl. June and July.   H. ☉.

Claffis XVI.

MONADELPHIA

PENTANDRIA.

WALTHERIA. *Gen. pl.* 827.

Monogyna. *Capf.* unilocularis, 2-valvis, 1-fperma.

1. W. foliis ovalibus plicatis ferrato-dentatis tomento-   *america-*
    fis, capitulis pedunculatis. *Syft. veget.* 610.   *na.*
American Waltheria.
Nat. of South America.
Cult. 1691, in the Royal Garden at Hampton-court.
    Pluk. phyt. t. 150. *f.* 6.
Fl. Moft part of the Year.   S. ♂.

HERMANNIA. *Gen. pl.* 828.

Pentagyna. *Capf.* 5-locularis. *Petala* bafi femitubu-
           lata, obliqua.

1. H. foliis ovatis crenatis plicatis tomentofis, calycibus   *althæifo-*
    florentibus campanulatis angulatis, ftipulis oblongis   *lia.*
    foliaceis.
Hermannia althæifolia. *Sp. pl.* 941.
Marfh-mallow-leav'd Hermannia.
Nat. of the Cape of Good Hope.
Cult. 1728, by Mr. Philip Miller. *R. S. n.* 326.
Fl. March——July.   G. H. ♄.

2. H. foliis cordato-ovatis denticulatis plicatis tomento-   *plicata.*
    fis, calycibus florentibus oblongo-ovatis fubcylin-
    draceis.
                                          Plaited-

Plaited-leav'd Hermannia.

Nat. of the Cape of Good Hope. Mr. *Fr. Maſſon.*

Introd. 1774.

Fl. November and December. G. H. ♄.

candi- cans. 3. H. foliis ovatis ſubcordatis obtuſis tomentoſis, calycibus florentibus patulis ſubangulatis, ſtipulis ſubu-
latis.

White Hermannia.

Nat. of the Cape of Good Hope. Mr. *Fr. Maſſon.*

Introd. 1774.

Fl. April——June. G. H. ♄.

alnifolia, 4. H. foliis cuneiformibus lineatis crenato-emarginatis.
Syſt. veget. 610.

Alder-leav'd Hermannia.

Nat. of the Cape of Good Hope.

Cult. 1728, by Mr. Philip Miller. *R. S. n.* 327.

Fl. February——May. G. H. ♄.

odorata. 5. H. foliis oblongo-lanceolatis tomentoſis ſubtus rugo-
ſis: inferioribus truncato-denticulatis, calycibus urceolatis anguloſis.

Sweet-ſcented Hermannia.

Nat. of the Cape of Good Hope. Mr. *William Pa-
terſon.*

Introd. 1780, by the Counteſs of Strathmore.

Fl. Moſt part of the Year. G. H. ♄.

byſſopifo- lia. 6. H. foliis lanceolatis obtuſis ferratis. *Sp. pl.* 940.

Hyſſop-leav'd Hermannia.

Nat. of the Cape of Good Hope.

Cult. 1725, by Mr. Philip Miller. *R. S. n.* 172.

Fl. April——June. G. H. ♄.

7. H.

7. H. foliis lanceolatis obtusis integerrimis. *Sp. pl.* 942.   *lavendu-*
   Lavender-leav'd Hermannia.   *lifolia.*
   Nat. of the Cape of Good Hope.
   Cult. 1732, by James Sherard, M. D. *Dill. elth.* 179.
     t. 147. *f.* 176.
   Fl. May——September.   G. H. ♄ .

8. H. foliis lanceolatis acutis lævibus superne serratis.   *denudata.*
     Linn. suppl. 301. *Cavan. diss.* 6. *p.* 329. *t.* 181. *f.* 1.
   Smooth Hermannia.
   Nat. of the Cape of Good Hope.   Mr. *Fr. Masson.*
   Introd. 1774.
   Fl. May —— July.   G. H. ♄ .

   M E L O C H I A.   *Gen. pl.* 829.

     Pentagyna.   *Caps.* 5-locularis, 1-sperma.

1. M. floribus umbellatis, capsulis pyramidatis pentago-   *pyrami-*
     nis : angulis mucronatis, foliis nudis. *Syst. veget.*   *data.*
     611. *Jacqu. hort.* 1. *p.* 11. *t.* 30.
   Pyramidal Melochia.
   Nat. of Brazil.
   Cult. 1768, by Mr. Philip Miller. *Mill. dict. edit.* 8.
   Fl. July.   S. ♄ .

2. M. floribus capitatis sessilibus, capsulis subrotundis,   *corchori-*
     foliis subcordatis sublobatis. *Syst. veget.* 611.   *folia.*
   Red Melochia.
   Nat. of the East Indies.
   Cult. 1732, by James Sherard, M. D. *Dill. elth.* 221.
     t. 176. *f.* 217.
   Fl. July and August.   S. ☉ .

                    E R O D I U M.

ERODIUM. *L'Herit. Geran.*

Cal. 5-phyllus. *Cor.* 5-petala. *Nect. Squamulæ* 5, cum filamentis alternantes ; et *Glandulæ* melliferæ, bafi ftaminum infidentes. *Fructus* 5-coccus, roftratus : *roftra* fpiralia, introrfum-barbata.

* *Foliis compofitis, aut pinnatifidis.*

craffifolium.
  1. E. umbellis multifloris, foliis pinnatifido-laciniatis craffis : laciniis-linearibus. *L'Herit. n.* 5.
    Upright Crane's-bill.
    Nat. of the Ifland of Cyprus. *John Sibthorp*, M. D.
    Introd. 1788.
    Fl. April and May.   G. H. ♃.

romanum.
  2. E. acaule, fcapis radicalibus multifloris, foliis pinnatis : foliolis pinnatifidis. *L'Herit. n.* 11.
    Geranium romanum. *Sp. pl.* 951.
    Roman Crane's-bill.
    Nat. of Rome.
    Cult. 1724, in Chelfea Garden. *R. S. n.* 130.
    Fl. May and June.   H. ☉.

cicutarium.
  3. E. pedunculis multifloris, foliis pinnatis : foliolis feffilibus pinnatifidis. *L'Herit. n.* 12.
    Geranium cicutarium. *Sp. pl.* 951. *Curtis lond.*
    Hemlock-leav'd Crane's-bill.
    Nat. of Britain.
    Fl. April——September.   H. ☉.

mofchatum.
  4. E. pedunculis multifloris, foliis pinnatis : foliolis fubpetiolatis inæqualiter incifis. *L'Herit. n.* 13.
    Geranium mofchatum. *Sp. pl.* 951. *Jacqu. hort.* 1.
    p. 22. *tab.* 55.
    Mufk Crane's-bill.

                                   Nat.

Nat. of England.
Fl. May——July. H. ☉.

5. E. pedunculis fubmultifloris, foliis ternatis crenato-   *gruinum.*
dentatis : extimo pinnatifide lobato. *L'Herit. n.* 15.
Geranium gruinum. *Sp. pl.* 952.
Broad-leav'd Annual Crane's-bill.
Nat. of the Ifland of Candia.
Cult. 1596, by Mr. John Gerard. *Hort. Ger.*
Fl. June and July. H. ☉.

6. E. pedunculis multifloris biflorifve, foliis interrupte   *ciconium.*
bipinnatifidis laciniatis, caule ftrigofo. *L'Herit.*
n. 16.
α Geranium ciconium. *Sp. pl.* 952. *Jacqu. hort.* 1.
p. 7. *tab.* 18.
Long-beak'd Crane's-bill.
β Geranium botrys. *Cavan. diff.* 4. *p.* 218. *tab.* 90. *f.* 2.
Hairy Crane's-bill.
Nat. of the South of Europe.
Cult. 1724, in Chelfea Garden. *R. S. n.* 132.
Fl. June and July. H. ☉.

7. E. pedunculis paucifloris, foliis tripartitis ternatifve   *incarna-*
trifidis fcabris, caule fruticulofo. *L'Herit. n.* 21.   *tum.*
tab. 5.
Geranium incarnatum. *Linn. fuppl.* 306.
Flefh-colour'd Crane's-bill.
Nat. of the Cape of Good Hope.
Introd. 1787, by Mr. Francis Maffon.
Fl. July. G. H. ♄.

** *Foliis lobatis, aut indivifis.*

8. E. pedunculis multifloris, foliis cordatis trilobis : lo-   *malacoi-*
bis lobatis obtufis obfolete dentatis. *L'Herit. n.* 22.   *des.*
α Geranium malacoides. *Sp. pl.* 952.

Mallow-

Mallow-leav'd Crane's-bill.

β Geranium chium. *Sp. pl.* 951.

Various-leav'd Crane's-bill.

Nat. of the South of Europe.

Cult. 1596, by Mr. John Gerard. *Hort. Ger.*

Fl. May——July.   H. ☉.

glauco-   9. E. pedunculis multifloris, foliis oblongis obfolete cre-
phyllum.   natis glaucis, roftris plumofis. *L'Herit. n.* 25.

Geranium glaucophyllum. *Sp. pl.* 952.

Glaucous Crane's-bill.

Nat. of Egypt.

Cult. 1732, by James Sherard, M. D. *Dill. elth.* 150.
    t. 124. *f.* 150.

Fl. July and Auguft.   G. H. ☉.

mariti-   10. E. pedunculis fubtrifloris, foliis cordatis incifis crena-
mum.   tis fcabris, caulibus depreffis. *L'Herit. n.* 29.

Geranium maritimum. *Sp. pl.* 951.

Sea Crane's-bill.

Nat. of England.

Fl. May——July.   H. ♃.

chamæ-   11. E. fubacaule, pedunculis unifloris, foliis cordatis ob-
dryoides.   tufis crenatis. *L'Herit. n.* 31. *tab.* 6.

Geranium chamædryoides. *Cavan. diff.* 4. *p.* 197.
    tab. 76. *f.* 2.

Geranium parvulum. *Scop. infubr.* 1. *p.* 8. *tab.* 3.

Geranium Reichardi. *Murray in commentat. gotting.*
    1780. *p.* 11. *tab.* 3. *Curtis magaz.* 18.

Geranium æftivum, minimum, fupinum, alpinum,
    chamædryoides, flore albo variegato. *Bocc. muf.*
    160. *tab.* 128.

Dwarf Crane's-bill.

Nat. of Minorca and Corfica.

Fl. April——September.   G. H. ♃.

HEPTAN-

HEPTANDRIA.

PELARGONIUM. *L'Herit. Geran.*

Cal. 5-partitus: lacinia suprema definente in tubulum capillarem, nectariferum, fecus pedunculum decurrentem. *Cor.* 5-petala, irregularis. *Filam.* 10, inæqualia: quorum 3 (raro 5) castrata. *Fructus* 5-coccus, rostratus: *rostra* spiralia, introrsum barbata.

* *Acaulia, radice rapacea, umbellis compositis.*

1. P. acaule, umbella composita, foliis obovatis lanceo-   *hirfutum.*
latifve integerrimis pinnatifidisve hirtis ciliatis.
Geranium hirsutum. *Burm. geran. n.* 68. *tab.* 2.
   Cavan. diff. 4. *p.* 247. *tab.* 101. *f.* 2.
Geranium lobatum β. hirsutum. *Sp. pl.* 950.
Various-leav'd Crane's-bill.
Nat. of the Cape of Good Hope.
Introd. 1788, by Mr. Francis Masson.
Fl. March.   G. H. ♃.

2. P. acaule, umbella subcomposita, foliis pinnatis: fo- *pinnatum.*
liolis subrotundo-ovatis indivisis utrinque hirsutis.
L'Herit. n. 14.
Geranium pinnatum. *Sp. pl. ed.* 1. *p.* 677.
Geranium prolificum γ. pinnatum. *Sp. pl. ed.* 2.
   p. 950.
Geranium astragalifolium. *Cavan. diff.* 4. *p.* 257.
   tab. 104. *f.* 2.
Geranium africanum, astragali folio. *Commel. præl.*
   53. *tab.* 3.
Pinnated Crane's-bill.

Nat. of the Cape of Good Hope.
Introd. 1788, by Mr. Francis Maſſon,
Fl. April.   G. H. ♃.

rapace-   3. P. acaule, umbella compoſita, foliis dècompoſite laci-
um.   niatis villoſis. *L'Herit. n.* 17.
              Geranium rapaceum. *Syſt. nat. ed.* 10. *p.* 1141.
              Geranium prolificum *a.* *Sp. pl.* 949.
              Geranium africanum myrrhidis folio, flore albicante,
                radice rapacea. *Commel. hort.* 2. *p.* 125. *tab.* 63.
              Caraway-leav'd Crane's-bill.
              Nat. of the Cape of Good Hope.
              Introd. 1788, by Mr. Francis Maſſon.
              Fl. April.   G. H. ♃.

              ** *Subacaulia, radice tuberoſa.*

lobatum.   4. P. acaule, umbella compoſita, foliis ternatis quinatiſve
              lobatis tomentoſis. *L'Herit. n.* 22.
              Geranium lobatum *a.* *Sp. pl.* 950.
              Vine-leav'd Crane's-bill.
              Nat. of the Cape of Good Hope.
              Cult. 1739, in Chelſea Garden. *Rand. chel. n.* 65.
              Fl. July and Auguſt.   G. H. ♃.

triſte.   5. P. ſubcauleſcens, umbella ſimplici, foliis multifido-
              laciniatis villoſis: laciniis lanceolatis. *L'Herit.n.* 23.
              Geranium triſte. *Sp. pl.* 950.
              Night ſmelling Crane's-bill.
              Nat. of the Cape of Good Hope.
              Introd. before 1632, by Mr. John Tradeſcant, Sen.
                Ger. emac. 948.
              Fl. Moſt part of the Summer.   G. H. ♃.

flavum.   6. P. ſubcauleſcens, umbellis ſimplicibus, foliis decompo-
              ſite laciniatis hirſutis: laciniis linearibus. *L'Herit.*
              n. 24.
                                                        Geranium

Geranium flavum. *Linn. mant.* 257.

Geranium daucifolium. *Murray in commentat. gotting.*
1780. *p.* 13. *tab.* 4.

Carrot-leav'd Crane's-bill.

Nat. of the Cape of Good Hope.

Cult. 1724, in Chelsea Garden. *R. S. n.* 135.

Fl. July——September.   G. H. ♃.

* * * *Herbacea, aut suffruticosa.*

7. P. pedunculis paucifloris, foliis subrotundo-cordatis   *tabulare.*
quinquelobis obtusis, caulibus decumbentibus pilo-
sis. *L'Herit. n.* 25. *tab.* 9.

Geranium tabulare. *Sp. pl.* 947.

Geranium elongatum. *Cavan. diff.* 4. *p.* 233. *tab.* 101.
f. 3.

Rough-stalked Crane's-bill.

Nat. of the Cape of Good Hope.

Introd. 1775, by Mr. Francis Masson.

Fl. Most part of the Summer.   G. H. ♃.

8. P. pedunculis subquadrifloris, foliis orbiculatis pal-   *alchimil-*
mato-incisis pilosissimis, caule herbaceo decumbente,   *loides.*
stigmatibus sessilibus. *L'Herit. n.* 26.

Geranium alchimilloides. *Sp. pl.* 948.

Lady's Mantle-leav'd Crane's-bill.

Nat. of the Cape of Good Hope.

Cult. 1693, by Mr. Jacob Bobart. *Br. Muf. Sloan.*
mff. 3343.

Fl. May——October.   G. H. ♃.

9. P. pedunculis subquinquefloris, foliis subrotundo-cor-   *odoratissi-*
datis mollissimis. *L'Herit. n.* 28.   *mum.*

Geranium odoratissimum. *Sp. pl.* 948.

Sweet-scented Crane's-bill.

Nat. of the Cape of Good Hope.

Cult,

Cult. 1724, in Chelsea Garden.   *R. S. n.* 143.
Fl. Most part of the Summer.   G. H. ♃.

grossula-   10. P. pedunculis subbifloris filiformibus, foliis cordatis
riodes.   subrotundis incisis dentatis, caulibus glaberrimis.
                    L'Herit. n. 30.
                    Geranium grossularioides.   *Sp. pl.* 948.
                    Gooseberry-leav'd Crane's-bill.
                    Nat. of the Cape of Good Hope.
                    Cult. 1731, by Mr. Philip Miller.   *Mill. dict. edit.* 1.
                    n. 39.
                    Fl. April——August.   G. H. ♃.

anceps.   11. P. umbellis multifloris, floribus subcapitatis, foliis
                    cordato-subrotundis obsolete lobatis, caulibus tri-
                    quetro-ancipitibus.   *L'Herit. n.* 31.
                    Angular-stalked Crane's-bill.
                    Nat. of the Cape of Good Hope.   Mr. *Fr. Masson.*
                    Introd. 1788.
                    Fl. May.   G. H. ♃.

althæ-   12. P. pedunculis multifloris, foliis cordato-ovatis sinua-
oides.   tis dentatis: summis pinnatifidis, petalis calyci
                    æqualibus.   *L'Herit. n.* 39. *tab.* 10.
                    Geranium althæoides.   *Sp. pl.* 949.
                    Althæa-leav'd Crane's-bill.
                    Nat. of the Cape of Good Hope.
                    Cult. 1724, in Chelsea Garden.   *R. S. n.* 139.
                    Fl.   G. H. ♂.

senecioi-   13. P. pedunculis trifloris, involucris calycibusque obtu-
des.   sis, foliis bipinnatifido-laciniatis, caule herbaceo.
                    L'Herit. n. 42. *tab.* 11.
                    Small white-flower'd Crane's-bill.
                    Nat. of the Cape of Good Hope.   Mr. *Fr. Masson.*
                    Introd.

Introd. 1775.
Fl. June and July. G. H. ⊙.

14. P. pedunculis fubtrifloris, corollis fubtetrapetalis, *corian-*
foliis bipinnatis linearibus, caule herbaceo læviuf- *drifolium.*
culo. *L'Herit. n.* 43.
Geranium coriandrifolium. *Sp. pl.* 949.
Coriander-leav'd Crane's-bill.
Nat. of the Cape of Good Hope.
Cult. 1724, in Chelfea Garden. *R. S. n.* 136.
Fl. March——September. G. H. ♂.

15. P. pedunculis fubtrifloris, corollis fubtetrapetalis, *myrrhifo-*
foliis bipinnatifidis: inferioribus cordatis lobatis, *lium.*
caule ftrigofiufculo. *L'Herit. n.* 44.
Geranium myrrhifolium. *Sp. pl.* 949.
Myrrh-leav'd Crane's-bill.
Nat. of the Cape of Good Hope.
Cult. 1731, by Mr. Philip Miller. *Mill. dict. edit.* 1.
n. 35.
Fl. May——Auguft. G. H. ♄.

**** *Fruticofa, caule carnofo aut craffo.*

16. P. umbellis multifloris, foliis decompofite pinnatis *tenuifoli-*
multifidis linearibus hirfutis, caule carnofo, ramis *um.*
floriferis gracilibus. *L'Herit. n.* 48. *tab.* 12.
Fine-leav'd Crane's-bill.
Nat. of the Cape of Good Hope.
Introd. 1768, by Mr. William Malcolm.
Fl. Moft part of the Summer. G. H. ♄.

17. P. umbellis multifloris, foliis pinnatifidis laciniatis, *carnofum.*
petalis linearibus, articulis carnofo-gibbofis.
L'Herit. n. 49.
Geranium carnofum. *Sp. pl.* 946.
Flefhy-ftalked Crane's-bill.

E e 3 *Nat.*

Nat. of the Cape of Good Hope.
Cult. 1724, in Chelfea Garden.   R. S. n. 137.
Fl. June——Auguft.   G. H. ♄.

cerato-   18. P. umbellis multifloris, foliis remote pinnatis carno-
phyllum.   fis teretibus, laciniis canaliculatis obfolete trifidis.
                L'Herit. n. 50. tab. 13.
                Horn-leav'd Crane's-bill.
                Nat. of the South-weft Coaft of Africa.
                Introd. 1786, by Mr. Anthony Hove.
                Fl. May.   G. H. ♄.

gibbofum. 19. P. umbellis multifloris, foliis pinnatis apice pinnati-
                fido - confluentibus, geniculis carnofis gibbofis.
                L'Herit. n. 51.
                Geranium gibbofum.   Sp. pl. 946.
                Gouty Crane's-bill.
                Nat. of the Cape of Good Hope.
                Cult. 1712. Philofoph. tranfact. n. 333. p. 420. n. 74.
                Fl. Moft part of the Summer.   G. H. ♄.

fulgidum. 20. P. umbella gemina, foliis tripartitis pinnatifido-
                incifis : lacinia intermedia maxima.   L'Herit.
                n. 53.
                Geranium fulgidum.   Sp. pl. 945.
                Celandine-leav'd Crane's-bill.
                Nat. of the Cape of Good Hope.
                Cult. 1732, by James Sherard, M.D.   Dill. elth.
                156. t. 130. f. 157.
                Fl. Moft part of the Summer.   G. H. ♄.

quercifo- 21. P. umbellis fubmultifloris, foliis cordatis pinnatifidis
lium.   crenatis: finubus rotundatis, filamentis apice ad-
                fcendentibus.   L'Herit. n. 54. tab. 14.
                Geranium quercifolium.   Linn. fuppl. 306. Cavan.
                diff. 4. p. 246. tab. 119. f. 1.
                                                Geranium

Geranium terebinthinaceum. *Murray in commentat. gotting.* 7. (1785) *p.* 88. *tab.* 4.
Great oak-leav'd Crane's-bill.
β foliis duplicato-pinnatifidis. *L'Herit. tab.* 15.
Small oak-leav'd Crane's-bill.
Nat. of the Cape of Good Hope.
Introd. 1774, by Mr. Francis Maſſon.
Fl. March——Auguſt.   G. H. ♄.

22. P. umbellis paucifloris, foliis pinnatifido-lacinia- *Radula.*
tis ſcabris margine revolutis : laciniis linearibus.
L'Herit. n. 55. *tab.* 16.
Geranium Radula. *Cavan. diſſ.* 4. *p.* 262. *tab.* 101.
f. 1.
Geranium revolutum. *Jacqu. ic. collect.* 1. *p.* 84.
Multifid-leav'd Crane's-bill.
Nat. of the Cape of Good Hope. Mr. *Fr. Maſſon.*
Introd. 1774.
Fl. March——July.   G. H. ♄.

23. P. umbellis multifloris ſubcapitatis, foliis palmatis *graveo-*
ſeptemlobatis : laciniis oblongis obtuſis ; margi- *lens.*
nibus revolutis. *L'Herit. n.* 56. *tab.* 17.
Geranium terebinthinaceum. *Cavan. diſſ.* 4. *p.* 250.
tab. 114. *f.* 1.
Strong-ſcented Crane's-bill.
Nat. of the Cape of Good Hope. Mr. *Fr. Maſſon.*
Introd. 1774.
Fl. March——July.   G. H. ♄.

24. P. umbellis multifloris, foliis ſubrotundo-cordatis *papilio-*
angulatis, corollis papilionaceis : alis carinaque *naceum.*
minutis. *L'Herit. n.* 58.
Geranium papilionaceum. *Sp. pl.* 945.
Butterfly Crane's-bill.
Nat. of the Cape of Good Hope.

Cult. 1724, in Chelſea Garden.   *R. S. n.* 146.

Fl. April——July.   G. H. ♄.

*inqui-   25. P. umbellis multifloris, foliis orbiculato-reniformibus
nans.*   ſubindiviſis crenatis tomentoſo-viſcidis.   *L' Herit.*
                      n. 59.

Geranium inquinans.   *Sp. pl.* 945.

Scarlet-flower'd Crane's-bill.

Nat. of the Cape of Good Hope.

Cult. 1714, by Biſhop Compton. *Philoſoph. tranſact.*
          n. 346. *p.* 362. *n.* 121.

Fl. May——September.   G. H. ♄.

hybridum. 26. P. umbellis multifloris, foliis obovatis crenatis gla-
                      bris carnoſis, petalis linearibus.   *L' Herit. n.* 60.

Geranium hybridum.   *Linn. mant.* 97.

Baſtard Crane's-bill.

Nat. of the Cape of Good Hope.

Cult. 1732, by James Sherard, M. D.   *Dill. elth.*
          152. *t.* 125. *f.* 152.

Fl. May——September.   G. H. ♄.

zonale.   27. P. umbellis multifloris, foliis cordato-orbiculatis vix
                      lobatis dentatis zonatis.   *L' Herit. n.* 61.

Geranium zonale.   *Sp. pl.* 947.

Common Horſe-ſhoe Crane's-bill.

Nat. of the Cape of Good Hope.

Cult. 1710, by the Dutcheſs of Beaufort.   *Br. Muſ.*
          H. S. 135. *fol.* 6.

Fl. April——November.   G. H. ♄.

*heteroga-   28. P. umbellis multifloris, foliis ſuborbiculatis inciſo-
mum.*   lobatis dentatis, caule erecto fruticoſo.   *L' Herit.*
                      n. 62. *tab.* 18.

                                                              Red-

Red-flower'd Crane's-bill.
Nat.
Cult. 1786, by Meſſrs. Lee and Kennedy.
Fl. Moſt part of the Summer.   G. H. ♄.

29. P. foliis orbiculato-reniformibus obſolete lobatis *mon-*
complicatis criſpis. *L'Herit. n.* 63.   *ſtrum.*
Cluſter-leav'd Crane's-bill.
Nat.
Cult. 1784, by Mrs. Norman.
Fl.   G. H. ♄.

30. P. umbellis multifloris, foliis ternatifidis lobatis den- *bicolor.*
tatis undulatis villoſis. *L'Herit. n.* 64.
Geranium bicolor. *Jacqu. hort.* 3. *p.* 23. *tab.* 39.
Cavan. diſſ. 4. *p.* 248. *tab.* 111. *f.* 1.
Two-colour'd Crane's-bill.
Nat.
Introd. 1778, by John Earl of Bute.
Fl. July and Auguſt.   G. H. ♄.

31. P. floribus capitatis, foliis cordatis trilobis ſcabriuſ- *viti-*
culis, caulibus erectis. *L'Herit. n.* 65.   *folium.*
Geranium vitifolium. *Sp. pl.* 947.
Balm-ſcented Crane's-bill.
Nat. of the Cape of Good Hope.
Cult. 1724, in Chelſea Garden. *R. S. n.* 145.
Fl. April——Auguſt.   G. H. ♄.

32. P. floribus capitatis, foliis cordatis lobatis undatis *capita-*
mollibus, caulibus diffuſis. *L'Herit. n.* 66.   *tum.*
Geranium capitatum. *Sp. pl.* 947.
Roſe-ſcented Crane's-bill.
Nat. of the Cape of Good Hope.
                              Introd.

Introd. 1690, by Mr. Bentick. *Br. Muf. Sloan. mff.*
3370.
Fl. April——Auguft.   G. H. ♄.

glutino-
fum.
33. P. umbellis paucifloris, foliis cordatis haftato-quin-
quangulis vifcofis. *L'Herit. n.* 67. *tab.* 20.
Geranium glutinofum. *Jacqu. ic. collect.* 1. *p.* 85.
Geranium vifcofum. *Cavan. diff.* 4. *p.* 246. *tab.* 108.
f. 2. *Scop. infubr.* 2. *p.* 27. *tab.* 14.
Clammy Crane's-bill.
Nat. of the Cape of Good Hope.
Introd. about 1777, by Meffrs. Kennedy and Lee.
Fl. May and June.   G. H. ♄.

cuculla-
tum.
34. P. umbellis fubmultifloris, foliis reniformibus cucullatis dentatis. *L'Herit. n.* 68.
Geranium cucullatum. *Sp. pl.* 946. (exclufis fynonymis Dillenii et Martyni, quæ fequentis.)
Hooded Crane's-bill.
Nat. of the Cape of Good Hope.
Introd. 1690, by Mr. Bentick. *Br. Muf. Sloan.*
mff. 3370.
Fl. Moft part of the Summer.   G. H. ♄.

angulo-
fum.
35. P. umbellis multifloris, foliis rotundatis cucullatis
angulofis dentatis. *L'Herit. n.* 69.
Geranium angulofum. *Mill. dict.*
Geranium acerifolium. *Cavan. diff.* 4. *p.* 243. *tab.*
112. *f.* 2.
Geranium africanum arborefcens, foliis cucullatis
angulofis. *Dill. elth.* 155. *tab.* 129. *f.* 156.
Geranium africanum arborefcens, Ibifci folio angu-
lofo, floribus amplis purpureis. *Martyn dec.* 3.
p. 28. *tab.* 28.
Marfh-mallow-leav'd Crane's-bill.

Nat.

Nat. of the Cape of Good Hope.
Cult. 1724, in Chelsea Garden. *R. S. n.* 150.
Fl. July and August. G. H. ♄.

36. P. umbellis subquinquefloris, foliis palmato-quinque- *aceri-*
   lobis serratis inferne cuneatis indivisis. *L'Herit.* *folium.*
   n. 70. *tab.* 21.
   Maple-leav'd Crane's-bill.
   Nat. of the Cape of Good Hope.
   Cult. 1784, by Mr. Archibald Thompson.
   Fl. April and May. G. H. ♄.

37. P. umbellis multifloris, foliis cordatis acutis denta- *cordatum.*
   tis, petalis inferis linearibus acutis.
   Geranium cordifolium. *Cavan. diss.* 4. *p.* 240. *tab.*
   117. *f.* 3.
α foliis indivisis mollissimis planis.
   Heart-leav'd Crane's-bill.
β foliis laciniatis crispis.
   Curl'd heart-leav'd Crane's-bill.
   Nat. of the Cape of Good Hope. Mr. *Fr. Masson.*
   Introd. 1774.
   Fl. March——July. G. H. ♄.

38. P. pedunculis bifloris, ramis tetragonis carnosis, *tetrago-*
   corollis tetrapetalis. *L'Herit. n.* 72. *tab.* 23. *num.*
   Geranium tetragonum. *Linn. suppl.* 305 *Jacqu.*
   ic. collect. 1. *p.* 92. *Scop. insubr.* 1. *p.* 12. *tab.* 5.
   Cavan. diss. 4. *p.* 231. *tab.* 99. *f.* 2.
   Square-stalked Crane's-bill.
   Nat. of the Cape of Good Hope.
   Introd. 1774, by Mr. Francis Masson.
   Fl. June——August. G. H. ♄.

39. P. umbellis paucifloris, foliis quinquelobis integerri- *peltatum.*
   mis

mis carnofis peltatis, ramis angulatis.   *L'Herit.*
n. 73.
Geranium peltatum.   *Sp. pl.* 947.   *Curtis magaz.* 20.
Peltated Crane's-bill.
Nat. of the Cape of Good Hope.
Cult. 1701, by the Duchefs of Beaufort.   *Raj. hift.* 3.
p. 514. *n.* 38.
Fl. June——Auguft.   G. H. ♃.

lateripes.   40. P. umbellis multifloris, foliis cordatis quinquelobis
fubdentatis carnofis, ramis teretibus.   *L'Herit.*
n. 74. *tab.* 24.
Ivy-leav'd Crane's-bill.
Nat. of the Cape of Good Hope.
Cult. 1787, by Meff. Grimwood and Barret.
Fl. Moft part of the Summer.   G. H. ♃.

cortufæ-   41. P. umbellis multifloris, foliis cordatis incifo-lobatis
folium.   undatis obtufe dentatis, ftipulis fubulatis.   *L'Herit.*
n. 75. *tab.* 25.
Cortufa-leav'd Crane's-bill.
Nat. of the South-weft Coaft of Africa.
Introd. 1786, by Mr. Anthony Hove.
Fl. July.   G. H. ♃.

craffi-   42. P. umbellis multifloris, foliis reniformibus obacumi-
caule.   natis, caule carnofo ramofo lævi.   *L'Herit. n.* 77.
tab. 26.
Thick-ftalk'd Crane's-bill.
Nat. of the South-weft Coaft of Africa.
Introd. 1786, by Mr. Anthony Hove.
Fl. July.   G. H. ♃.

cotyledo-   43. P. umbellis compofitis, foliis cordatis peltatis rugo-
nis.   fis, caule carnofo.   *L'Herit. n.* 81. *tab.* 27.
Geranium cotyledonis.   *Linn. mant.* 569.
                                              Hollyhock-

Hollyhock-leav'd Crane's-bill.
Nat. of the Ifland of St. Helena.
Introd. 1765, by Mr. John Bufh.
Fl. May——July.   G. H. ♄.

***** *Fruticofa, caule lignofo.*

44. P. umbellis paucifloris: pedicellis elongatis, foliis   *ovale.*
ellipticis dentatis, caulibus hirfutis. *L'Herit.*
n. 83. *tab.* 28.
Geranium ovale. *Burm. fl. cap.* 19.
Geranium ovatum. *Cavan. diff.* 4. *p.* 238. *tab.* 103.
f. 3.
Oval-leav'd Crane's-bill.
Nat. of the Cape of Good Hope. Mr. *Fr. Maffon.*
Introd. 1774.
Fl. May——July.   G. H. ♄.

45. P. umbellis paucifloris, foliis ovatis inæqualiter fer-   *betuli-*
ratis lævigatis. *L'Herit. n.* 84.   *num.*
Geranium betulinum. *Sp. pl.* 946.
Birch-leav'd Crane's-bill.
Nat. of the Cape of Good Hope.
Introd. 1786, by Mr. Francis Maffon.
Fl. Moft part of the Summer.   G. H. ♄.

46. P. pedunculis bifloris, foliis lanceolatis integerrimis *glaucum.*
acuminatis glaucis. *L'Herit. n.* 89. *tab.* 29.
Geranium glaucum. *Linn. fuppl.* 306.
Geranium lanceolatum. *Cavan. diff.* 4. *p.* 235.
tab. 102. *f.* 2. *Curtis magaz.* 56.
Spear-leav'd Crane's-bill.
Nat. of the Cape of Good Hope.
Introd. 1775, by Meffrs. Kennedy and Lee.
Fl. June——Auguft.   G. H. ♄.

47. P.

tricuspi-   47. P. pedunculis bifloris, foliis tricuspidibus : lobo in-
datum.   termedio productiore subserrato ; costa subtus mu-
         ricata. *L'Herit. n.* 90. *tab.* 30.
         Three-pointed Crane's-bill.
         Nat. of the Cape of Good Hope.
         Introd. 1780, by Messrs. Kennedy and Lee.
         Fl. May——August.   G. H. ♄.

acetosum.   48. P. umbellis paucifloris, foliis obovatis crenatis gla-
         bris carnosis, petalis linearibus. *L'Herit. n.* 97.
         Geranium acetosum. *Sp. pl.* 947.
         Sorrel Crane's-bill.
         Nat. of the Cape of Good Hope.
         Cult. 1724, in Chelsea Garden. *R. S. n.* 141.
         Fl. May——September.   G. H. ♄.

scabrum.   49. P. umbellis paucifloris, foliis cuneatis semitrifidis
         scabris : lobis lanceolatis laxe serratis. *L'Herit.*
         n. 99. *tab.* 31.
         Geranium scabrum. *Sp. pl.* 946.
         Rough-leav'd Crane's-bill.
         Nat. of the Cape of Good Hope.
         Introd. 1775, by Messrs. Kennedy and Lee.
         Fl. August——November.   G. H. ♄.

crispum.   50. P. pedunculis subbifloris, foliis distichis cordatis tri-
         lobis crispis muricatis. *L'Herit. n.* 100. *tab.* 32
         & 33.
         Geranium crispum. *Linn. mant.* 257.
         Geranium hermanniæfolium. *Linn. suppl.* 305. (non
         Linn. mant.),
         Curl'd-leav'd Crane's-bill.
         Nat. of the Cape of Good Hope.
         Introd. 1774, by Mr. Francis Masson.
         Fl. July——November.   G. H. ♄.

                                     51. P.

51. P. pedunculis fubbifloris, foliis cordatis trilobis un-   *adulter.-*
dulatis villofis mollibus. *L'Herit. n.* 103. *tab.* 34.   *num.*
Hoary trifid-leav'd Crane's-bill.
Nat. of the Cape of Good Hope.
Introd. 1785, by Mr. Archibald Thompfon.
Fl. April and May.   G. H. ♄ .

52. P. umbellis paucifloris, foliis cordatis tripartito-lo-   *exftipula-*
batis dentatis canis, ftipulis fubnullis. *L'Herit.*   *tum.*
n. 107. *tab.* 35.
Geranium exftipulatum.   *Cavan. diff.* 4. *p.* 253.
diff. 5. *p.* 271. *tab.* 123. *f.* 1.
Soft-leav'd trifid Crane's-bill.
Nat. of the Cape of Good Hope. Mr. *William*
Paterfon.
Introd. 1779, by the Countefs of Strathmore.
Fl. May——Auguft.   G. H. ♄ .

O C T A N D R I A.

A I T O N I A. *Linn. fuppl.* 49.

Monogyna. *Cal.* 4-partitus. *Cor.* 4-petala. *Bacca*
ficca, 4-angularis, 1-locularis, polyfperma.

1. AITONIA. *Linn. fuppl.* 303.   *capenfis.*
Cape Aitonia.
Nat. of the Cape of Good Hope.
Introd. 1774, by Mr. Francis Maffon.
Fl. Moft part of the Year.   G. H. ♄ .

DECANDRIA.

DECANDRIA.

GERANIUM. *L'Herit. Geran.*

Cal. 5-phyllus. *Cor.* 5-petala, regularis. *Nect.* glandulæ 5 melliferæ, basi longiorum filamentorum adnatæ. *Fructus* 5-coccus, rostratus: *rostra* simplicia, nuda (nec spiralia, nec barbata.)

* *Pedunculis unifloris.*

sibiricum.   1. G. pedunculis subuhifloris, foliis quinquepartitis acutis: foliolis pinnatifidis. *Sp. pl.* 957. *Jacqu. hort.* 1. *p.* 7. *tab.* 19.
Siberian Crane's-bill.
Nat. of Siberia.
Cult. 1768, by Mr. Philip Miller. *Mill. dict. edit.* 8.
Fl. June and July.   H. ♃.

sanguineum.   2. G. pedunculis unifloris, foliis quinquepartitis trifidis orbiculatis. *Sp. pl.* 958.
α Geranium sanguineum maximo flore. *Bauh. pin.* 318.
Common Bloody Crane's-bill.
β Geranium hæmatodes Lancastriense, flore eleganter striato. *Dill. elth.* 163. *t.* 136. *f.* 163.
Lancashire Crane's-bill.
Nat. of Britain.
Fl. June——August.   H. ♃.

* * *Pedunculis bifloris; fruticosa.*

anemone-folium.   3. G. foliis palmatis: foliolis pinnatifidis, caule fruticoso. *L'Herit. n.* 6. *tab.* 36.
Geranium palmatum. *Cavan. diff.* 4. *p.* 216. *tab.* 84. *f.* 2.

Smooth

Smooth Crane's-bill.
Nat. of Madeira. Mr. *Francis Maſſon.*
Introd. 1778.
Fl. May——Auguſt.   G. H. ♄.

4. G. calycibus inflatis, petalis integris, piſtillo longiſſi- *macro-*
   mo, ſcapo dichotomo. *Syſt. veget.* 616. *Jacqu. ic.* *rhizum.*
   collect. 1. *p.* 258.
   Long-rooted Crane's-bill.
   Nat. of Italy.
   Cult. 1658, in Oxford Garden. *Hort. oxon. edit.* 2.
      p. 69.
   Fl. May and June.   H. ♃.

      *** *Pedunculis bifloris; perennia.*
5. G. foliis ſubpeltatis quinquepartitis ſubtus canefcenti- *canefcens.*
   bus : lobis incifis, petalis emarginatis. *L'Herit.*
   n. 12. *tab.* 38.
   Silky-leav'd Crane's-bill.
   Nat. of the Cape of Good Hope. Mr. *Fr. Maſſon.*
   Introd. 1787.
   Fl. May and June.   G. H. ♃.

6. G. calycibus ariſtatis, petalis integris, arillis hirſutis, *incanum.*
   foliis ſubdigitatis pinnatifidis. *Sp. pl.* 957.
   Hoary Crane's-bill.
   Nat. of the Cape of Good Hope.
   Cult. 1704, by the Dutcheſs of Beaufort. *Raj. hiſt.*
      3. *p.* 513. *n.* 28.
   Fl.   G. H. ♃.

7. G. foliis multipartitis : laciniis linearibus ſubdiviſis *tubero-*
   obtuſis. *Sp. pl.* 953. *ſum.*
   Tuberous-rooted Crane's-bill.
   Nat. of Italy.

Cult. 1596, by Mr. John Gerard. *Hort. Ger.*
Fl. May.   H. ♃.

phæum.   8. G. pedunculis folitariis oppofitifoliis, ealycibus fub-
ariftatis, caule erecto, petalis undulatis. *Syft.*
veget. 616.
Dark-flower'd Crane's-bill.
Nat. of England.
Fl. April——June.   H. ♃.

reflexum.   9. G. pedunculis foliifque alternis, petalis reflexis laci-
niatis longitudine calycis mutici. *Linn. mant.* 257.
Purple-flower'd Crane's-bill.
Nat. of Italy.
Cult. 1758, by Mr. Philip Miller.
Fl. May——July.   H. ♃.

lividum.   10. G. foliis femifeptilobis incifis, calycibus fimplicibus
pilofis, petalis planis fubundulatis. *L'Herit. n.* 18.
tab. 39.
Geranium foliis femifeptilobis rugofis dentatis, pe-
talis planiffimis circumferratis indivifis. *Hall.*
hift. 935.
Wrinkled-leav'd Crane's-bill.
Nat. of Switzerland.
Introd. 1775, by the Doctors Pitcairn and Fothergill.
Fl. June.   H. ♃.

nodofum.   11. G. petalis emarginatis, foliis caulinis trilobis inte-
gris ferratis fubtus lucidis. *Syft. veget.* 616.
Knotty Crane's-bill.
Nat. of England.
Fl. May——September.   H. ♃.

ftriatum.   12. G. foliis quinquelobis : lobis medio dilatatis, petalis
bilobis venofo-reticulatis. *Sp. pl.* 953. *Curtis*
magaz. 55.

Streak'd

Streak'd Crane's-bill.
Nat. of Italy.
Cult. 1629. *Park. parad.* 227. *f. 7.*
Fl. May——October.   H. ♃.

13. G. caule dichotomo erecto, foliis quinquepartitis in-   *macula-*
    cifis: fummis feffilibus. *Sp. pl.* 955.   *tum.*
    Spotted Crane's-bill.
    Nat. of Carolina and Virginia.
    Cult. 1732, by James Sherard, M. D. *Dill elth.* 158.
    t. 132. *f.* 159.
    Fl. May——July.   H. ♃.

14. G. foliis fubpeltatis multipartitis rugofis acutis, pe-   *pratenfe.*
    talis integris. *Sp. pl.* 954. *Curtis lond.*
    Meadow Crane's-bill.
    Nat. of Britain.
    Fl. May and June.   H. ♃.

15. G. pedunculis longiffimis declinatis, foliis quinque-   *paluftre.*
    lobis incifis, petalis integris. *Sp. pl.* 954.
    Marfh Crane's-bill.
    Nat. of Germany.
    Cult. 1732, by James Sherard, M. D. *Dill. elth.*
    160. *t.* 134. *f.* 161.
    Fl. June——Auguft.   H. ♃.

16. G. foliis fubpeltatis feptempartitis, petalis integris   *aconiti-*
    venofo-lineatis. *L'Herit. n.* 26. *tab.* 40.   *folium.*
    Aconite-leav'd Crane's-bill.
    Nat. of Switzerland and Dauphiné.
    Introd. 1775, by the Doctors Pitcairn and Fothergill.
    Fl. May.   H. ♃.

17. G. foliis fubpeltatis quinquelobis incifo-ferratis,   *fylvati-*
    caule erecto, petalis emarginatis. *Sp. pl.* 954.   *cum.*

Wood

Wood Crane's-bill.
Nat. of England.
Fl. May and June.   H. ♃.

pyrenai-   18. G. foliis rotundatis femiquinquelobis trilobifve in-
cum.   cifis, petalis bilobis. *L'Herit. n.* 31.
               Geranium pyrenaicum.   *Linn. mant.* 97 & 257.
               Curtis lond.
               Geranium perenne.   *Hudf. angl. ed.* 1. *p.* 265.
               Mill. dict.
               Mountain Crane's-bill.
               Nat. of Britain.
               Fl. May——Auguft.   H. ♃.

               **** *Pedunculis bifloris* ; *annua.*

molle.   19. G. pedunculis foliifque floralibus alternis, petalis
                 bifidis, calycibus muticis, caule erectiufculo.. *Syft.*
                 veget. 617. *Curtis lond.*
                 Common Crane's-bill, or Dove's-foot.
                 Nat. of Britain.
                 Fl. May and June.   H. ☉.

rotundi-   20. G. petalis fubintegris calyci fubariftato æqualibus,
folium.   foliis reniformibus lobatis incifis : finubus glandu-
                   lofis. *L'Herit. n.* 33.
                 Geranium rotundifolium. *Sp. pl.* 957.
                 Round-leav'd Crane's-bill.
                 Nat. of England.
                 Fl. July.   H. ☉.

lucidum.   21. G. calycibus pyramidatis angulatis elevato-rugofis,
                   foliis quinquelobis rotundatis. *Sp. pl.* 955.
                 Shining Crane's-bill, or Dove's-foot.
                 Nat. of Britain.
                 Fl. June and July.   H. ☉.

                                                            22. G.

22. G. pedunculis folio longioribus, foliis quinqueparti-   *columbi-*
 to-multifidis, calycibus pentagonis, arillis glabris. *num.*
 L'Herit. n. 35.
 Geranium colombinum. *Sp. pl.* 956.
 Long-ftalk'd Crane's-bill.
 Nat. of Britain.
 Fl. July.   H. ⊙.

23. G. pedunculis folio brevioribus, foliis quinqueparti-   *diffectum.*
 to-trifidis multifidifque, petalis emarginatis, arillis
 villofis. *L'Herit. n.* 36.
 Geranium diffectum. *Sp. pl.* 956.
 Jagged Crane's-bill.
 Nat. of Britain.
 Fl. May——July.   H. ⊙.

24. G. foliis quinquepartitis incifis, calycibus ariftatis,   *caroli-*
 petalis emarginatis, arillis hirfutis. *L'Herit. n.* 38. *nianum.*
 Geranium carolinianum. *Sp. pl.* 956.
 Carolina Crane's-bill.
 Nat. of North America.
 Cult. 1725, in Chelfea Garden. *R. S. n.* 171.
 Fl. July.   H. ⊙.

25. G. petalis emarginatis, arillis hirtis, cotyledonibus   *bohemi-*
 trifidis medio truncatis. *Sp. pl.* 955. *cum.*
 Bohemian Crane's-bill.
 Nat.
 Cult. 1683, by Mr. James Sutherland. *Sutherl. hort.*
 edin. 129. *n.* 3.
 Fl. June——Auguft.   H. ⊙.

26. G. foliis quinatis ternatifque incifis, calycibus de-   *robertia-*
 cemangulatis. *L'Herit. n.* 42. *num.*
 Geranium robertianum. *Sp. pl.* 955. *Curtis lond.*

    F f 3   *a* Geranium

α Geranium robertianum primum. *Bauh. pin.* 319.
  Stinking Crane's-bill, or Herb Robert.
β Geranium lucidum faxatile, foliis Geranii robertiani.
     Raj. fyn. 358.
  Shining ftinking Crane's-bill.
  Nat. of Britain.
  Fl. April——Auguft.   H. ⊙.

DODECANDRIA.

PENTAPETES. *Gen. pl.* 834.

Cal. 5-partitus.   *Stam.* 20: horum 5 caftrata, longa.
        Capf. 5-locularis, polyfperma.

phœnicea. 1. P. foliis haftato-lanceolatis ferratis.   *Sp. pl.* 958.
             Scarlet-flower'd Pentapetes.
             Nat. of India.
             Cult. 1690, in the Royal Garden at Hampton-court.
                Catal. mff.
             Fl. July.   S. ⊙.

Erythro-   2. P. foliis cordatis fubcrenatis fubtus tomentofis rugo-
xylon.   fo-reticulatis, floribus fubumbellatis decandris.
             Alcea arborea populi nigræ foliis prona parte albican-
                tibus, flore ampliffimo rubicundo.   *Pluk. mant.*
                p. 6. *t.* 333. *f.* 1. *Amman in comm. petrop.* 8. *p.* 217.
             St. Helena Red-wood.
             Nat. of the Ifland of St. Helena.
             Introd. 1772, by Sir Jofeph Banks, Bart.
             Fl.   S. ♄.

POLYAN-

POLYANDRIA.

ADANSONIA. *Gen. pl.* 836.

Cal. fimplex, deciduus. *Stylus* longiffimus : *Stigmata* plura. *Capf.* lignofa, 10-locularis pulpa farinacea, polyfperma.

1. ADANSONIA. *Syft. veget.* 620.   *digitata.*
Ethiopian Sour Gourd, or Monkies-bread.
Nat. of Senegal and Egypt.
Introd. 1724, by William Sherard, Efq. *Mill. dict. edit.* 8.
Fl.   S. ♄.

BUTONICA. (Barringtonia. *Linn. fuppl.* 50.)

Cal. fimplex, 2-phyllus, fuperus, perfiftens. *Drupa* ficca tetragona : *Nuce* 1-4-loculari.

1. BUTONICA. *Rumph. amb.* 3. *p.* 179. *tab.* 114. *De* *fpeciofa.*
Lamarck encycl. 1. *p.* 521.
Barringtonia fpeciofa. *Forft. nov. gen.* 38. *fl. auftr.* 47.
J. F. Mill. ic. 7. *Linn. fuppl.* 312. *Thunb. nov. gen.* 2. *p.* 47. *Cook's voyage, vol.* 1. *tab.* 24.
Commerçona. *Sonnerat iter nov. guin. p.* 14. *tab.* 8. 9.
Mammea afiatica. *Ofb. it.* 278. *Sp. pl.* 731.
Laurel-leav'd Butonica.
Nat. of the Eaft Indies, and the South Sea Iflands.
Introd. 1786, by Mr. Anthony Hove.
Fl.   S. ♄.

CAROLINEA.   *Linn. fuppl.* 51.

Cal. fimplex, tubulofus, truncatus.   *Petala* enfiformia.
Pomum 5-fulcatum, 2-loculare.

princeps.   1. CAROLINEA.   *Linn. fuppl.* 314.
Digitated Carolinea.
Nat. of the Weft Indies.
Introd. 1787, by Mr. Alexander Anderfon.
Fl,   S. ♄.

B O M B A X.   *Gen. pl.* 835.

Cal. 5-fidus. *Stam.* 5 f. multa. *Capf.* lignofa, 5-locula-
ris, 5-valvis. *Sem.* lanata. *Recept.* 5-gonum.

pentan-   1. B. floribus pentandris, foliis feptenatis. *Syft. veget.*
drum.   620.
Seven-leav'd Silk Cotton Tree.
Nat. of the Eaft Indies.
Cult. 1739, by Mr. Philip Miller. *Rand. chel.* Gof-
fypium 3.
Fl.   S. ♄.

Ceiba.   2. B. floribus polyandris, foliis quinatis. *Sp. pl.* 959.
Five-leav'd Silk Cotton Tree.
Nat. of both Indies.
Cult. 1692, in the Royal Garden at Hampton-court.
Br. Muf. Sloan. mff. 3343.
Fl.   S. ♄.

S I D A.

S I D A. *Gen. pl.* 837.

Cal. fimplex, angulatus. *Stylus* multipartitus. *Capf.* plures, 1-fpermæ.

* *Quinque-decem-capfulares.*

1. S. foliis cordato-oblongis ferratis, ftipulis fetaceis,   *fpinofa.* axillis fubfpinofis. *Sp. pl.* 960.
Prickly Sida.
Nat. of the Eaft Indies.
Introd. 1771, by Monf. Richard.
Fl. July——September.   S. ⊙.

2. S. foliis lanceolato-rhomboidibus ferratis, axillis fub-   *rhombifo-* fpinofis. *Sp. pl.* 961.   *lia.*
Rhombus-leav'd Sida.
Nat. of both Indies.
Cult. 1732, by James Sherard, M.D. *Dill. elth.* 216. t. 172. *f.* 212.
Fl. June and Auguft.   S. ♂.

3. S. foliis orbiculatis plicatis ferratis. *Sp. pl.* 961.   *alnifolia.*
Alder-leav'd Sida.
Nat. of the Eaft Indies.
Cult. 1732, by James Sherard, M.D. *Dill. elth.* 215. t. 172. *f.* 211.
Fl. July——September.   S. ⊙.

4. S. foliis cordatis ferratis fubtomentofis, ramis trique-   *triquetra.* tris. *Sp. pl.* 962. *Jacqu. hort.* 2. *p.* 54. *t.* 118.
Triangular-ftalk'd Sida.
Nat. of the Weft Indies.
Introd. 1775, by Jofeph Nicholas de Jacquin, M.D.
Fl. July and Auguft,   S. ♄.

carpini-
folia.

5. S. foliis bifariis ovato-lanceolatis ferratis : ferraturis
arctatis, umbellis axillaribus.   *Linn. fuppl.* 307.
Jacqu. ic. collect. 1. *p.* 38. *Cavan. diff.* 5. *p.* 274.
tab. 134. *f.* 1.
Hornbeam-leav'd Sida.
Nat.——(Cultivated in Madeira.)
Introd. 1774, by Mr. Francis Maffon.
Fl. Moft part of the Summer.   G. H. ♄.

cordifo-
lia.

6. S. foliis cordatis fubangulatis ferratis villofis. *Sp. pl.*
961.
Heart-leav'd Sida.
Nat. of India and the Cape of Good Hope.
Cult. 1732, by James Sherard, M.D. *Dill. elth.* 211.
t. 171. *f.* 209.
Fl. July——September.   S. ☉.

periplo-
cifolia.

7. S. foliis cordato-lanceolatis integerrimis, caule pani-
culato. *Syft. veget.* 622.
Great Bind-weed-leav'd Sida.
Nat. of both Indies.
Cult. 1691, in the Royal Garden at Hampton-court.
Pluk. phyt. t. 74. *f.* 7.
Fl. July and Auguft.   S. ♂.

arborea.

8. S. foliis cordatis ovatis acuminatis crenatis quinque-
nerviis molliffime tomentofis, pedunculis axillari-
bus unifloris. *Linn. fuppl.* 307. *L'Herit. ftirp.*
nov. p. 131. *tab.* 63.
Sida peruviana. *Cavan. diff.* 5. *p.* 276. *tab.* 130.
Great-flower'd Sida.
Nat. of Peru.
Introd. about 1772.
Fl. July and Auguft.   S. ♄.

9. S.

9. S. foliis palmatis peltatis: laciniis lanceolatis runci- *jatrophoi-*
nato-lobatis. *L'Herit. ſtirp. nov. p.* 117. *tab.* 56. *des.*
Sida palmata. *Cavan. diſſ.* 1. *p.* 20. *diſſ.* 5. *p.* 274.
tab. 131. *f.* 3.
Mulberry-leav'd Sida.
Nat. of South America.
Introd. 1787, by Monſ. Thouin.
Fl. Auguſt. S. ☉.

** *Multicapſulares.*

10. S. foliis cordatis ſublobatis, ſtipulis patentibus, pedun- *occiden-*
culis petiolo brevioribus, capſulis multilocularibus. *talis.*
Syſt. veget. 622.
Downy Sida.
Nat. of America.
Cult. 1732, by James Sherard, M.D. *Dill. elth.* 7.
t. 6. *f.* 6.
Fl. July and Auguſt. S. ☉.

11. S. foliis cordatis oblongis indiviſis, capſulis multilo- *america-*
cularibus longitudine calycis : loculis lanceolatis. *na.*
Sp. pl. 963.
Woolly Sida.
Nat. of Jamaica.
Cult. 1759, by Mr. Ph. Miller. *Mill. dict. edit.* 7. *n.* 16.
Fl. July and Auguſt. S. ☉.

12. S. foliis ſubrotundo-cordatis indiviſis, pedunculis folio *Abutilon.*
brevioribus, capſulis multilocularibus : corniculis
bifidis. *Sp. pl.* 963.
Broad-leav'd Sida.
Nat. of both Indies.
Cult. 1596, by Mr. John Gerard. *Hort. Ger.*
Fl. June——Auguſt. S. ☉.

13. S.

afiatica. 13. S. foliis cordatis indivifis, ftipulis reflexis, pedunculis longioribus, capfulis multilocularibus hirfutis calyce brevioribus. *Sp. pl.* 964.
Small-flower'd Indian Mallow.
Nat. of the Eaft Indies.
Introd. about 1768.
Fl. July and Auguft. S. ☉.

indica. 14. S. foliis cordatis fublobatis, ftipulis reflexis, pedunculis petiolo longioribus, capfulis multilocularibus fcabris calyce longioribus. *Syft. veget.* 623.
Rough capful'd Sida.
Nat. of India.
Cult. 1739, in Chelfea Garden. *Rand. chel.* Abutilon 3.
Fl. July and Auguft. S. ☉.

crifpa. 15. S. foliis cordatis fublobatis crenatis tomentofis, capfulis cernuis inflatis multilocularibus crenatis repandis. *Sp. pl.* 964.
Curl'd Sida.
Nat. of Carolina, and of the Bahama Iflands.
Cult. 1726, by James Sherard, M.D. *Dill. elth.* 6. *t.* 5. *f.* 5.
Fl. July and Auguft. S. ☉.

criftata. 16. S. foliis angulatis: inferioribus cordatis; fuperioribus panduriformibus, capfulis multilocularibus. *Sp. pl.* 964.
Crefted Sida.
Nat. of Mexico.
Cult. 1725, by James Sherard, M.D. *Dill. elth.* 3. *t.* 2, *f.* 2.
Fl. June——Oſtober. S. ☉.

MALACHRA.

M A L A C H R A. *Linn. mant.* 13.

Cal. commun. 3-phyllus, multiflorus, major. *Arilli* 5, monofpermi.

1. M. capitulis pedunculatis triphyllis feptemfloris. *Syft.*   *capitata.*
veget. 624.
Heart-leav'd Malachra.
Nat. of the Weft Indies.
Introd. 1782, by Mr. Francis Maffon.
Fl. Auguft and September.   S. ☉.

A L T H Æ A.   *Gen. pl.* 839.

Cal. duplex: exterior 9-fidus. *Arilli* plurimi, monofpermi.

1. A. foliis fimplicibus tomentofis. *Sp. pl.* 966.   *officinalis.*
Common Marfh-mallow.
Nat. of Britain.
Fl. July——September.   H. ♃.

2. A. foliis inferioribus palmatis; fuperioribus digitatis. *cannabi-*
Sp. pl. 966. *Jacqu. auftr.* 2. *p.* 1. *t.* 101.   *na.*
Hemp-leav'd Marfh-mallow.
Nat. of the South of Europe.
Cult. 1597, by Mr. John Gerard. *Ger. herb.* 789.
Fl. June and July.   H. ♃.

3. A. foliis utrinque tomentofis: inferioribus quinque- *narbonen-*
lobis; fuperioribus trilobis, pedunculis folitariis uni- *fis.*
floris. *Jacqu. ic.*
Althæa narbonenfis. *Cavan. diff.* 2. *p.* 94. *tab.* 29.
f. 2. *Pourret in act. tolof.* 3. *p.* 307.
Narbonne Althæa.
                       Nat.

Nat. of the South of France.
Introd. 1780, by Monf. Thouin.
Fl. Auguft and September.   H. ♃.

birfuta.   4. A. foliis trifidis pilofo-hifpidis fupra glabris, peduncu-
lis folitariis unifloris. *Syft. veget.*624. *Jacqu. auftr.*2.
p. 44. *t.* 170.
Hairy Marfh-mallow.
Nat. of the South of Europe.
Cult. 1683, by Mr. James Sutherland.   *Sutherl. hort.*
edin. 12. *n.* 4.
Fl. June and July.   H. ♂.

A L C E A.   *Gen. pl.* 840.

Cal. duplex: exterior 6-fidus.   *Arilli* plures, monofpermi.

rofea.   1. A. foliis finuato-angulofis.   *Sp. pl.* 966.
Common Holly-hock.
Nat. of China.
Cult. 1597.   *Ger. herb.* 782. *f.* 1.
Fl. July——September.   H. ♂.

ficifolia.   2. A. foliis palmatis.   *Sp. pl.* 967.
Fig-leav'd Holly-hock.
Nat. of the Levant.
Cult. 1597.   *Ger. herb.* 782. *f.* 2.
Fl. June——September.   H. ♂.

M A L V A.   *Gen. pl.* 841.

Cal. duplex: exterior 3-phyllus.   *Arilli* plurimi, monofpermi.

* *Foliis indivifis.*

fpicata.   1. M. foliis cordatis crenatis tomentofis, fpicis oblongis
hirtis. *Sp. pl.* 967.

                                       Spiked

Spiked Mallow.
Nat. of Jamaica.
Cult. 1726, by Mr. Philip Miller. *R. S. n.* 242.
Fl. September and October. S. ♄.

2. M. foliis ovatis crenato-ferratis, floribus axillaribus *fcoparia.*
    confertis, caule fruticofo, ramis virgatis. *L'Herit.*
    ftirp. nov. p. 53. *tab.* 27.
Malva fcoparia. *Cavan. diff.* 2. *p.* 65. *tab.* 21. *f.* 4.
Small yellow-flower'd upright Mallow.
Nat. of Peru.
Introd. 1782, by Monf. Thouin.
Fl. Auguft and September. S. ♄.

3. M. foliis lanceolatis, floribus axillaribus binis, pedun- *angufti-*
    culis petiolo brevioribus, calyce exteriori fetaceo *folia.*
    deciduo.
Malva anguftifolia. *Cavan. diff.* 2. *p.* 64. *tab.* 20. *f.* 3.
Narrow-leav'd Mallow.
Nat. of Mexico.
Introd. 1780, by Benjamin Bewick, Efq.
Fl. Auguft. G. H. ♄.

4. M. foliis cordatis crenatis, floribus lateralibus folita- *america-*
    riis; terminalibus fpicatis. *Sp. pl.* 968. *na.*
American Mallow.
Nat. of North America.
Cult. 1759, by Mr. Ph. Miller. *Mill. dict. edit.* 7. *n.* 15.
Fl. June and July. H. ☉.

        ** *Foliis angulatis.*

5. M. caule erecto herbaceo, foliis palmatis, fpicis fecun- *peruvia-*
    dis axillaribus, feminibus denticulatis. *Syft. veget.* *na.*
    625. *Jacqu. hort.* 2. *p.* 73. *t.* 156.
Peruvian Mallow.
Nat. of Peru.

                         Cult.

Cult. 1759, by Mr. Ph. Miller. *Mill. dict. edit.* 7. *n.* 7.
Fl. June——Auguſt.   H. ⊙.

limenſis.   6. M. caule erecto herbaceo, foliis lobatis, ſpicis ſecundis
axillaribus, ſeminibus lævibus. *Sp. pl.* 968. *Jacqu.*
hort. 2. *p.* 66. 141.
Blue-flower'd Mallow.
Nat. of Peru.
Introd. 1768, by Monſ. Richard.
Fl. July.   H. ⊙.

bryonifo-   7. M. caule fruticoſo tomentoſo, foliis palmatis ſcabris,
lia.   pedunculis multifloris. *Sp. pl.* 968.
Bryony-leav'd Mallow.
Nat. of the Cape of Good Hope.
Introd. 1774, by Mr. Francis Maſſon.
Fl. July and Auguſt.   G. H. ♄.

lactea.   8. M. frutescens, foliis angulatis acutis cordatis villoſis,
petalis obcordatis calyce brevioribus, pedunculis
paniculatis.
Panicled Mallow.
Nat.
Introd. 1780, by Benjamin Bewick, Eſq.
Fl. January and February.   S. ♄.

capenſis.   9. M. foliis cordatis quinquelobis, caule arboreſcente.
Syſt. veget. 625.
Gooſeberry-leav'd Mallow.
Nat. of the Cape of Good Hope.
Cult. 1713, in Chelſea Garden. *Philoſ. tranſact. n.* 337.
p. 197. *n.* 68.
Fl. Moſt part of the year.   G. H. ♄.

carolinia-   10. M. caule repente, foliis multifidis. *Sp. pl.* 969.
na.   Creeping

Creeping Mallow.
Nat. of Carolina.
Cult. 1723, in Chelſea Garden. *R. S. n.* 51.
Fl. June and July. H. ☉.

11. M. caule proſtrato, foliis cordato-orbiculatis obſo- *rotundi-*
    lete quinquelobis, pedunculis fructiferis declinatis. *folia.*
    Sp. pl. 969. *Curtis lond.*
    Round-leav'd, or Dwarf Mallow.
    Nat. of Britain.
    Fl. June——Auguſt. H. ☉.

12. M. caule erecto herbaceo, foliis ſeptemlobatis acutis, *ſylveſtris.*
    pedunculis petiolifque piloſis. *Sp. pl.* 969. *Curtis*
    lond.
    Common Mallow.
    Nat. of England.
    Fl. May——October. H. ♂.

13. M. caule erecto herbaceo, foliis quinquelobatis ob- *mauri-*
    tuſis, pedunculis petiolifque glabriuſculis. *Sp.* *tiana.*
    pl. 970.
    Ivy-leav'd Mallow.
    Nat. of the South of Europe.
    Introd. 1768, by John Earl of Bute.
    Fl. June and July. H. ☉.

14. M. caule erecto, foliis angulatis, floribus axillaribus *verticil-*
    glomeratis ſeſſilibus, calycibus ſcabris. *Sp. pl.* 970. *lata.*
    Jacqu. hort. 1. *p.* 15. *t.* 40.
    Whorl-flower'd Mallow.
    Nat. of China.
    Cult. 1683, by Mr. James Sutherland. *Sutherl. hort.*
    edin. 220. *n.* 1.
    Fl. June and July. H. ☉.

VOL. II. G g 15. M.

crifpa.   15. M. caule erecto, foliis angulatis crifpis, floribus
            axillaribus glomeratis.   *Sp. pl.* 970.
            Curl'd Mallow.
            Nat. of Syria.
            Cult. 1596, by Mr. John Gerard.   *Hort. Ger.*
            Fl. June——Auguft.   H. ☉.

Alcea.   16. M. caule erecto, foliis multipartitis fcabriufculis.
            Sp. pl. 971.
            Vervain Mallow.
            Nat. of England.
            Fl. July——October.   H. ♃.

mofchata. 17. M. caule erecto, foliis radicalibus reniformibus in-
            cifis ; caulinis quinquepartitis pinnato-multifidis.
            Syft. veget. 626.   *Curtis lond.*
            Mufk Mallow.
            Nat. of Britain.
            Fl. Auguft.   H. ♃.

ægyptia. 18. M. caule erecto, foliis palmatis dentatis, corollis ca-
            lyce minoribus.   *Syft. veget.* 626.   *Jacqu. hort.* 1.
            p. 27. *t.* 65.
            Palmated Mallow.
            Nat. of Egypt.
            Cult. 1739, by Mr. Ph. Miller.   *Rand. chel.* Alcea 9.
            Fl. June and July.   G. H. ♃.

            L A V A T E R A.   *Gen. pl.* 842.

            Cal. duplex : exterior 3-fidus. *Arilli* plurimi, mono-
            fpermi.

arborea. 1, L caule arboreo, foliis feptemangularibus tomentofis
            plicatis, pedunculis confertis unifloris axillaribus.
            Sp. pl. 972.
                                                            Tree

Tree Lavatera, or Mallow.
Nat. of Britain.
Fl. Auguft.   H. ♂.

2. L. caule fruticofo, foliis quinquelobo-haftatis, floribus   *olbia.*
    folitariis. *Sp. pl.* 972. *Jacqu. hort.* 1. *p.* 30. *t.* 73.
Downy-leav'd Lavatera.
Nat. of the South of France.
Cult. 1570, by Mr. Hugh Morgan.   *Lobel. adv.* 294.
Fl. June——October.   G. H. ♄.

3. L. caule fruticofo, foliis fubcordatis fubtrilobis rotun-   *triloba.*
    datis crenatis, ftipulis cordatis, pedunculis unifloris
    aggregatis. *Sp. pl.* 972. *Jacqu. hort.* 1. *p.* 30. *t.* 74.
Three-lob'd Lavatera.
Nat. of Spain.
Cult. 1759, by Mr. Ph. Miller. *Mill. dict. edit.* 7. *n.* 7.
Fl. June and July.   G. H. ♄.

4. L. caule fruticofo, foliis feptemangularibus tomentofis   *lufitanica.*
    plicatis, racemis terminalibus. *Sp. pl.* 973.
Portugal Lavatera.
Nat. of Portugal.
Cult. 1748, by Mr. Philip Miller.   *Mill. dict. edit.* 5.
    Althæa 2.
Fl. Auguft and September.   G. H. ♄.

5. L. caule herbaceo, fructibus denudatis, calycibus in-   *thurin-*
    cifis. *Sp. pl.* 973. *Jacqu. auftr.* 4. *p.* 6. *t.* 311.   *giaca.*
Great-flower'd Lavatera.
Nat. of Sweden, Germany, and Hungary.
Cult. 1732, by James Sherard, M. D.   *Dill. elth.* 9.
    t. 8. *f.* 8.
Fl. July——September.   H. ♄.

eretica.   6. L. caule erecto, ramis inferioribus diffusis, pedunculis confertis unifloris, foliis lobatis : superioribus acutis. *Sp. pl.* 973. *Jacqu. hort.* 1. *p.* 15. *t.* 41.
Cretan Lavatera.
Nat. of Candia.
Cult. 1768, by Mr. Philip Miller. *Mill. dict. edit.* 8.
   Malva 7.
Fl. July.   H. ☉.

trimestris.   7. L. caule scabro herbaceo, foliis glabris, pedunculis unifloris, fructibus orbiculo tectis. *Sp. pl.* 974.
Jacqu. hort. 1. *p.* 29. *t.* 72.
Common annual Lavatera.
Nat. of the South of Europe and the Levant.
Cult. 1640. *Park. theat.* 300. *f.* 5.
Fl. July——September.   H. ☉.

U R E N A.   *Gen. pl.* 844.

Cal. duplex : exterior 5-fidus. *Caps.* 5-locularis, echinata, monosperma.

lobata.   1. U. foliis angulatis. *Sp. pl.* 974.
Angular-leav'd Urena.
Nat. of China.
Cult. 1732, by James Sherard, M. D. *Dill. elth.*
   430. *t.* 319. *f.* 412.
Fl. July and August.   S. ♄.

sinuata.   2. U. foliis sinuato-palmatis : sinubus obtusis. *Sp. pl.* 974.
Cut-leav'd Urena.
Nat. of the East Indies.
Cult. 1759, by Mr. Ph. Miller. *Mill. dict. edit.* 7. *n.* 3.
Fl. July and August.   S. ♄.

GOSSY-

GOSSYPIUM. *Gen. pl.* 845.

Cal. duplex: exterior 3-fidus. *Capf.* 4-locularis. *Sem.*
lana obvoluta.

1. G. foliis quinquelobis fubtus eglandulofis, caule herba- *herba-*
ceo. *Syft. veget.* 628. *ceum.*
Common Cotton.
Nat. of the Eaft Indies.
Cult. 1594, by Mr. John Gerard. *Ger. herb.* 753.
Fl. July. S. ☉.

2. G. foliis palmatis: lobis lanceolatis, caule fruticofo. *arboreum.*
Sp. pl. 975.
Tree Cotton.
Nat. of the Eaft Indies.
Cult. 1731, by Mr. Philip Miller. *Mill. dict. edit.* 1.
Xylon 5.
Fl. S. ♄.

3. G. foliis trilobis acutis fubtus uniglandulofis, ramulis *religio-*
nigro-punctatis. *Syft. veget.* 628. *fum.*
Spotted-bark'd Cotton-Tree.
Nat. of India.
Introd. 1777, by Daniel Charles Solander, LL. D.
Fl. July. S. ♄.

4. G. foliis trilobis integerrimis fubtus triglandulofis. *barba-*
Sp. pl. 975. *denfe.*
Barbadoes Cotton-Tree.
Nat. of Barbadoes.
Cult. 1759, by Mr. Ph. Miller. *Mill. dict. edit.* 7. *n.* 2.
Fl. September. S. ♂.

HIBISCUS. *Gen. pl.* 846.

Cal. duplex: exterior polyphyllus. *Capf.* 5-locularis, polyfperma.

paluftris.   1. H. caule herbaceo fimpliciffimo, foliis ovatis fubtri-
lobis fubtus tomentofis, floribus axillaribus. *Sp,*
pl. 976.
Marfh Hibifcus.
Nat. of Virginia and Canada.
Cult. 1759, by Mr. Ph. Miller. *Mill. dict. edit.* 7. *n.* 18.
Fl. July and Auguft.   H. ♃.

præmor-   2. H. foliis fubrotundis dentato-ferratis retufis pubefcen-
fus.   tibus.
Hibifcus præmorfus. *Linn. fuppl.* 309. *Jacqu. ic.*
collect. 1. *p.* 81.
Pavonia cuneifolia. *Cavan. diff.* 3. *p.* 139. *tab.* 45. *f.* 1.
Urena præmorfa. *L'Herit. ftirp. nov. tom.* 2. *tab.* 51.
Round-leav'd fhrubby Hibifcus.
Nat. of the Cape of Good Hope. Mr. *Fr. Maffon.*
Introd. 1774.
Fl. June——Auguft.   G. H. ♄.

populneus.   3. H. foliis cordatis integerrimis, caule arboreo. *Syft.*
veget. 629.
Poplar-leav'd Hibifcus.
Nat. of the Eaft Indies and the South Sea Iflands.
Introd. 1770, by Monf. Richard.
Fl.   S. ♄.

tiliaceus.   4. H. foliis cordatis fubrotundis indivifis acuminatis cre-
natis, caule arboreo, calyce exteriore decemdentato.
Syft. veget. 629.
Lime-tree-leav'd Hibifcus.
                                                       Nat.

Nat. of the Eaft Indies.

Cult. 1739, by Mr. Philip Miller. *Mill. dict. vol.* 2. Ketmia 7.

Fl.   S. ♄ .

5. H. foliis ovatis acuminatis ferratis, caule arboreo.   *Rofa* *Syft. veget.* 629.   *finenfis.*

α flore fimplici.

China Rofe.

β flore pleno.

Double China Rofe.

Nat. of India.

Cult. 1731, by Mr. Philip Miller. *Mill. dict. edit.* 1. Ketmia 6.

Fl. July and Auguft.   S. ♄ .

6. H. foliis cordato-quinquangularibus obfolete ferratis,   *mutabilis.* caule arboreo. *Sp. pl.* 977.

α flore fimplici.

Changeable Rofe.

β flore pleno.

Double-flower'd changeable Rofe.

Nat. of the Eaft Indies.

Introd. 1690, by Mr. Bentick. *Br. Muf. Sloan. mff.* 3370.

Fl. November and December.   S. ♄ .

7. H. foliis cordatis crenatis indivifis, capfulis fpinis ex-   *fpinifex.* tantibus. *Sp. pl.* 978. *Jacqu. hort.* 2. *p.* 46. *t.* 103.

Prickly-fruited Hibifcus.

Nat. of the Weft Indies.

Introd. 1778, by William Wright. M. D.

Fl. July.   S. ♄ .

8, H. foliis fubcordatis tricufpidatis ferratis, calycibus   *Solandra.* ecalyculatis. *L'Herit. ftirp. nov. p.* 103. *tab.* 49.

                                 Solandra

Solandra lobata. *Murray in commentat. gotting.* 6.
    (1784) *p.* 20. *tab.* I.
Triguera acerifolia. *Cavan. diff.* I. *p.* 41. *tab.* II.
Maple-leav'd Hibifcus.
Nat. of the Ifland of Bourbon.
Introd. 1786, by Mr. Zier.
Fl. July and Auguft.   S. ☉.

fyriacus.   9. H. foliis cuneiformi-ovatis : fuperne incifo-dentatis,
                  caule arboreo. *Sp. pl.* 978. *Curtis magaz.* 83.

purpu-   α floribus purpureis.
reus.   Purple Althæa-frutex.

ruber.   β floribus rubris.
               Red Althæa-frutex.

albus.   γ floribus albis.
               White Althæa-frutex.

variega-   δ floribus variegatis.
tus.   Striped-flower'd Althæa-frutex.
               Nat. of Syria.
               Cult. 1629.   *Park. parad.* 367. *f.* 5.
               Fl. Auguft and September.   H. ♄.

ficulneus.   10. H. foliis quinquefido-palmatis, caule aculeato, flori-
                    bus pedunculatis. *Sp. pl.* 978.
                 Fig-leav'd Hibifcus.
                 Nat. of Ceylon.
                 Cult. 1732, by James Sherard, M. D.   *Dill. elth.*
                    190. *t.* 157. *f.* 190.
                 Fl. June and July.   S. ♄.

fpeciofus.   11. H. foliis palmatis glabris : laciniis lanceolatis ferra-
                    tis, caule pedunculis calycibufque lævibus.
                 Smooth Hibifcus.
                 Nat. of South Carolina.
                 Cult. 1778, by John Fothergill, M. D.

                                                            Fl.

Fl. September. G. H. ♃ .

OBS. Valde affinis Hibifco lævi *Scop. infubr.* 3. *p.* 35.
tab. 17; fed in hoc folia omnia quinqueloba,
quæ triloba defcribit Scopoli.

12. H. foliis ferratis : inferioribus ovatis indivifis ; fupe- *Sabda-*
rioribus tripartitis, caule inermi, floribus feffilibus. *riffa.*
Sp. pl. 978.
Various-leav'd Hibifcus.
Nat. of India.
Cult. 1596, by Mr. John Gerard. *Hort. Ger.*
Fl. June——September. S. ⊙.

13. H. recurvato-aculeatus, foliis quinquelobis, caly- *furatten-*
cibus exterioribus appendiculatis, ftipulis femicor- *fis.*
datis, floribus pedunculatis. *Syft. veget.* 630.
Prickly-ftalk'd Hibifcus.
Nat. of the Eaft Indies.
Cult. 1768, by Mr. Philip Miller. *Mill. dict. edit.* 8.
Fl. July. S. ⊙.

14. H. foliis palmato-digitatis feptempartitis, caule pe- *Manihot.*
tiolifque inermibus. *Syft. veget.* 630.
Palmated-leav'd Hibifcus.
Nat. of China and Japan.
Cult. 1712, in Chelfea Garden. *Philofoph. tranfact.*
n. 333. *p.* 417. *n.* 64.
Fl. Auguft. S. ♄ .

15. H. foliis fubpeltato-cordatis feptemangularibus ferra- *Abel-*
tis, caule hifpido. *Syft. veget.* 630. *mofchus.*
Target-leav'd Hibifcus.
Nat. of both Indies.

            Cult.

Cult. 1656, by Mr. John Tradeſcant, Jun. *Muſ.*
    Trad. 73.
Fl. July and Auguſt.   S. ♃.

eſculen- 16. H. foliis quinquepartito-pedatis, calycibus interiori-
tus.   bus latere rumpentibus. *Sp. pl.* 980.
           Eatable Hibiſcus.
           Nat. of the Weſt Indies.
           Cult. 1692, in the Royal Garden at Hampton-court.
               Br. Muſ. Sloan. mſſ. 3343.
           Fl. June and July.   S. ⊙.

æthiopi- 17. H. foliis ſubcuneatis ſubtridentatis : ſummis oppoſi-
cus.   tis, floribus terminalibus. *Linn. mant.* 258.
           Dwarf wedge-leav'd Hibiſcus.
           Nat. of the Cape of Good Hope.
           Introd. 1774, by Mr. Francis Maſſon.
           Fl. Auguſt.   G. H. ♄.

Trionum. 18. H. foliis tripartitis inciſis, calycibus inflatis. *Sp.*
               pl. 981.
           α Alcea veſicaria. *Bauh. pin.* 317.
           Common Bladder Hibiſcus.
           β Hibiſcus *africanus,* foliis tripartitis dentatis, lobis an-
               guſtioribus, caule hirſuto, calycibus inflatis. *Mill.*
               dict.

           Narrow-leav'd African Bladder Hibiſcus.
           γ Hibiſcus *hiſpidus,* foliis inferioribus trilobis ſummis
               quinquepartitis obtuſis crenatis, calycibus inflatis,
               caule hiſpido. *Mill. dict.*
           Broad-leav'd African Bladder Hibiſcus.
           Nat. α of Italy. β γ of the Cape of Good Hope.
           Cult. 1596, by Mr. John Gerard. *Hort. Ger.*
           Fl. June——September.   H. ⊙.

                              ACHANIA.

A C H A N I A. *Swartz prodr.*

Cal. duplex : exterior polyphyllus. *Cor.* convoluta.
Bacca 5-fperma.

1. A. foliis fcabriufculis acuminatis, foliolis calycis ex- *Malva-*
  terioris erectis. *Swartz prodr.* 102.   *vifcus.*
  Hibifcus Malvavifcus. *Sp. pl.* 978.
  Scarlet Achania, or Baftard Hibifcus.
  Nat. of Jamaica.
  Cult. 1714, by the Dutchefs of Beaufort. *Br. Muf.*
  H. S. 131. *fol.* 57.
  Fl. Moft part of the Year.   S. ♄ .

2. A. foliis tomentofis, foliolis calycis exterioris patulis. *mollis.*
  Woolly Achania.
  Nat. of America.
  Introd. 1780, by Benjamin Bewick, Efq.
  Fl. Auguft and September.   S. ♄ .

3. A. foliis pilofis obtufis acutifque. *Swartz prodr.* 102. *pilofa.*
  Hairy Achania.
  Nat. of Jamaica.
  Introd. 1780, by Mr. Gilbert Alexander.
  Fl. November.   S. ♂ .

S T U A R T I A. *Gen. pl.* 847.

Cal. fimplex. *Stylus* fimplex, ftigmate 5-fido. *Po-*
 mum exfuccum, 5-lobum, 1-fpermum, 5-fariam
 diffiliens.

1. STUARTIA. *Sp. pl.* 982.
  Common Stuartia.   *Malaco-*
  Nat. of Carolina and Virginia.   *dendron.*

           Cult.

Cult. 1752, by Mr. Philip Miller.   *Mill. dict. edit.* 6.
Appendix.

Fl. July and August.   H. ♄.

C A M E L L I A. *Gen. pl.* 848.

Cal. imbricatus, polyphyllus: foliolis interioribus
majoribus.

japonica.   1. CAMELLIA. *Sp. pl.* 982. *Curtis magaz.* 42. *Jacqu.*
ic. vol. 2. *collect.* 1. *p.* 117.
Japan Rose.
Nat. of China and Japan.
Cult. before 1742, by Robert James Lord Petre.
Edwards's birds, vol. 2. *p.* 67.
Fl. January——May.   G. H. ♄.

END OF THE SECOND VOLUME

Printed in the United States
By Bookmasters